Oligosaccharides of Chitin and Chitosan

Liming Zhao
Editor

Oligosaccharides of Chitin and Chitosan

Bio-manufacture and Applications

Springer

Editor
Liming Zhao
State Key Laboratory of Bioreactor Engineering,
School of Biotechnology
East China University of Science and Technology
Shanghai, China

ISBN 978-981-13-9404-1 ISBN 978-981-13-9402-7 (eBook)
https://doi.org/10.1007/978-981-13-9402-7

This Springer imprint is published by the registered company Springer Nature Singapore Pte Ltd.
The registered company address is: 152 Beach Road, #21-01/04 Gateway East, Singapore 189721, Singapore

Preface

Saccharides, also known as carbohydrates, are a group of important active molecules which have great structural and functional significance for human life. However, the research of saccharides had lagged far behind that of nucleic acids and proteins for a long time. The biological functions of bioactive oligosaccharide and polysaccharide were not identified until the 1980s, as well as the profound impact of saccharides on daily life. Since then, people started to realize that carbohydrates were as fundamental as proteins and nucleic acids for life and the basis, development, and application of saccharides had become a hot spot of international scientific research. At present, the researches on saccharides have received unprecedented attentions from scientists not only in the fields of basic molecular biology, biochemistry, or synthetic biology but also in applied medicine, pharmacy, and food science. In 1988, Raymond Dwek published a review entitled "Glycobiology," marking the birth of this frontier field "Glycoengineering." Now, glycoengineering has become another important scientific field in life sciences after genetic engineering and protein engineering. Preparation, separation, and purification of carbohydrates, as well as modification and functional exploration toward them, are the core contents of glycoengineering.

Aminosaccharides include chitin and its derivatives, which are mainly composed of chitin and chitosan, chitooligosaccharides, chitin oligosaccharides, and glucosamine or N-acetyl-D-glucosamine. As the second most abundant natural polysaccharides after cellulose, the biological yield of aminosaccharides is estimated to be about ten billion tons/year. Meanwhile, aminosaccharides are also the most abundant natural organic macromolecule compounds containing nitrogen besides proteins. Aminosaccharides have unique physical, chemical, and biological properties such as cationic polyelectrolytes, multifunctional group reactivity, antimicrobial activity, biocompatibility, biodegradability, etc. These amazing properties have attracted the attention of consumers, industries, and academic researchers in food, biomedical, agricultural, and other fields.

In this book, the history and recent research development of the chitooligosaccharides/chitin oligosaccharides and its monomers are summarized, as well as their applications and bio-manufacture. Most importantly, the bio-manufacturing techniques for production, purification, characterization, modification of chitooligosaccharides/chitin oligosaccharides, and their monomers are addressed in detail. Meanwhile, according to their bioactivities and biomaterial properties, the potential applications in food, biomedical, and agricultural industries were focused. Several new and cutting-edge insights here may help to solve problems in related industries and encourage further academic investigations, such as relying on synthetic biology to construct "intelligent" microorganisms which convert glucose into chitin/chitosan polymers or oligomers with high efficiency. To guide and facilitate tracking of the important progress in the chitooligosaccharides/chitin oligosaccharides and its monomers, the book *Oligosaccharides of Chitin and Chitosan: Bio-manufacture and Applications* was thus proposed by Prof. Dr. Zhao Liming (East China University of Science and Technology, China).

In this book, Chap. 1 introduces the discovery and development of chitin/chitosan and their degradation products, while Chap. 2 presents the current studies of the oligosaccharides and its monomers. These two chapters provide important background information, as well as the development and application of related technology. In Chaps. 3, 4, 5, 6 and 7, we introduce the preparation (Chaps. 3 and 4), separation, purification and detection (Chap. 5), clean production (Chap. 6), and modification (Chap. 7) of chitooligosaccharides/chitin oligosaccharides and its monomers. As the most important contents, these chapters focus on fermentation or enzyme catalysis-based bio-manufacturing techniques of chitooligosaccharides/chitin oligosaccharides and their monomers. Recent developments and prospects are described, including microbial strain screening and construction, improvement and optimization of fermentation as well as enzymatic catalysis process, and product downstream processing (especially the separation and purification procedures). In Chaps. 8, 9, 10 and 11, the application of chitooligosaccharides/chitin oligosaccharides and their monomer is addressed, including biological activities and potential application in food industry (Chap. 8), in biomedical materials (Chap. 9), in agriculture (Chap. 10), and in breeding industry (Chap. 11).

The authors of this book are from East China University of Science and Technology (China), Jiangnan University (China), China Ocean University (China), Chinese Academy of Science Dalian Physical and Chemistry Institute (China), Dalian Minzu University, and Chinese Academy of Sciences Institute of Process Engineering, all well known for their achievements in the field of chitin and its derivatives, enzyme and fermentation technology in functional carbohydrates, plant protection, and breeding. We are particularly grateful to all the authors for their indispensable contributions.

Shanghai, China Liming Zhao

Contents

Part I Introductory and Background Information

1 **The History of Chito/Chitin Oligosaccharides and Its Monomer** 3
Zhen Qin and Liming Zhao

2 **Current Studies of Chito/Chitin Oligosaccharides and Its Monomer** 15
Qiming Chen and Liming Zhao

Part II Bio-Manufacturing of Chito/Chitin Oligosaccharides and Its Monomer

3 **Preparation of Chitooligosaccharides and Its Monomer** 29
Weizhi Liu, Qianqian Lyu, and Zhen Qin

4 **Preparation of Chitin Oligosaccharides and Its Monomer** 55
Shaoqing Yang, Zhengqiang Jiang, Yihao Liu, and Shuai Ma

5 **Detection and Separation of Chito/Chitin Oligosaccharides** 83
Zhen Qin, Yanying Hou, Waheed Ahamed, Yun Li, and Liming Zhao

6 **Cleaner Production Guide of Chito/Chitin Oligosaccharides and Its Monomer** 107
Xiangzhao Mao, Na Guo, and Jianan Sun

7 **Modification of Chitosan/Chitin and Its Oligosaccharides** 129
Xiaoli Liu and Qixing Jiang

Part III The Application of Chito/Chitin Oligosaccharides and Its Monomer

8 **Biological Activities and Potential Application in Food Industry** 163
Mengyao Zhao and Liming Zhao

9 **The Application of Chitooligosaccharides on Biomaterials** 275
Qiming Chen and Liming Zhao

10 **The Application of Chito/Chitin Oligosaccharides as Plant Vaccines** 289
Xiaoming Zhao, Mengyu Wang, Wenxia Wang, Qishun Liu, Jianguo Li, and Heng Yin

11 **The Application of Oligosaccharides in Breeding Industry** 325
Qingsong Xu, Lunxue Wang, and Yuguang Du

Editor and Contributors

About the Editor

Liming Zhao, Ph.D., is a professor of Food Biotechnology at School of Biotechnology, East China University of Science and Technology (ECUST). He received his B.S., M.S., and Ph.D. degrees in Food Science from Jiangnan University (*former Wuxi University of Light Industry*), and he was visiting professor at the University of California, Davis, from March 2013 to March 2014. Before he became a teacher and researcher at ECUST in 2010, he had ever worked for domestic or international companies as engineer or general manager for over 8 years. His research interests include the development and application research on chitin/chitosan and its derivatives, membrane and chromatography separation and purification technologies and their industrial-scale application, and the research and development on biomaterial and bio-based materials like chitosan and polyamides. He has published more than 110 original scientific papers in peer-reviewed journals and 10 popular articles, as author or coauthor. He published his first monograph of *the Application of Membrane Technology in Food and Fermentation Industries* in July 2011, and he is also the editor in chief of a teaching book *Principle of Food Engineering* published in September 2013.

Dr. Zhao has contributed significantly to the fields of aminosaccharides biomanufacture and separation technology in glycoengineering and food engineering fields, more than 25 patents had been authorized, and most of them had already successfully industrial-scale applied. He was awarded *Outstanding Youth Award* of the Chinese Institute of Food Science and Technology (CIFST) in 2013, *Distinguished Young Scholars of Lun Shiyi Education Fund* in 2018, and *Excellent Teachers of Baosteel Education Fund* in 2017. He has been conferred adjunct/visiting professorships from four universities, namely, Chengdu University, Shandong University of Technology, Jinan University, and Shanxi University.

Dr. Zhao currently serves as director of CIFST and vice chairman of its Youth Work Committee, director of China Association of Biology and Fermentation Industry, and director of Shanghai Food Institute. He currently serves on the editorial board for *Food Science & Nutrition* (Wiley), *Current Biotechnology*, *China Brewing*, and *Science and Technology of Food Industry*.

Contributors

Waheed Ahamed State Key Laboratory of Bioreactor Engineering, School of Biotechnology, East China University of Science and Technology, Shanghai, China

Qiming Chen State Key Laboratory of Bioreactor Engineering, School of Biotechnology, East China University of Science and Technology, Shanghai, China

Yuguang Du Institute of Process Engineering, Chinese Academy of Sciences, Beijing, China

Na Guo College of Food Science and Engineering, Ocean University of China, Qingdao, China

Yanying Hou State Key Laboratory of Bioreactor Engineering, School of Biotechnology, East China University of Science and Technology, Shanghai, China

Qixing Jiang State Key Laboratory of Food Science and Technology, School of Food Science and Technology, Collaborative Innovation Center of Food Safety and Quality Control in Jiangsu Province, Jiangnan University, Wuxi, China

Zhengqiang Jiang College of Food Science and Nutritional Engineering, China Agricultural University, Beijing, China

Jianguo Li Agricultural Technology Promotion Center of Ankang City, Ankang, Shaanxi, China

Yun Li State Key Laboratory of Bioreactor Engineering, School of Biotechnology, East China University of Science and Technology, Shanghai, China

Qishun Liu Dalian Institute of Chemical Physics, Chinese Academy of Sciences, Dalian, China

Weizhi Liu MOE Key Laboratory of Marine Genetics and Breeding, College of Marine Life Sciences, Ocean University of China, Qingdao, China

Xiaoli Liu State Key Laboratory of Food Science and Technology, School of Food Science and Technology, Collaborative Innovation Center of Food Safety and Quality Control in Jiangsu Province, Jiangnan University, Wuxi, China

Yihao Liu College of Food Science and Nutritional Engineering, China Agricultural University, Beijing, China

Qianqian Lyu MOE Key Laboratory of Marine Genetics and Breeding, College of Marine Life Sciences, Ocean University of China, Qingdao, China

Shuai Ma College of Food Science and Nutritional Engineering, China Agricultural University, Beijing, China

Xiangzhao Mao College of Food Science and Engineering, Ocean University of China, Qingdao, China

Laboratory for Marine Drugs and Bioproducts of Qingdao National Laboratory for Marine Science and Technology, Qingdao, China

Zhen Qin State Key Laboratory of Bioreactor Engineering, School of Biotechnology, East China University of Science and Technology, Shanghai, China

Jianan Sun College of Food Science and Engineering, Ocean University of China, Qingdao, China

Lunxue Wang Dalian Institute of Chemical Physics, Chinese Academy of Sciences, Dalian, China

Mengyu Wang Dalian Maritime University, Dalian, China

Wenxia Wang Dalian Institute of Chemical Physics, Chinese Academy of Sciences, Dalian, China

Qingsong Xu Key Laboratory of Biotechnology and Bioresources Utilization of Ministry of Education, Dalian Minzu University, Dalian, China

Shaoqing Yang College of Food Science and Nutritional Engineering, China Agricultural University, Beijing, China

Heng Yin Dalian Institute of Chemical Physics, Chinese Academy of Sciences, Dalian, China

Liming Zhao State Key Laboratory of Bioreactor Engineering, School of Biotechnology, East China University of Science and Technology, Shanghai, China

Mengyao Zhao State Key Laboratory of Bioreactor Engineering, School of Biotechnology, East China University of Science and Technology, Shanghai, China

Xiaoming Zhao Dalian Institute of Chemical Physics, Chinese Academy of Sciences, Dalian, China

Abbreviations

ACh	Acetylcholine
AChE	Acetylcholinesterase
AD	Alzheimer's disease
ADFI	Average daily feed intake
ADG	Average daily gain
AE	Aminoethyl
AEC	Anion-exchange chromatography
AEM	Anion-exchange membrane
Akt	Protein kinase B
ALP	Alkaline phosphatase
AMC	Affinity membrane chromatography
AMV	Alfalfa mosaic virus
Aβ	β-amyloid
Bak	Bcl-2 antagonist/killer 1
Bax	Bcl-2associated X protein
Bcl-2	B-cell lymphoma 2
Bcl-XL	B-cell lymphoma-extra large
BMAL1	Brain muscle Arnt-like 1
BMP	Bone morphogenetic protein
BMVIT	Ministry for Transport, Innovation, and Technology
BUN	Blood urea nitrogen
BWG	Body weight gain
CaMKKβ	Calmodulin-dependent protein kinase kinase-beta
CaSR	Calcium-sensing receptor
CaSR-PLC-IP3	Calcium-sensing receptor-phospholipase C-IP3
CAT	Catalase
CBM	Carbohydrate-binding module
CD36	Cluster of differentiation 36
CE	Capillary electrophoresis
CEC	Cation-exchange chromatography
CEM	Cationic exchange membrane
ChOS	Chitin oligosaccharides
Ci	Intercellular CO_2 concentration
CLOCK	Circadian locomotor output cycles kaput

CMAPS	Compound muscle action potentials
CMV	Cucumber mosaic virus
COS	Chitooligosaccharides
COS-Cr	COS chelating chromium
COS-V	COS chelating vanadium
COX-2	Cyclooxygenase 2
CP	Crude protein
CPT1A	Carnitine palmitoyltransferase I
CSA	Cross-sectional area
CSE	Colistin sulfate
Csx A	Exo-chitosanase
CVD	Continuous volumetric diafiltration
CYP7A1	Cholesterol 7α-hydroxylase
Cyto c	Cytochrome C
DA	Degree of acetylation
DD	Degree of deacetylation
DEAE	Diethylaminoethyl
DEGs	Differentially expressed genes
DGAT2	Diacylglycerol acyltransferase 2
DHB	2,5-Dihydroxybenzoic acid
DMAC	Dimethylacetamide
DMEM	Dimethylaminoethyl
DMF	Dimethyl formamide
DMI	Dry matter intake
DMSO	Dimethylsulfoxide
DON	Deoxynivalenol
DP	Degree of polymerization
DPPH	2,2-Diphenyl-1-picrylhydrazyl
DSS	Dextran sulfate sodium
EAAU	Experimental autoimmune anterior uveitis
EAT	Ehrlich ascites tumor
ED	Electrodialysis
ELSD	Evaporative light-scattering detector
ERK1/2	Extracellular signal regulated kinase 1/2
FABP4	Fatty acid-binding protein 4
FAS	Fas cell surface death receptor
FCR	Feed conversion ratio
FI	Feed intake
FoxO3a	Forkhead box O3
FSMP	Food for special medical purposes
FTIR	Fourier transform infrared spectroscopy
G6Pase	Glucose-6-phosphatase
GC-MS	Gas chromatography-mass spectrometer
GH	Glycoside hydrolase (Chapters 3 and 4)
GI	Gastrointestinal tracts

GLC	Glucosamine
GlcN	Glucosamine
GlcNAc	N-acetyl-D-glucosamine
GlmA	Exo-D-glucosaminidase
GLUT	Glucose transporter
GnT-V	N-acetylglucosaminyltransferase V
GPx1	Glutathione peroxidase 1
Gs	Stomatal conductance
GSH	Glutathione
GSH-Px	Glutathione peroxidase
H/L	Heterophil/lymphocyte
HDL	High-density lipoprotein
HDLC	High-density lipoprotein cholesterol
HFD	High-fat diet
HILIC	Hydrophilic interaction liquid chromatography column
HIMC	Hydrophobic interaction membrane chromatography
HIV-1	Human immunodeficiency virus type 1
HMGCR	3-Hydroxy-3-methyl-glutaryl-CoA reductase
HNF	Hepatocyte nuclear factor
HO-1	Hemeoxygenase1
HPAEC	High-performance anion-exchange chromatography
HPLC	High-performance liquid chromatography
HR	Hypersensitive responses
IEC	Ion-exchange chromatography
IGF	Insulin-like growth factor
IL	Interleukin
IMC	Ion-exchange membrane chromatography
iNOS	Inducible nitric oxide synthase
IR	Infrared spectrometry
JA/ET	Jasmonic acid/ethylene
JAK	Janus tyrosine kinase
JNK	c-Jun N-terminal kinase
LC-MS	Liquid chromatograph-mass spectrometry
LCOs	Lipo-chitooligosaccharides
LDL	Low-density lipoprotein
LDLC	Low-density lipoprotein-cholesterol
LDLR	Low-density lipoprotein receptor
LHD	Lactate dehydrogenase
LMCS	Low-molecular-weight chitosan
LOX	Lipoxygenases
LPS	Lipopolysaccharide
Ls	Stomatal limitation
LSC	Liquid-solid chromatography
LXRα	Liver X receptor alpha
MALDI-TOF	Matrix-assisted laser desorption ionization

MAPK	Mitogen-activated protein kinase
MDA	Malondialdehyde
MMC	Multistage membrane chromatography
MMP	Matrix metalloproteinase
MPO	Myeloperoxidase
MS	Mass spectrometry
MW	Molecular weight
N-acetyl COSs	Chitin oligosaccharides
NAFLD	Nonalcoholic fatty liver disease
NDF	Neutral detergent fiber
NF	Nanofiltration
NFM	Nanofiltration membrane
NF-κB	Nuclear factor-kappa B
NMR	Nuclear magnetic resonance
NO	Nitric oxide
NOS	Nitric oxide synthase
NR	Nitrate reductase
Nrf2	Nuclear factor erythroid-2-related factor 2
OPG	Osteoprotegerin
PAD	Pulsed amperometric detection
PAL	Phenylalanine ammonia lyase
PAMAM	Polyamidoamine
PCSK9	Proprotein convertase subtilisin/kexin type 9
PDB	Protein data bank
PEPCK-C	Phosphoenolpyruvate carboxykinase-cytoplasm
PEPCK-M	Phosphoenolpyruvate carboxykinase-mitochondria
PGE2	Prostaglandin
PI3K	Phosphoinositide 3-kinase
PMA	Phorbol 12-myristate 13-acetate
PMO	Porcine milk oligosaccharides
Pn	Net photosynthetic rate
POD	Peroxidase
PPAR	Peroxisome proliferator-activated receptors
PPO	Polyphenol oxidase
PR-1a	Pathogenesis-related protein 1a
PSV	Peanut stunt virus
PXR	Pregnenolone X receptor
RANKL	Receptor activator of NF-κB ligand
RID	Refractive index detector
RO	Reverse osmosis
ROS	Reactive oxygen species
RTIL	Room temperature ionic liquid
SA	Salicylic acid
SCFA	Short-chain fatty acid
SCOS	Sulfated COS

SEC	Size exclusion chromatography
SGLT1	Sodium-glucose co-transporters
SOD	Superoxide dismutase
SRBSDV	Southern rice black-streaked dwarf virus
SREBP-2	Sterol regulatory element-binding protein
STAT3	Signal transducer and activator of transcription 3
STZ	Streptozocin
T-AOC	Total antioxidant capacity
TC	Total cholesterol
TG	Triglycerides
TGA	Thermal gravimetric analysis
TGF-β	Transforming growth factor-β
TLC	Thin layer chromatography
TLR4	Toll-like receptor 4
TMV	Tobacco mosaic virus
TNF-α	Tumor necrosis factor alpha
TNV	Tobacco necrosis virus
Tr	Transpiration rate
UF	Ultrafiltration
UNIDO	United Nations Industrial Development Organization
VEGF	Vascular endothelial growth factor
VEGFA	Vascular endothelial growth factor A
WBC	White blood cell

Part I

Introductory and Background Information

1 The History of Chito/Chitin Oligosaccharides and Its Monomer

Zhen Qin and Liming Zhao

Abstract

Chitooligosaccharides and chitin oligosaccharides are collectively referred to as aminooligosaccharides. The monomer of chito/chitin oligosaccharides are N-acetyl-D-glucosamine and glucosamine, two of the few nitrogenous sugars in nature. Chito/chitin oligosaccharides are mainly produced by hydrolysis from natural aminopolysaccharides (chitosan and chitin), which are extracted from crab and shrimp shells resource. Chitin and chitosan are well known as natural polysaccharides with abundant biological activities. Similarly, chito/chitin oligosaccharides also have a number of commercial uses as their biological activities. This chapter summarizes the history and chemical-based topics like definition, origin, structure, molecular weight, classification of chitin, chitosan, chito/chitin oligosaccharides and their monomers (N-acetyl-D-glucosamine and glucosamine). Due to the broad research interests and market demands, the discovery and development of biological activities of chito/chitin oligosaccharides are summarized as argument.

1.1 Introduction

1.1.1 Chitin

Chitin $(C_8H_{13}O_5N)_n$ is a natural polysaccharide composed of β-1,4-linked N-acetyl-D-glucosamine (GlcNAc, NAG), chemical construction is shown in Fig. 1.1. Chitin is an abundant biopolymer on earth next to cellulose. Professor Henri Braconnot

Z. Qin · L. Zhao (✉)
State Key Laboratory of Bioreactor Engineering, School of Biotechnology, East China University of Science and Technology, Shanghai, China
e-mail: zhaoliming@ecust.edu.cn

L. Zhao (ed.), *Oligosaccharides of Chitin and Chitosan*,
https://doi.org/10.1007/978-981-13-9402-7_1

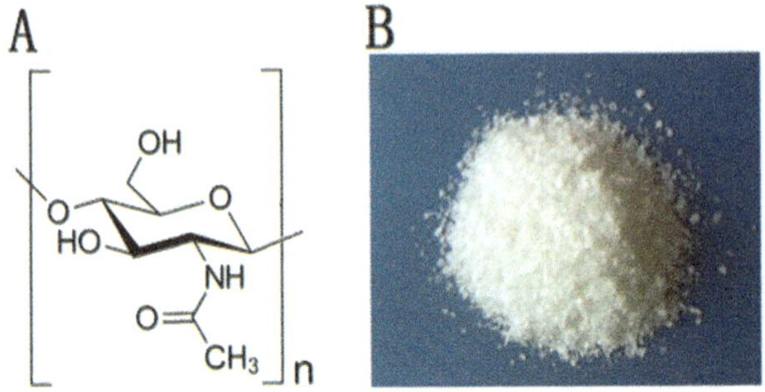

Fig. 1.1 Chitin. (**a**) Chemical construction of chitin; (**b**) Picture of chitin powder

firstly discovered and isolated chitin from mushrooms in 1811. After that, Odier found the same polysaccharide in the insects' exoskeleton and named it chitin in 1823. Chitin can be obtained from a wild range of sources. Exoskeletons of arthropods (crustaceans, insects and arachnids) and mollusks (beaks and endoskeleton of cephalopods) are main sources of chitin. And various microorganisms are also the sources of chitin, such as the cell walls polysaccharides of yeasts and fungi, and the spines of diatoms. However, commercially chitin product is mainly obtained from marine sources, e.g., crustacean shells from crabs and shrimps at present.

Like another common polysaccharide – cellulose, the structure of chitin often forms crystalline nanofibrils or whiskers. In terms of physiological function, chitin may be compared to the animal protein – keratin. Chitin has been proved to be used in a variety of industrial, pharmaceutical and biotechnology fields. Chitin and cellulose are polysaccharides similar on the structure; chitin is a linear polymer consists of β-(1,4)-N-acetyl-D-glucosamine. As the different crystalline microfibrils order, natural chitin occurs in three polymorphic forms: α-chitin, β-chitin and γ-chitin. α-Chitin is arranged in anti-parallel strands, and it is the most abundant and stable form found in nature. It acts as shell for insects, crabs, lobsters and shrimps, as well as structural skeletons in the cell walls of fungi and yeast. The β form is arranged in parallel chains. Consequently, β-Chitin is less stable than the α-Chitin. The source of different chitin is also not the same. The β-chitin has been found in the pens of squids, the extracellular fibers of diatoms and the spines and chaetae of certain annelids. γ-chitin is a mixture composed of α and β structures, which is the least common form. It has been found in the squid's stomach and in the cocoons of two beetles.

Chitin is a nitrogenous polysaccharide which has an acetamide group ($NH\text{-}CO\text{-}CH_3$) at C-2 in place of the hydroxyl group in cellulose. The pure chitin is white or yellowish, odorless and tasteless. Chitin exhibits excellent biodegradability and biocompatibility that is potential to be used in food, medicine or material industries. However, chitin is highly hydrophobic; thus, it is insoluble in water and even in most organic solvents. The water insolubility of chitin is one of the main limitations in the large-scale application. Therefore, the research about water-soluble derivatives of chitin has become one of the research hotpots.

1.1.2 Chitosan

Chitosan is the product of the deacetylation reaction of chitin, which is a linear polysaccharide composed of β-1,4-linked D-glucosamine (GlcN) and a small amount of N-acetyl-D-glucosamine (Fig. 1.2). Chitosan is generally produced by a deacetylation reaction of treating the chitin shells of shrimp and other crustaceans with a basic substance such as sodium hydroxide (Fig. 1.3). Chitosan was discovered as a transformation of chitin in water soluble form after chemical manipulation by Roughet in 1859. Later, in 1870, this transformation of chitin was named chitosan. The nature sources of chitin include shells of crabs and shrimps. Unlike chitin, natural chitosan not exist in animal species, and even is rarely found in nature, except fungi. Both chitosan and chitin present in the cell walls synthesis of fungi, e.g., Zygomycetes. Moreover, chitosan has two distinct forms in the fungal cell wall: free form of chitosan and chitosan bound to β-glucan.

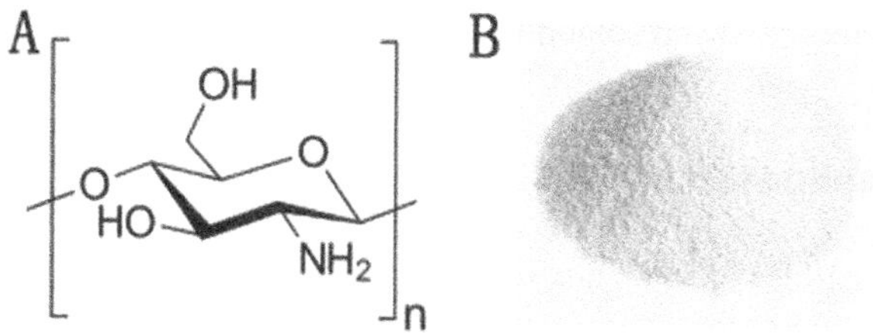

Fig. 1.2 Chitosan. (**a**) Chemical construction of Chitosan; (**b**) Picture of chitosan powder

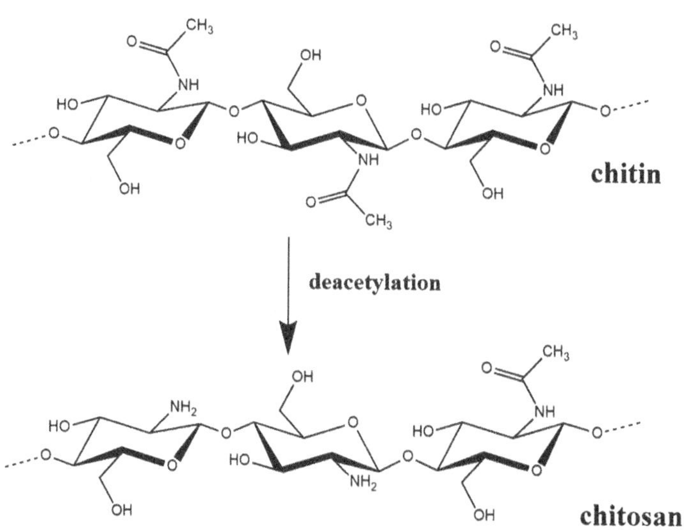

Fig. 1.3 Forming chitosan by partial deacetylation of chitin

Commercial chitosan product is mostly obtained from the deacetylation of natural chitin, which is completely or partially deacetylated by treatment with acid or alkali solution. Structurally, chitosan is a linear polysaccharide containing copolymer of β-1,4-linked D-glucosamine and N-acetyl-D-glucosamine. The degree of deacetylation (DD) is generally defined as the glucosamine/N-acetyl-D-glucosamine ratio, which goes up as chitin is converted to chitosan. Therefore, chitin and chitosan can be distinguished by the ratio above. When the content of N-acetyl-D-glucosamine is higher than that of glucosamine, the biopolymer is often called chitin; otherwise, the compound is called chitosan.

The physical and chemical properties of chitosan are similar to those of chitin as mentioned above. Due to the positive charges of the amino groups in glucosamine, chitosan is the only water-soluble cationic polysaccharide commercially available. Because of its distinctive physicochemical characteristics such as biocompatibility, biodegradability and low toxicity, chitosan is considered having great potential for the applications in various industries. However, the higher molecular weight and high viscosity of chitosan restrict its specific applications. Fortunately, chitosan can be further hydrolyzed into its low molecular weight (MW) derivatives, which exhibit more biologically active and solubility than chitosan.

1.1.3 Chitooligosaccharides

Chitooligosaccharides (COS) are the degradation products of chitosan, which exhibit various biological activity, such as antitumor, antioxidant, immunostimulatory, antimicrobial and elicitors of plant immunity. Since the discovery of chitosan in the late 1870s, many scientists have put a lot of efforts on research of its physicochemical properties. Under different reaction conditions, chitooligosaccharides can be prepared with different physicochemical properties. According to the academic definition of the oligosaccharides, the degrees of polymerization (DP) of an oligosaccharide should be between 2 and 10. However, in industrial production and commercial application, chitosan with DP degrees of polymerization (DP) less than 20 and an average molecular weight less than 3900 Da are also classified named chitooligosaccharides (or chitosan oligosaccharides, chitosan oligomer). The DD of chitooligosaccharides is generally higher than 75%. Chitooligosaccharides has three types of reactive functional groups: an amino/acetamido group, a primary hydroxyl group, and a secondary hydroxyl group at the C-2, C-3, and C-6 positions, respectively. Chitooligosaccharides and modified chitooligosaccharides are both found to be much easier to use in large-scale applications (Fig. 1.4).

The most important factors that affect the physicochemical properties of chitooligosaccharides are DP and DD, which are also correlated with their biological actives. The DP, especially the corresponding molecular weight, contributes to the change in the physicochemical properties of chitooligosaccharides. Low DP chitooligosaccharides is more soluble than chitosan (often with relatively high DP). Compared with chitosan, chitooligosaccharides is more easily absorbed in the intestine and rapidly enters to blood circulation. Thus, molecular weight is considered to

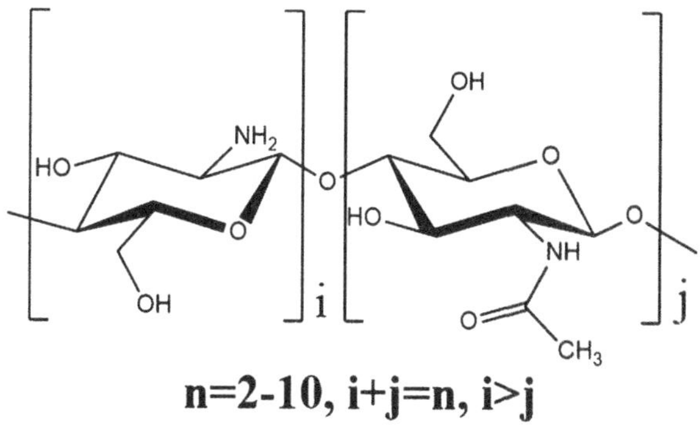

Fig. 1.4 Chemical construction of chitooligosaccharide

Fig. 1.5 Chemical construction of chitin oligosaccharides

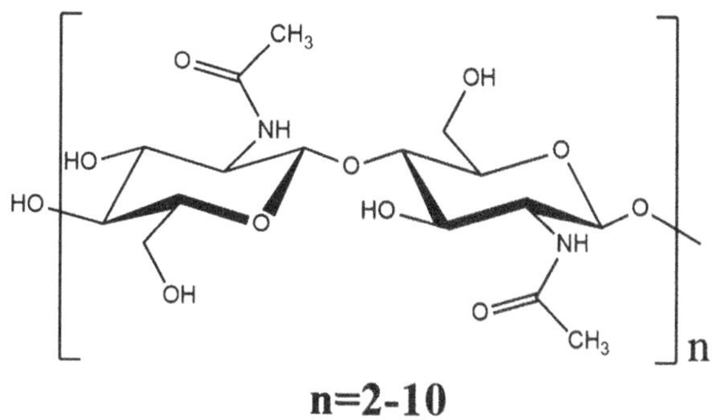

be a major feature of chitooligosaccharides and is highly correlated with their biological activity. DD is also one of the important parameters of chitooligosaccharides. The higher DD chitooligosaccharides have more free amino groups and more positive charges in the solution. DD is the most important parameter determining the solubility of chitooligosaccharides. The rate and extent of biodegradation of chitooligosaccharides in organisms are related to DD, and the decrease in degradation rate leads to an increase in DD (Fig. 1.5).

1.1.4 Chitin Oligosaccharides

Chitin oligosaccharides (N-acetyl COS, ChOS) are the partially degraded products of chitin by chemical, physical or enzymatic methods. In general, he degree of polymerization (DP) of chitinoligosaccharides is 2–10. The indicator of chitinoligosaccharides and chitooligosaccharides is the degree of deacetylation. The degree of acetylation of chitinoligosaccharides is generally high (>75%). Chitin oligosaccharides has attracted considerable attention in recent years due to their water-solubility and temperature and pH stability, as well as excellent biological activities, such as antimicrobial, anti-tumor and hypoglycemic activities. In addition to these

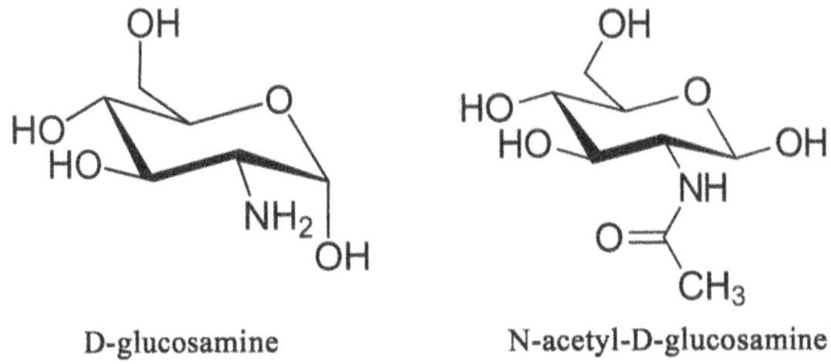

Fig. 1.6 Chemical construction of glucosamine and N-acetyl-D-glucosamine

biological activities, ChOS are also considered as effective plant elicitors and potential angiogenic inhibitor, involving in the promotion of bifidus proliferation, the characteristics of which could be widely used in agricultural, food and pharmaceutical industrials. Thus, production of ChOS is a promising way for the disposal and recycling of enormous waste chitin materials (Fig. 1.6).

1.1.5 Glucosamine and N-acetyl-D-Glucosamine

Glucosamine (GlcN) and N-acetyl-D-glucosamine (GlcNAc) are the monomers of chito/chitin oligosaccharides. Glucosamine and N-acetyl-D-glucosamine are natural amino monosaccharides and derivatives from glucose. Glucosamine was first prepared by the hydrolysis of chitin with concentrated hydrochloric acid by Georg Ledderhose in 1876. And the chemical construction of glucosamine was determined by Walter Haworth in 1939. Glucosamine/N-acetyl-D-glucosamine is not only part of the constitutional unit of the structural polysaccharides (chitosan and chitin), but also one of the most abundant bioactive monosaccharides with significant biological activity. Since glucosamine is the precursor for glycosaminoglycan, which is a major component of cartilage. Research has focused on the potential for supplemental glucosamine to beneficially influence cartilage structure and alleviate arthritis. In addition, N-acetyl-D-glucosamine is a major component of the bacterial and fungal cell walls, which has a significant effect on the treatment of inflammatory diseases. Thus, glucosamine/N-acetyl-D-glucosamine have been widely used as dietary supplement and pharmaceutical drug for decades. In the United States, glucosamine/N-acetyl-D-glucosamine is one of the most common dietary supplements used by adults.

For a long time, glucosamine/N-acetyl-D-glucosamine is produced commercially by the enzymatic/chemical hydrolysis of crustacean exoskeletons (Fig. 1.7). The crab shell is first deproteinized and decalcified by acid-base treatment to obtain chitin. Chitin is then processed by acid hydrolysis or enzymatic hydrolysis to obtain N-acetyl-D-glucosamine. N-acetyl-D-glucosamine are finally deacetylated by concentrated alkali or enzyme to obtain glucosamine. Hydrolysis process has many potential problems such as environmental pollution, product allergy and shortage of raw material supply. Although the use of enzymes alleviated the above problems in

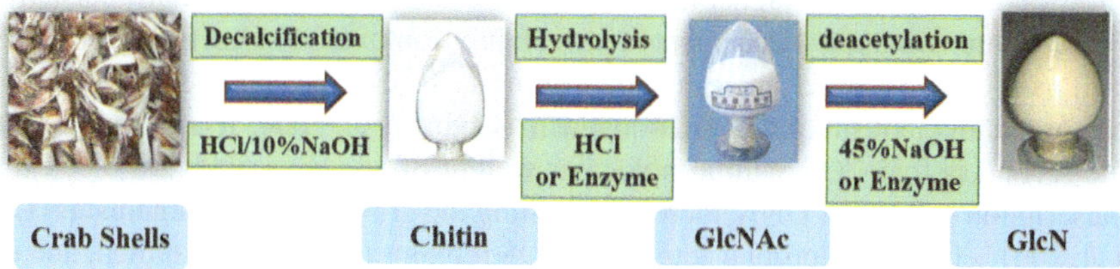

Fig. 1.7 Preparation of glucosamine/N-acetyl-D-glucosamine by hydrolysis

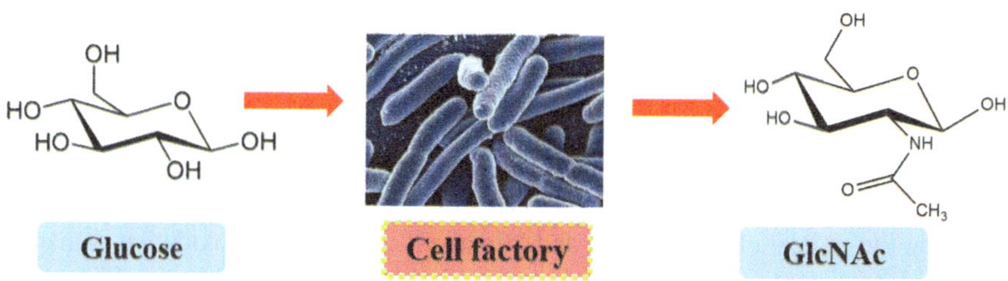

Fig. 1.8 Preparation of glucosamine/N-acetyl-D-glucosamine by fermentation

hydrolysis process, cost and production efficiency still need to be resolved. Synthetic biology is a newly research field with potential to product a certain substance in cell by optimizing intracellular inheritance and regulating metabolic processes. Thus glucosamine/N-acetyl-D-glucosamine also can be produced by microbial fermentation using recombinant microorganism such as *Escherichia coli* and *Bacillus subtilis*. Fermentation method as an environmentally friendly process is attracting increasing attention. In addition, N-acetyl-D-glucosamine deacetylase has been discovered in recent years and has potential for industrial preparation of glucosamine by enzymatic deacetylation. The enzymatic deacetylation combined fermentation will completely solve the pollution and the source of raw materials problem of glucosamine preparation (Fig. 1.8).

1.2 Biological Activities of Chito/Chitin Oligosaccharides

1.2.1 Antimicrobial Activity

COS has antimicrobial activity and is a natural antimicrobial agent that inhibits the growth of various pathogenic microorganisms. COS and its derivatives were reported to have antimicrobial activity in the 1980s (Allan and Hadwiger 1979). Since then COS has received much attention due to its antimicrobial activity. The antimicrobial activity of COS is mainly related to its Mw, DP, DD and chemical modification. The antimicrobial activity of COS is superior to that of chitosan and chitin, and the COS with higher molecular weight shows higher antimicrobial activity. There are two main mechanisms for the inhibition of COS. One is that the COS itself has positive charges and can adhere to the surface of the cell, thereby affecting the normal metabolism of the microorganism (Chung et al. 2004; Je and Kim 2006). The second is that COS with a molecular weight of less than 5000 Da block the transcription process of DNA, thereby inhibiting the growth of microorganisms (Liu et al. 2001). The antimicrobial activity of COS remains to be further studied, and its research is of great significance. The development of new antimicrobial drugs using COS has great potential.

1.2.2 Antioxidant Activity

COS and its derivatives have strong reducing ability and can effectively scavenge hydroxyl radicals and superoxide anions. Because COS has positive charges, it provides positrons to free radical, which makes the free radical in a stable state, thus blocking the radical chain reaction and effectively reducing the oxidative damage. COS and its derivatives have stronger antioxidant activity than chitosan. The antioxidant activity of COS is related to its DD, DP and MW. The smaller the molecular weight of COS is, the stronger its antioxidant activity will be, and the COS with low MW (5 kDa) has highest antioxidant capacity (Mengibar et al. 2013). The low DP of COS shows good antioxidant activity, and the COS of DP between 10 and 12 has the strongest scavenging ability (Li et al. 2012). Because of its strong antioxidant activity, COS can effectively remove free radicals from the body, which has very important research and application value.

1.2.3 Anti-inflammatory Activity

The anti-inflammatory activity of COS depends on their physicochemical properties. The MW and DP of COS have great effects on their anti-inflammatory ability. COS with MW lower than 10 kDa and low DP have better anti-inflammatory activity, and their anti-inflammatory activity is directly related to the dose used. Studies have shown that Similar proportions of monoacetylated and deacetylated oligomers is necessary for the mixtures of chitooligosaccharides to achieve anti-inflammatory effects, and it directly depends on the preparation method to which chitosan was submitted (Sanchez et al. 2018). The use of COS to develop new anti-inflammatory agents has great potential in biological and medical applications.

1.2.4 Anti-tumor/Anticancer Activity

Studies in the twentieth century have shown that COS or can effectively inhibit the growth of cancer cells. Highly charged COS effectively inhibits the growth of cancer cells (Huang et al. 2006). Mw of COS plays a key role in its anti-tumor activity, and its tumor suppressive effect is negatively correlated with MW (Salah et al. 2013). The COS with low molecular weight has higher antitumor activity, and as the MW decreases, the tumor suppressing ability of COS is obviously enhanced. Since COS has significant antitumor activity and has the potential to treat cancer, the development of new anticancer agents using COS is of great significance. At present, the anticancer activity and its mechanism of COS are not fully understood. Therefore, whether COS can be used as an anticancer agent requires further research.

1.2.5 Immunostimulatory Activities

Studies have shown that COS has immunomodulatory activities and its immunological activity is related to its molecular weight. COS promotes the secretion of NO and pro-inflammatory cytokines through the PI3K-Akt and the MAPK pathway as well as NF-κB activation (Yang et al. 2017). NO is involved in the clearance of invading pathogens by macrophages, and proinflammatory cytokines are also involved in the destruction of pathogens. Low molecular weight chitosan can induce the expression and secretion of NO and proinflammatory cytokines in cells. After treatment with COS, the amount of NO and pro-inflammatory cytokines in the cells increased significantly, and the amount increased was positively correlated with the dose of COS used. The immunomodulatory activity of COS makes it be of great application value in medicine.

1.2.6 Wound Healing and Tissue Regeneration Properties

Both chitin and chitosan have functions to promote wound healing and tissue regeneration, which is related to their immunostimulatory activity (Okamoto et al. 2003). Compared with chitin, Chitinoligosaccharides and N-acetyl-D-glucosamine, chitosan with higher deacetylation degree, chitooligosaccharides and glucosamine have stronger effects on fibroblast activation and can significantly stimulate the production of fibroblasts (Minagawa et al. 2007), thereby accelerating wound healing. Studies have shown that COS-modified membranes can accelerate wound healing while improving antibacterial ability (Luo et al. 2016). The osteogenesis effect of the COS membrane is comparable to that of the gelatin membrane and is considered to be a superior material for osteogenic differentiation of stem cells (Ratanavaraporn et al. 2009). COS has great application prospects in medicine, due to its ability to promote wound healing and tissue regeneration.

1.2.7 Hypocholesterolemic Activity

Several studies have attributed some metabolic syndrome to a high level of cholesterol, researchers are also extensively examining the hypocholesterolemic activity of COS and its derivatives. COS has shown angiotensin converting enzyme (ACE) inhibiting effects because of its positive charge and its affinity towards the ACE functional site. Pyo-Jam et al. (Park et al. 2003) studied the ACE inhibitory activity of three different degrees of deacetylation COS mixtures, indicating that the relatively lowest degree of deacetylation COS, which was prepared from 50% deacetylated chitosan, exhibited the highest ACE inhibitory activity. Zhang et al. (Zhang et al. 2012) investigated the hypolipidemic activities of different MW chitosan in rats fed high-fat diets. The studies showed that hypolipidemic activity of lower MW of chitosan was better than higher MW of chitosan which might be partially attributed to the increase of serum and liver LPL activities.

1.2.8 Elicitors of Plant Immunity

COS is an essential and natural plant immunity elicitor. Many studies reported that N-acetyl COS was the potent elicitor for rice cells, it induced a set of defense reactions including depolarization of ion fluxes, membrane potential, phytoalexin synthesis at extremely low concentration et al. The effects of COS depend on their concentration or dosage, DP, application ways, application time, and growth period of plants (Liaqat and Eltem 2017). Previous research has found that the defense reactions of N-acetyl COS are dependent on their physicochemical properties. N-acetyl COS which has higher DP (DP>3) show stronger defense activities than the lower DP or the deacetylated COS (Yamada et al. 1993). Lan et al. used COS as an elicitor to enhance barley germination for improving the quality of malt (Lan et al. 2016). The study showed that malt quality was significantly improved by COS in seed priming at 1 mg/L. In addition, much previous studies have indicated that COS has great potential to provide biopesticides and biofertilizers (Zong et al. 2017; Zou et al. 2016).

1.3 Future Perspectives and Limitations

Aminooligosaccharides (chito/chitin oligosaccharides) have broad application prospects as bioactive compounds in many fields. Therefore, exploring the clear structure-activity relationship of aminooligosaccharides, preparing aminooligosaccharides products with specific DP or DD are the future development trends. Aminooligosaccharides have many possibilities for application in foods, agriculture, cosmetics and medicine. It is essential to develop novel methods for large scale production of aminooligosaccharides products with desired characteristics. In addition, the pure standards of aminooligosaccharides are quite expensive, in fact, aminooligosaccharides standards with DP higher than 6 are difficult to obtain in large-scale at present. For this reason, it is still a hard work to correctly analyses the chito/chitin oligosaccharides products mixtures with broad DP. Although aminooligosaccharides related research has made great progress, it is still a difficult task to produce purified products with good characteristics. Therefore, the large-scale production of purified aminooligosaccharides, proper characterization of the aminooligosaccharides products, the reduction in production costs and knowing the detail mechanism of action of aminooligosaccharides is valuable for further study.

The biopolymers, chitosan and chitin, still have a lot of usefulness and many application properties in the application of biological materials. However, due to their properties of solubility, adsorption and viscosity, the wide application of these glycans is limited. At the same time, chitinoligosaccharides, chitooligosaccharides and their derivatives exhibit their versatility, complete solubility and low viscosity properties, which are superior to their polymers, chitosan and chitin, in these respects; so COS have received widespread attention. Especially in biomedicine and pharmaceutical applications. This is closely related to its degree of deacetylation, molecular weight, degree of polymerization and cationic nature. Although COS have great application prospects, it is not currently possible to mass-produce chitooligosaccharides and single COS, and the study on the physiological activities of chitooligosaccharides is also limited. A novel and efficient method for the preparation of COS, the separation of chitooligosaccharides and the mechanism of action of COS should be the focus of further research.

References

Allan CR, Hadwiger LA (1979) Fungicidal effect of chitosan on fungi of varying cell wall composition[J]. Exp Mycol 3(3):285–287

Chung YC, Su YP, Chen CC et al (2004) Relationship between antibacterial activity of chitosan and surface characteristics of cell wall.[J]. Acta Pharmacol Sin 25(7):932–936

Huang R, Mendis E, Rajapakse N et al (2006) Strong electronic charge as an important factor for anticancer activity of chitooligosaccharides (COS)[J]. Life Sci 78(20):0–2408

Je JY, Kim SK (2006) Chitosan derivatives killed Bacteria by disrupting the outer and inner membrane[J]. J Agric Food Chem 54(18):6629–6633

Lan W, Wang W, Yu Z et al (2016) Enhanced germination of barley (Hordeum vulgareL.) using chitooligosaccharide as an elicitor in seed priming to improve malt quality[J]. Biotechnol Lett 38(11):1935–1940

Li K, Xing R, Liu S et al (2012) Separation of chito-oligomers with several degrees of polymerization and study of their antioxidant activity[J]. Carbohydr Polym 88(3):896–903

Liaqat F, Eltem R (2017) Chitooligosaccharides and their biological activities: a comprehensive review[J]. Carbohydr Polym 7:5113. S0144861717314789

Liu XF, Guan YL, Yang DZ et al (2001) Antibacterial action of chitosan and carboxymethylated chitosan[J]. J Appl Polym Sci 79(7):12

Luo C, Liu W, Luo B et al (2016) Antibacterial activity and cytocompatibility of chitooligosaccharide-modified polyurethane membrane via polydopamine adhesive layer[J]. Carbohydr Polym. S0144861716310906

Mengíbar M, Mateos-Aparicio I, Miralles B et al (2013) Influence of the physico-chemical characteristics of chito-oligosaccharides (COS) on antioxidant activity[J]. Carbohydr Polym 97(2):776–782

Minagawa T, Okamura Y, Shigemasa Y et al (2007) Effects of molecular weight and deacetylation degree of chitin/chitosan on wound healing[J]. Carbohydr Polym 67(4):640–644

Okamoto Y, Yano R, Miyatake K et al (2003) Effects of chitin and chitosan on blood coagulation[J]. Carbohydr Polym 53(3):337–342

Park PJ, Je JY, Kim SK (2003) Angiotensin I Converting Enzyme (ACE) inhibitory activity of hetero-Chitooligosaccharides prepared from partially different Deacetylated Chitosans[J]. J Agric Food Chem 51(17):4930–4934

Ratanavaraporn J, Kanokpanont S, Tabata Y et al (2009) Growth and osteogenic differentiation of adipose-derived and bone marrow-derived stem cells on chitosan and chitooligosaccharide films[J]. Carbohydr Polym 78(4):873–878

Salah R, Michaud P, Mati F et al (2013) Anticancer activity of chemically prepared shrimp low molecular weight chitin evaluation with the human monocyte leukaemia cell line, THP-1[J]. Int J Biol Macromol 52(none):333–339

Sanchez A, Mengibar M, Fernandez et al (2018) Influence of preparation methods of Chitooligosaccharides on their physicochemical properties and their anti-inflammatory effects in mice and in RAW264.7 macrophages[J]. Mar Drugs 16:430

Yamada A, Shibuya N, Kodama O et al (1993) Induction of phytoalexin formation in suspension-cultured rice cells by N-Acetylchitooligosaccharides[J]. Biosci Biotech Bioch 57(3):405–409

Yang Y, Xing R, Liu S et al (2017) Immunostimulatory effects of Chitooligosaccharides on RAW 264.7 mouse macrophages via regulation of the MAPK and PI3 K/Akt signaling pathway[J]. Int J Biol Macromol 108

Zhang J, Zhang W, Mamadouba B et al (2012) A comparative study on hypo-lipidemic activities of high and low molecular weight chitosan in rats[J]. Int J Biol Macromol 51(4):504–508

Zong H, Li K, Liu S et al (2017) Improvement in cadmium tolerance of edible rape (*Brassica rapa*, L.) with exogenous application of chitooligosaccharide[J]. Chemosphere 181:92–100

Zou P, Li K, Liu S et al (2016) Effect of sulfated Chitooligosaccharides on wheat seedlings (Triticum aestivum L.) under salt stress[J]. J Agric Food Chem 64(14):2815–2821

Current Studies of Chito/Chitin Oligosaccharides and Its Monomer

2

Qiming Chen and Liming Zhao

Abstract

In order to get an overview of the worldwide advance of research on chito/chitin oligosaccharides and its monomers, the open published literatures on Web of Science and Wanfang data were retrieved and analysed. It is obvious that the amount of studies on chitooligosaccharides rapidly grows after it overcame problems of large-scale production, and China shares the main parts of the industrial yield and academic research of chitooligosaccharides. However, there are still many crisis problems in industrial-scale preparation and green manufacture. The predominant application fields of chitooligosaccharides is agriculture as fertilizer. Nevertheless, the literatures about chitinoligosaccharides are far more less than that of chitooligosaccharides, and among which most focused on preparation study. It shows that the preparation technolgy of chitinoligosaccharides is still the bottleneck, there might be an explosive growth of research when a breakthrough is made in the preparation technology.

Keywords

Chitooligosaccharides · Chitin oligosaccharides · Research overview · Web of Science · Wanfang data

Q. Chen · L. Zhao (✉)
State Key Laboratory of Bioreactor Engineering, School of Biotechnology, East China University of Science and Technology, Shanghai, China
e-mail: zhaoliming@ecust.edu.cn

L. Zhao (ed.), *Oligosaccharides of Chitin and Chitosan*,
https://doi.org/10.1007/978-981-13-9402-7_2

2.1 Introduction

Chitooligosaccharides (COS) and chitin oligosaccharides (ChOS, N-acetyl COS) are the derivatives of chitin, which is a long-chain polymer of N-acetyl-D-glucosamine existed in invertebrates and fungi (Tang et al. 2015). In 1957 (Horowitz et al. 1957) and 1965 (Powning and Irzykiewicz 1965), first studies about COS and ChOS were reported, respectively. Since then, there are more and more studies on them. As the only natural cationic polymer on earth, which has abundant biological activities, the research and application development on chitin and its' derivatives became to be a hot spot of international scientific research. Since China is the main production country of the chitin and its derivatives in the world, we retrieved and analyzed not only English literatures but also Chinese ones. In this chapter, we mainly analyzed the open published records concerning COS and ChOS as well as their monomers, the data retrieved from Web of Science (http://www.isiknowledge.com/) and Wanfang data (http://www.wanfangdata.com). Chitosan oligosaccharide/oligochitosan/chitooligosaccharides/chito oligosaccharides and chitin oligosaccharide/chitinoligosaccharides were selected as title keywords. Web of Science (previously known as Web of Knowledge) is an online subscription-based scientific citation indexing service originally produced by Clarivate Analytics. It provides over 90 million records in the world from 1900 to present. Wanfang data is one of the biggest databases provided journals, dissertations, academic conference proceedings, Chinese companies and their products, Chinese patents and so on. The statistical data of Wanfang data reflected the research level of China on chito/chitin oligosaccharides and its monomers.

Although the data does not cover completely, the result could represent the real situation. This data would benefit for the improvement of this field.

2.2 The Predominant Countries and Areas Research on Chito/Chitin Oligosaccharides

China, South Korea, USA, Japan and Canada provide most literatures on COS (Fig. 2.1a). It is notable that nearly 58% of records on Web of Science originated from China, result in that almost over 90% of the chitin and its derivatives of the world are manufactured in China. The background is that traditional preparation methods of chitin and chitosan are chemical extraction from shells of shrimp and craps as byproducts of aquatic product processing industry, which discharged large amount of wastewater at the early years, a lot of manufactures alone the seashore of China set up to produce the raw material from the byproducts. Obviously, China has become the most important research center of COS in the world now. As shown in Fig. 2.1b, the top five countries contribute to the research on ChOS are Netherlands, USA, Japan, China and Mexico in turn. However, there isn't a dominant country for ChOS like that for COS. Research on ChOS now is still in its initial stage. It is estimated that, the total annual yield of COS and ChOS is about 2000 tons in the world,

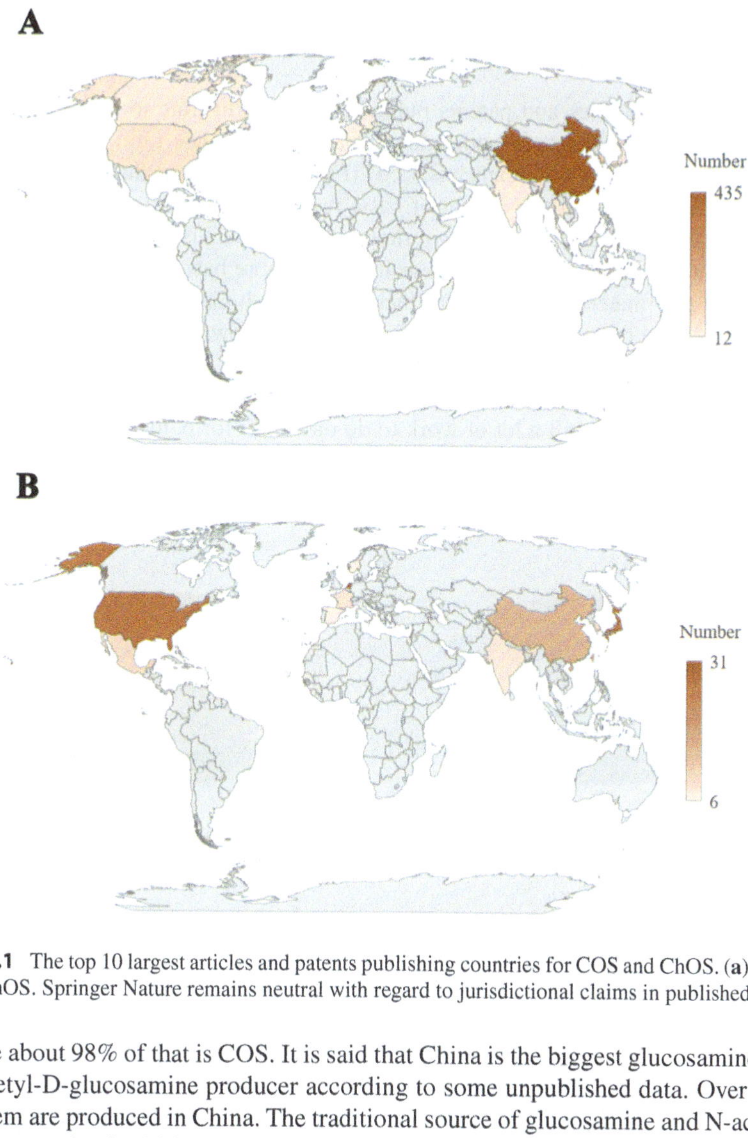

Fig. 2.1 The top 10 largest articles and patents publishing countries for COS and ChOS. (**a**) COS; (**b**) ChOS. Springer Nature remains neutral with regard to jurisdictional claims in published maps

while about 98% of that is COS. It is said that China is the biggest glucosamine and N-acetyl-D-glucosamine producer according to some unpublished data. Over 70% of them are produced in China. The traditional source of glucosamine and N-acetyl-D-glucosamine is chitin.

According to the literature records from Web of Science, Nevertheless, China, USA and Japan contribute mainly to the research on both COS and ChOS. However, as the second active region of the world economy, Chinese researchers and companies paid more and more attention on the studies of chito/chitin oligosaccharides and its monomers, this can be concluded from the data retrieved from both Wanfang data and Web of Science. The top ten hot spot countries concerning COS and ChOS studies are shown in Fig. 2.1.

2.3 The Comparation of Articles and Patents Published About COS and ChOS

The number of articles and patents published might illustrate to some extent the abundance of basic research and industrial application. The ratio of articles and patents for COS on Web of Science is 33:67 (Fig. 2.2a), it means that the application concern is almost two times of that of basic research, which might lead to mature industrialization. The corresponding ratio on Wanfang data is 60:40 (Fig. 2.2b), and the ratio indicated that China still needs more and more basic research in this field, other than raw material manufacture merely, although China has already become the predominant research center of COS in the world now.

The ratio of articles and patents number for ChOS is 46:54 on Web of Science (Fig. 2.3a) and 48:52 on Wanfang data (Fig. 2.3b). Considering the less amount of total studies, there is still a lot of work to do on ChOS for both basic research and application.

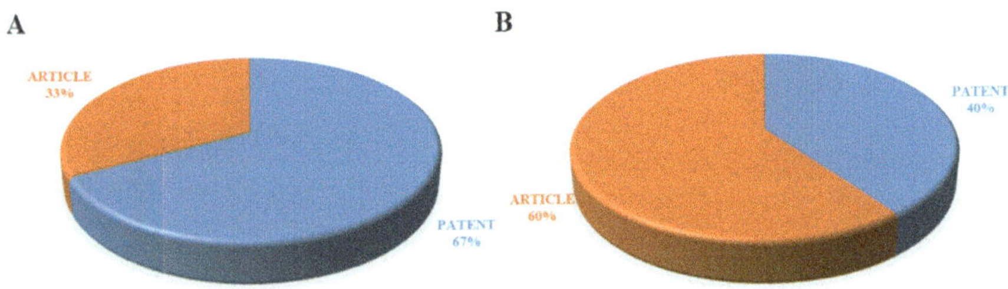

Fig. 2.2 The ratio of articles and patents number concerning COS on Web of Science and Wanfang data. (**a**) Web of Science. (**b**) Wanfang data

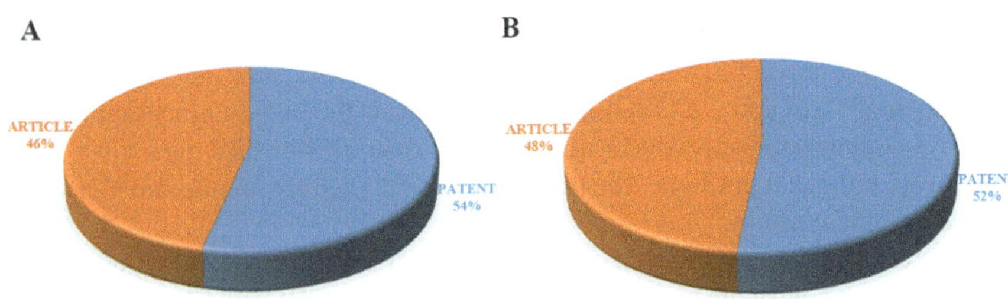

Fig. 2.3 The ratio of articles and patents number concerning ChOS on Web of Science and Wanfang data. (**a**) Web of Science. (**b**) Wanfang data

2.4 The Popularity Changes of Research on COS and ChOS Over Time

The studies had an explosive growth in 2012 for COS. After then, the amount of reports increased rapidly. The reason for the increase of studies amount in 2012 is that there were many patents published. The popularity changes trends of research on COS on Web of Science over time are showed in Fig. 2.4a. On Wanfang data, a

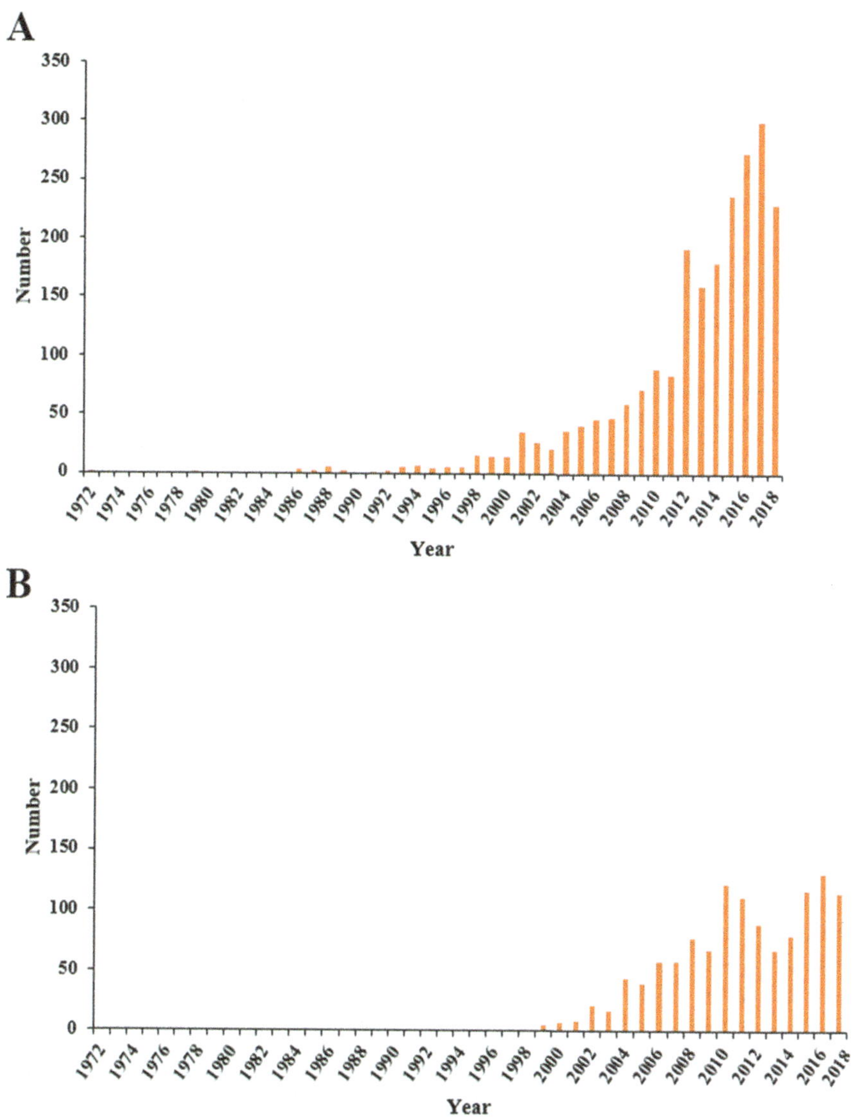

Fig. 2.4 Popularity change of research COS on Web of Science and Wanfang data over time. (**a**) Web of Science. (**b**) Wanfang data

peak appeared in 2011 (Fig. 2.4b). In this year, a company named Hainan Zhengye Zhongnong High Technology applied and published 37 pieces of patents. All of them focused on the application in plants. This phenomina also indicates that the application of COS in agriculture as plant immunity elictor, pepcide or fertilizer increased repaidly in China.

However, reports on ChOS are always tepid, and the total amount is still little on both Web of Science and Wanfang data since ChOS was firstly reported in 1965 (Fig. 2.5). As mentioned above, there is still a lot of work to do on the preparation and application development of ChOS.

2.5 The Future Research Trends on COS and ChOS

COS has various functions such as anti-bacteria (Benhabiles et al. 2012), anti-tumor (Chokradjaroen et al. 2018), blood lipid (Liu et al. 2018) and nerve repair (Wang et al. 2016). It is widely applied in agriculture, industry, medicine and even environment conservation. The main application of COS might be in plants field (usually as fertilizer) according to the data from both Web of Science and Wanfang data (Fig. 2.6). It is usually added in soils or to plant tissues such as leaves, roots and seeds, as fertilizer additives. The application of COS in fertilizer can contribute to its various functions on plants such as enhancement of virus resistance (Jia et al. 2016; Zhang et al. 2019), cold resistance (Kuang et al. 2009), antimicrobial infection (Jia et al. 2018; Ma et al. 2013), promoting germination and seedling growth (Li et al. 2019). It is said that ChOS might have more intensive effect on plants than COS (Yin et al. 2016). However, the reports of ChOS are significantly less than that of COS on both Web of Science and Wanfang data (Fig. 2.7). Preparation, modification and characterization of COS also occupy a big share on both Web of Science and Wanfang data (Fig. 2.6).

Almost most of the published records referred to the COS with mixture degree of polymerization (DP), the preparation and application of pure monomers with specific DP were seldom reported. This because there still lack of the monomer preparation methods including enzymatic hydrolysis and separation and purification technology, what's more, the research and industrial-scale application about clean or green manufacture are seriously insufficient. There are only 178 literature records about COS monomers, while most of them are almost about chitobiose (115/178), on Web of Science. This might partly due to the easy preparation of low DP COS, because the present specific chitosanase developed to hydrolysis chitosan and chitin can obtain mix DP $COS_{2\text{-}20}$ while only $ChOS_2$, while the completely enzymatic hydrolysis products of chitosan are mainly COS_2 (chitobiose). Another possible reason might contribute to the significant bioactivities of the chitobiose. Incredibly, there are nearly no reports on chitooctaose, chitononaose and chitodecaose. As to Wanfang data, there is only 17 records about COS monomers published, among that more than a third of the records are about chitohexose. Anyhow, industrial-scale preparation of monomers still unavailable, a long time might be necessary, especially COS with high DP from 6 to 10. It is important to develop

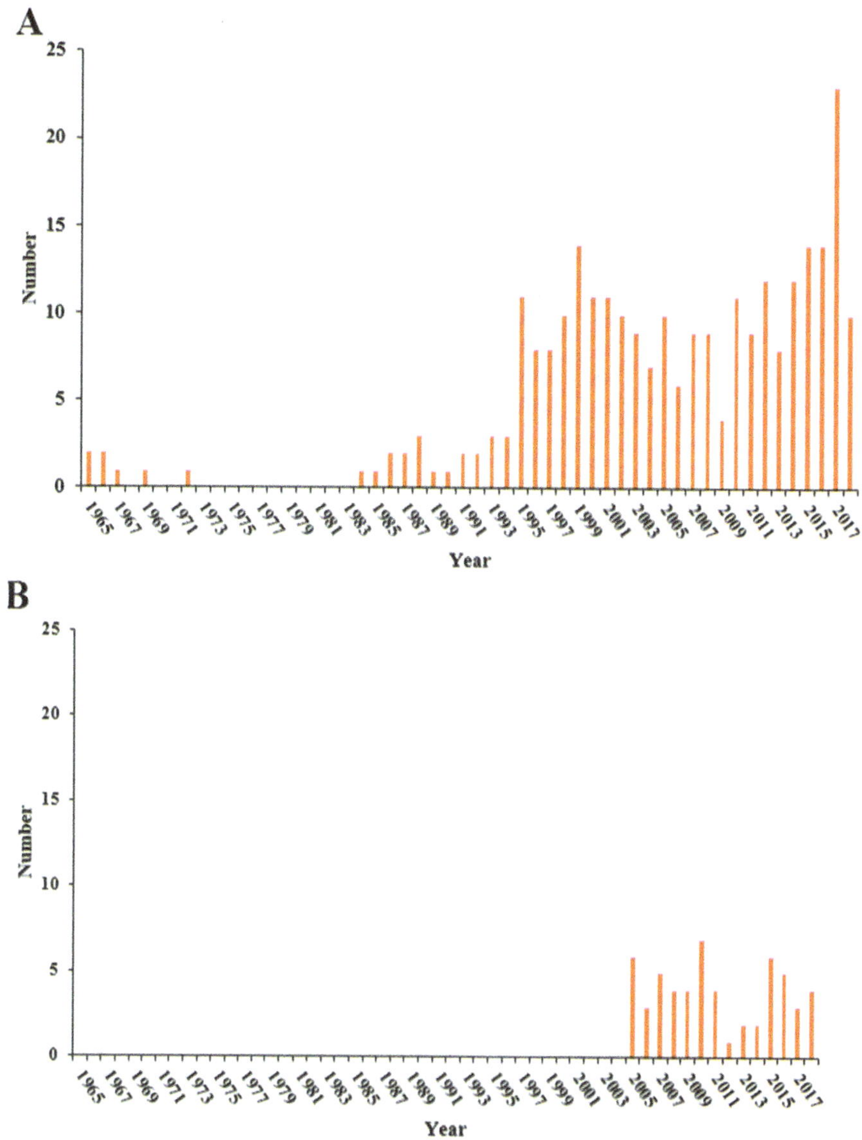

Fig. 2.5 Popularity change of research on ChOS on Web of Science and Wanfang data over time. (**a**) Web of Science. (**b**) Wanfang data

low-cost, environment friendly preparation methods for COS monomer considering research and market demands.

For ChOS, there are much fewer reports. The most records of ChOS on both Web of Science and Wanfang data are mainly concerning preparation methods (Fig. 2.7). It seems that the preparation technology is still the key problem and this greatly retart the development and application of ChOS. Fortunately, it would be efficiently

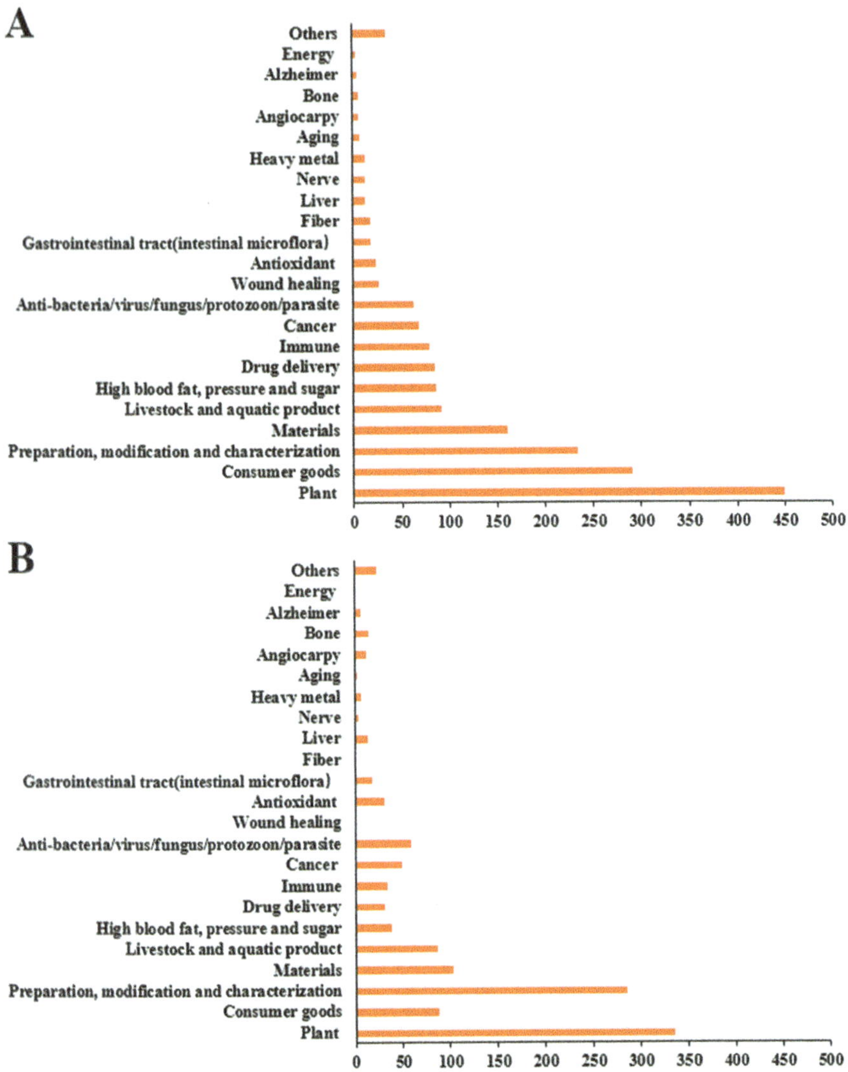

Fig. 2.6 The research trends of COS on Web of Science and Wanfang data. (**a**) Web of Science. (**b**) Wanfang data

resolved as the improvement of synthetic biology technology and bioengineering technique, "intelligent" microorganisms would be constructed to transform biomass like glucose to ChOS. According to the data, it might be that the biological activities of ChOS are higher than that of COS to some extent, which mainly result in the exist of acetylamino in ChOS while it is amino group in COS at the same site. More research should be done in future to improve the preparation technology and reveal the mechanisms of the bioactivities.

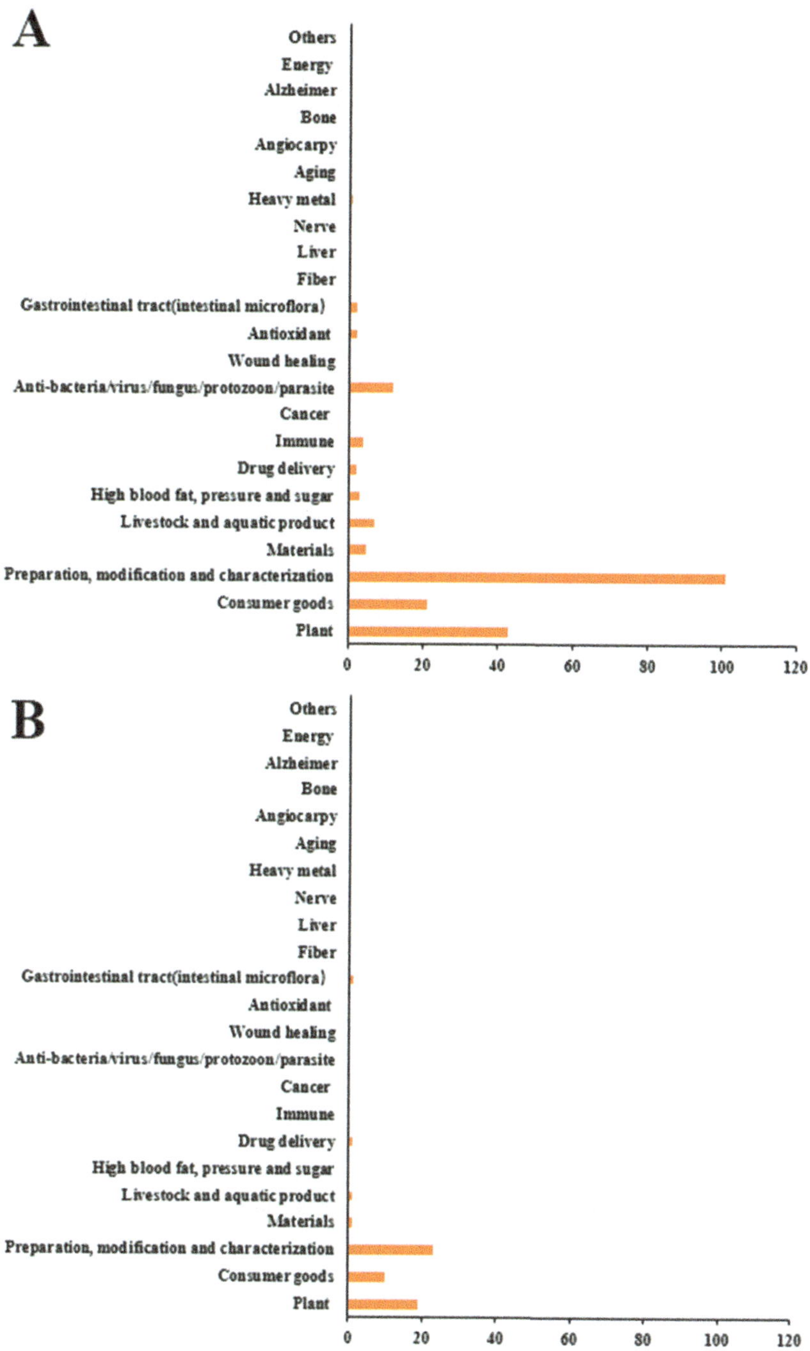

Fig. 2.7 The research trend of ChOS on Web of Science and Wanfang data. (a) Web of Science. (b) Wanfang data

Although chitin is the second most abundant biomass in the world (Ringo et al. 2012), most of them cannot be commercially developed, and the appropriate resource can be used to prepare high quality COS and ChOS is quite limit. Besides the traditional extraction from shells of shrimp and craps by chemical method, direct synthesis by synthetic biology technology and enzymatic synthesis are the optimum options in the recent future. There were some reports related to this aspects such as glycosidic synthase method (Martinez et al. 2012; Midori 2008; Slámová et al. 2015). Chemical enzymatic synthesis based on glycosyl transferases is also a possible way according to chemical enzymatic synthesis of heparin oligosaccharides. Although the yield and efficiency cannot satisfy the industrial-scale demands by now, it has great prospect and will be the future of the glycoengineering.

2.6 Conclusion

As the glycoengineering becomes a hot spot of scientific research from 1980s, the research on chito/chitin oligosaccharides continually increased, especially from this century, and the research and application development fields cover quite wide, application in plants, consumer goods, preparation, modification and characterization are the top three aspects. China is the predominant producer of chitin and its derivatives in the world, and contribute to more than 70% of the world market. Nevertheless, China, USA, Japan and South Korea contribute mainly to the research on both COS and ChOS. Because the traditional preparation methods of chitin and chitosan are chemical extraction from shells of shrimp and craps as byproducts of aquatic product processing industry, which would discharge large amount of pollutes, biomanufacturer and clean manufacture are necessary and even crucial for this industry in the future research and industrial-scale application, high quality COS or ChOS monomers are needed to developed, thus, synthetic biology, enzymatic transformation, and high performance separation and purification technologies are the keys. Then, there will be an explosive growth on function and application research and development when the bottlenecks mentioned above are broken.

References

Benhabiles MS, Salah R, Lounici H, Drouiche N, Goosen MFA, Mameri N (2012) Antibacterial activity of chitin, chitosan and its oligomers prepared from shrimp shell waste. Food Hydrocoll *29*(1):48–56. https://doi.org/10.1016/j.foodhyd.2012.02.013

Chokradjaroen C, Theeramunkong S, Yui H, Saito N, Rujiravanit R (2018) Cytotoxicity against cancer cells of chitosan oligosaccharides prepared from chitosan powder degraded by electrical discharge plasma. Carbohydr Polym. https://doi.org/10.1016/j.carbpol.2018.08.037

Horowitz ST, Roseman S, Blumenthal HJ (1957) The preparation of glucosamine oligosaccharides. I. Separation 1,2. J Am Chem Soc *79*(18):5046–5049. https://doi.org/10.1021/ja01575a059

Jia X, Meng Q, Zeng H, Wang W, Yin H (2016) Chitosan oligosaccharide induces resistance to Tobacco mosaic virus in Arabidopsis via the salicylic acid-mediated signalling pathway. Sci Rep 6:26144

Jia XC, Zeng HH, Wang WX, Zhang FY, Yin H (2018) Chitosan oligosaccharide induces resistance to pseudomonas syringae pv. tomato DC3000 in Arabidopsis thaliana by activating both salicylic acid- and jasmonic acid-mediated pathways. Mol Plant Microbe Interact 31(12):1271–1279

Kuang YJ, Peng HE, Gui-Ping YE, Qin CQ (2009) Effects of oligochitosan on cold resistance of eggplant seedlings. North Hortic 9:14–17

Li RX, He JX, Xie HG, Wang WX, Bose SK, Sun YQ, Hu JN, Yin H (2019) Effects of chitosan nanoparticles on seed germination and seedling growth of wheat (Triticum aestivum L.). Int J Biol Macromol 126:91–100

Liu SH, Chiu CY, Shi CM, Chiang MT (2018) Functional comparison of high and low molecular weight chitosan on lipid metabolism and signals in high-fat diet-fed rats. Mar Drugs. https://doi.org/10.3390/md16080251

Ma Z, Yang L, Yan H, Kennedy JF, Meng X (2013) Chitosan and oligochitosan enhance the resistance of peach fruit to brown rot. Carbohydr Polym 94(1):272–277

Martinez EA, Boer H, Koivula A, Samain E, Driguez H, Armand S, Cottaz S (2012) Engineering chitinases for the synthesis of chitin oligosaccharides: catalytic amino acid mutations convert the GH-18 family glycoside hydrolases into transglycosylases. J Mol Catal B-Enzym 74(1-2):89–96

Midori U (2008) Mutants of Mucor hiemalis endo-beta-N-acetylglucosaminidase show enhanced transglycosylation and glycosynthase-like activities. J Biol Chem 283(8):4469–4479

Powning RF, Irzykiewicz H (1965) Detection of chitin oligosaccharides on paper chromatograms. J Chromatogr A *17*:621–623

Ringo E, Zhou Z, Olsen RE, Song SK (2012) Use of chitin and krill in aquaculture – the effect on gut microbiota and the immune system: a review. Aquac Nutr 18(2):117–131

Slámová K, Krejzová J, Marhol P, Kalachova L, Kulik N, Pelantová H, Cvačka J, Křen V (2015) Synthesis of serivatized chitooligomers using transglycosidases engineered from the fungal GH20 β-N-acetylhexosaminidase. Adv Synth Catal 357(8):1941–1950

Tang WJ, Fernandez JG, Sohn JJ, Amemiya CT (2015) Chitin is endogenously produced in vertebrates. Curr Biol 25(7):897–900. https://doi.org/10.1016/j.cub.2015.01.058

Wang Y, Zhao Y, Sun C, Hu W, Zhao J, Li G, Gu X (2016) Chitosan degradation products promote nerve regeneration by stimulating Schwann cell proliferation via miR-27a/FOXO1 axis. Mol Neurobiol *53*(1):28–39. https://doi.org/10.1007/s12035-014-8968-2

Yin H, Du Y, Dong Z (2016) Chitin oligosaccharide and chitosan oligosaccharide: two similar but different plant elicitors. Front Plant Sci 7

Zhang HY, Yin H, Jin GJ (2019) Function of nitric oxide in chitosan oligosaccharide-induced resistance to tobacco mosaic virus. Int J Agric Biol 21(1):85–92

Part II

Bio-Manufacturing of Chito/Chitin Oligosaccharides and Its Monomer

Preparation of Chitooligosaccharides and Its Monomer

3

Weizhi Liu, Qianqian Lyu, and Zhen Qin

Abstract

Due to the wide applications of chitooligosaccharides and its monomer in various fields, the approaches to produce the chitooligosaccharides and its monomer have attracted much attention. The chemical-based approach generated a series of environmental pollution problems. In contrast, the enzymatic-based approach is able to make up this deficiency. However, the cost of the enzymes is still relative high, and the lack of fundamental understanding of the enzyme catalytic mechanisms hampers the preparation of chitooligosaccharides and its monomer by enzymatic-based approach. This review not only covers the typical chemical-based approach, also pays more attention to the description of the enzymatic-based approach used for chitooligosaccharides and its monomer generation. Both the nonspecific enzymes and specific enzymes are able to be utilized. Recently, more and more investigations are carried out to reveal the underlying catalytic mechanisms for the chitosanase, especially for the chitosanase GH46. These findings significantly facilitate the applications of chitosanase and protein-engineering to improve the chitosanase degradation efficiency. In addition, the metabolic engineering of model microorganism for microbial production of GlcNAc/GlcN is summarized. This review provides the viewpoints for further bio-manufacturing to achieve the industrial production of chitooligosaccharides and its monomer.

W. Liu (✉) · Q. Lyu
MOE Key Laboratory of Marine Genetics and Breeding, College of Marine Life Sciences, Ocean University of China, Qingdao, China
e-mail: liuweizhi@ouc.edu.cn

Z. Qin
State Key Laboratory of Bioreactor Engineering, School of Biotechnology, East China University of Science and Technology, Shanghai, China

L. Zhao (ed.), *Oligosaccharides of Chitin and Chitosan*,
https://doi.org/10.1007/978-981-13-9402-7_3

Keywords
Chitooligosaccharides · Chitosanase · Preparation · Glucosamine · Bio-manufacturing

3.1 Introduction

It is well known that the chitooligosaccharides and its monomer have various biological activities. And further studies demonstrated that the biological activities correlated well with the chemical structures of chitooligosaccharides, including the DP (Degree of polymerization) and DA (Degree of acetylation). For example, recent studies demonstrated that only the chitooligosaccharides with the DP>5 exhibiting the antibacterial activity against *Staphylococcus aureus* (Li et al. 2016). Therefore, it is urgent to carry out more and more detailed investigations for the preparations of chitooligosaccharides, especially for the enzymatic-based green method. In this context, the chitosanases are becoming the bottle neck for well-defined chitooligosaccharides preparation. Our group determined the first chitosanase-substrate complex structure and thereby elucidated the "three-step" mechanism for the chitosan polymer degradation (Lyu et al. 2014, 2015). Through these fundamental studies on the chitosanase, our aim is to facilitate the applications of chitosanase in the well-defined chitooligosaccharides preparation.

In this chapter, we reviewed the conventional methods to generate the chitooligosaccharides and its monomer and compared the advantage and disadvantage of them. Furthermore, we focused on the review of the most recent fundamental studies of the chitosanases, including the resources, biochemical characterization, structure and function study.

3.2 Chemical Hydrolysis Methods to Generate the Oligomer

Generally, a series of chemical agents were applied for the chitosan depolymerization, including the hydrochloric acid, nitrous acid, fluorolysis in anhydrous hydrogen fluoride, and the hydrogen peroxide. For example, the concentrated HCl were used to depolymerize the chitoan to produce the low DP GlcN oligomer (Trombotto et al. 2008). In earlier days, the concentrated HCl was used to produce the chitosan oligomer. This method is relative easy to scale-up, however, the defects include the low yield of chitosan oligomer with high amounts of glucosamine HCl. Obviously, this is not an environmental friendly method as the existence of unwanted acid waste. In addition, the oxidative depolymerization method was conducted using either the $NaNO_2$ (Mao et al. 2004) or the hydrogen peroxide (Tian et al. 2003). The combination of hydrogen peroxide with $FeCl_3$ were used to randomly depolymerize the chitosan (Nordtveit et al. 1994). The specific molecular weight chitosan oligosaccharides were obtained by the combination of hydrogen peroxide depolymerization with selective ethanol precipitation (Tian et al. 2003). Recently, the Galina

(Tishchenko et al. 2011) used the hydrogen peroxide to depolymerize the chitosan followed by the ammonium hydroxide precipitation with centrifugation. However, it was found that the chitosan oligomer was easy to aggregate and it was a time-consuming method.

Basically, the chemical degradation method involves hydrogen peroxide oxidation, acid hydrolysis and is easy to handle. But the relative molecular weight of the degradation products is broadly distributed, and therefore it is difficult to separate and purify them. Besides, the amount of used reagents is considerable, and the post-treatment is highly complicated. O_3, $NaNO_2$ and H_2O_2 are often used to obtain chitooligosaccharide as oxidants through oxidative degradation. Wu et al. prepared chitooligosaccharides derived from cicada slough of *Cryptotympana atrata* Fabricius by hydrolysis using hydrogen peroxide (H_2O_2) (Wu et al. 2013). The optimum hydrolysis conditions were as follows: time, 4 h; temperature, 65 °C; amount of H_2O_2, 2% (v/v); and pH 5. Under these conditions, the average degree of polymerisation decreased to similar to 4.5. Ronge Xing et al. generated three types of chitooligosaccharides using traditional (ZCOS), microwave irradiation (WCOS), and enzymatic hydrolysis (YCOS), and their monomer compositions and in vitro and in vivo immunomodulatory activities were measured (Xing et al. 2017). The chitosan powder (3 g) was applied in an Erlenmeyer flask in the presence of 2% acetic acid (100 mL, that is 2 mL glacial acetic acid plus 98 mL distilled water). Then H_2O_2, with the final concentration of 3%, was added to the chitosan aqueous solution. ZCOS was comprised of monosaccharides to pentasaccharides, with mainly disaccharides and trisaccharides with or without acetylation. Hongkui Zhang et al. developed a glucose-based solid acid catalyst (GSA) and was synthesized by hydrothermal carbonization (Zhang et al. 2018). Spherical, glucose-derived solid acid (GSA) catalyst with the -SO3H, -OH and -COOH groups, showed to be an efficient catalyst system for the hydrolysis of chitosan to generate D-glucosamine. The experimental results demonstrated that the yield of targeted product D-glucosamine could reach as high as 98.1% under optimal conditions (temperature: 110 °C; time: 6 h). After six catalytic cycles, no obvious deactivation was observed, indicating the relative high stability of the solid acid catalyst. This might open a relatively green route for chitosan hydrolysis based on the heterogeneous catalysts, instead of conventional liquid acid catalysts.

3.3 Physical Methods Used to Enhance the Depolymerization Efficiency

It was reported that several typical physical methods were applied to depolymerize the chitosan, including the microwave, ultrasonication, ultraviolet light and γ-irradiation (Li et al. 2016). Previous studies demonstrated that the aid of microwave significantly enhances the depolymerize efficiency of chitosan by the hydrogen peroxide (Sun et al. 2007). At the same reaction conditions, the usage of microwave reduced the depolymerization time from 4 h to less than 1 h as shown in Fig. 3.1. It was assumed that the microwave was able to provide the heat and destroy

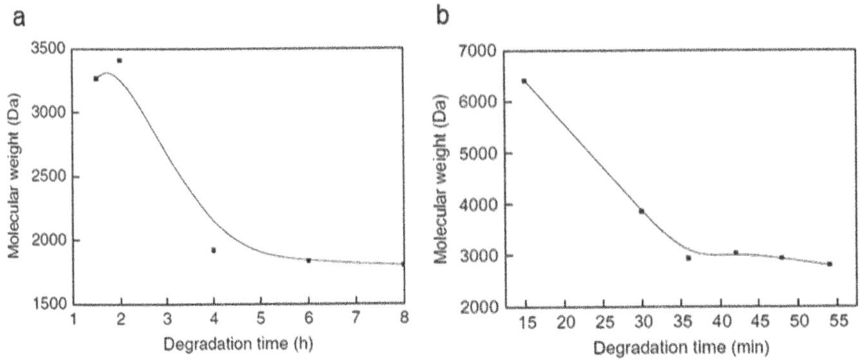

Fig. 3.1 Effect of degradation time on degradation of chitosan (**a**) or on degradation of chitosan under microwave radiation (**b**)

the chitosan polymer chain structure, therefore to be able to reduce the depolymeriztaion time.

Regarding to the ultrasonication, it was reported that two potentially mechanisms were evidenced for its enhancement of depolymerization efficiency. The first is a rapid and specific scission of chitosan chain and a lowering of their polydispersity, and the second is the generation of short polymer chains and oligomer with high polydispersity (Popa-Nita et al. 2009). During the depolymerization proceed, it was found that pH, ionic strength had no significant influence on the depolymerization. In contrast, the reactor geometry and acoustic parameters were determined to be highly relevant to the depolymerization. Also Xing et al. performed the salt-assisted acid depolymerization in the microwave irradiation field and it turned out that the depolymerization process was significantly accelerated by the addition of inorganic salts (Xing et al. 2005). In sum, considering the fact that chitosan exists in a very complicated structure, most detailed information of which remains unknown, therefore, extensive amounts of efforts need to put to explore the in-depth underlying mechanism for how various physical methods were able to enhance the chitosan depolymerization.

The purity of COS prepared by physical degradation is much easier to be controlled. Compared with conventional treatment methods used for COS preparation, microwave heating (MH) method can greatly shorten reaction time, enhance product yields and reduce the use of oxidative reagents, which attracted more interest in recent years. Also, physical degradation has been combined with chemical degradation method or enzymatic degradation method. To improve the enzymolysis efficiency of the immobilized cellulase, PAMAM was introduced by grafting it onto the surface of silica (Su et al. 2013). Because the introduction of PAMAM helps to immobilize more cellulases, the enzymolysis efficiency was markedly improved by increasing PAMAM generations. Both the immobilized cellulase-G3 PAMAM-grafted silica and microwave radiation were used for chitosan degradation. A high efficiency of enzymatic hydrolysis was obtained for a pH 6.2, 7 mg/ml chitosan solution at 50 °C with 40 W microwave-assisted digestion for 20 min. In this degradation process, cellulases exhibited a remarkable higher efficiency than a conventional protocol. The method combines the advantages of PAMAM and microwave-assisted technology, which can be potentially applied to high-throughput enzyme analysis.

3.4 Enzymatic Methods Used to Generate the Chitooligosaccharide

With the rapid development of molecular biology, omics and structural biology, the application of the protein engineering with rational design significantly accelerates the usage of biocatalysts both in the laboratory and on the industrial scale (Bornscheuer et al. 2012). Of course, the biocatalysis based on diverse enzymes were extensively explored on the generation of chitosan oligomer. Compared to the chemical and physical methods, the enzymatic method was the environmental-friendly and the process was easily controlled. Moreover, the oligomers could be generated without any further modifications. The following section will cover the detailed description of the enzymes involved in chitosan depolymerization.

Key enzymes can convert chitin into bioactive compounds such as chitooligosaccharides and N-acetyl-d-glucosamine (GlcNAc), which are potential to be applied to the medicinal and biotechnological fields. Theses enzymes involving natural production of chitooligosaccharides can be classified into two types: enzymes exhibiting specific enzymatic hydrolysis (chitinase, chitosanase, glucanase, etc.), and enzymes exhibiting non-specific enzymatic hydrolysis (lysozyme, protease, lipase, amylase, cellulose, etc.). Compared with acid hydrolysis, enzymatic degradation is apparently easier to operate and monitor. More importantly, enzymatic degradation can generate products without additional modifications. However, the cost, availability and specificity of enzymes have limited their application and commercial use. To overcome these limitations, enzymes involved in chitosan hydrolysis have attracted much attention in recently years and chitosan hydrolysis by food grade enzymes has been evaluated.

3.4.1 Nonspecific Enzymes Used for the Generation of the Chitooligosaccharides

Non-specific activity of various enzymes on chitosan has been widely reported. Several types of non-specific proteolytic and polysaccharide hydrolyase were applied for chitosan oligomer generation. They include the papain, pepsin and pronase, different types of cellulose, lysozyme. For example, the mixture of pectinase from *Phizopus oryzae*, papain from papaya latex was added into the chitosan sample. The depolymerization reaction occurs at 39 °C for 24 h. It turned out that the chitosan oligomer was generated with 33% the content of oligomer (dimers-octamers) and 54% for dimers-tetramers (Tishchenko et al. 2011). Gregor explored the chitosan depolymerization using the Cellobiohydrolases (EC 3.2.1.91). The oligomer generated by this approach had a polymerization degree between three and six unites. In addition, it was observed that degree of acetylation (DA) varied depending on the individual enzymes used for the degradation. They found that the HjCBH-mediated chitosan degradation yielded oligomer with high DA, in contrast, the TrlCBH generated oligomer mostly without acetyl group. These differences indicate that each types of enzymes have the potential to generate different pools of oligomer with distinct properties. It is well known that the bioactivity is correlated well with the oligomer properties including the DP and DA, therefore this study provides a promising approach to generated chitosan oligomer with well-defined properties (Tegl et al. 2016). More recently, degradation process of chitosan by the pepsin was extensively optimized, and it was found that the optimized conditions were a 10.0 g/L chitosan concentration of pH 4.0 at 50 °C, a 110 mg/L pepsin concentration and depolymerization time of 70 min, and the overall depolymerization process of chitosan follows the pseudo-second order and Haldane models (Gohi et al. 2016). Vishu Kumar previously characterized the chitosan depolymerization by Papain (from papaya latex; EC 3.4.22.2) and Pronase (from *Streptomyces griseus*; EC 3.4.24.31). It was found that optimum chitosan depolymerization occurred at pH 3.5 and 37 °C, resulting in LMMC (low molecular mass chitosan) and chitooligosacchарideic-monomeric mixture. The yield of the latter was 14–16% and 14–19% respectively for papain- and Pronase-catalysed reactions. In Pronase catalysed chitosanolysis, the presence of chitooligosaccharides along with only GlcNAc (N-acetylglucosamine) indicated the action pattern of Pronase-catalysed reaction is different from that of papain (Kumar et al. 2005).

More recently, the degradation of chitosan by papain and its kinetics were characterized by A-Dan Pan. The optimal conditions were chitosan substrate concentration 7.98 g/L, pH 4.55, 44.42 °C and the ratio of enzyme to substrate was 0.1037. In addition, the kinetic data and the chitosan depolymerization behavior towards different initial chitosan concentrations in the presence of papain were investigated. Because of lower residual standard deviation, a pseudo-second-order model was fitted (Pan et al. 2016).

More interestingly, almost all of the cellulases produced by different kinds of microorganisms could be classified into biofunctional enzymes, because their enzymatic activities on both cellulose and chitosan. Sequence alignment indicated these enzymes mainly belong to glycoside hydrolase family 8 (GH-8), few belong to GH-5 and GH-7. Some bifunctional cellulase-chitosanases used same domain for binding with cellulose or chitosan, while some used different substrate binding domains (Xia et al. 2008).

3.4.2 Specific Enzymes Used for the Chitooligosaccharides Generation

Based on the cleavage mode, the chitosanase could be divided into two categories: exo- and endo-chitosanase. So far, most of characterized chitosanases belongs to endo-chitosoanase. The exo-chitosanase (exo-β-D-glucosaminidase) could degrade the chitosan into D-glucosamine monosaccharides. For example, it was found that an exo-chitosanase (Csx A) derived from *A. orientalis* belonging to family 2 glycoside hydrolase is able to work on the non-reducing end of chitosan oligosaccharides, which were generated by the Csn. Further detailed structural analysis demonstrates that the Csx A possesses unique active site architecture compared to other GH2 enzymes (PDB: 2VZS), which confers Csx A the capability to degrade the cationic substrate (van Bueren et al. 2009). More recently, the archaeal exo-D-glucosaminidase (GlmA) that hydrolyzes chitosan oligosaccharides into monomer glucosamine was discovered and characterized. GlmA, a novel enzyme, belongs to the glycosidase hydrolase (GH)-A superfamily-subfamily 35 based on its amino acid sequence. Structural comparison reveals that a monomeric form of GlmA (PDB: 5GSL) is a structural homolog to GH42-galactosidases. Also, the conformational change of the active site of a dimeric GlmA was demonstrated to enhance its hydrolysis ability towards smaller substrate, which is similar to that of homotrimeric GH42-galactosidase. And one charged residue of GlmA was responsible for differentiating glucosamine from galactose (Mine et al. 2017). In addition, an exo-D-glucosaminidase was discovered and characterized, which derived from the anentomopathogenic fungus, *Paecilomyces lilacinus*. Detailed biochemical analysis demonstrated that this exo-D-glucosaminidase is a retaining enzyme (Chao et al. 2013).

For the generation of chitooligosaccharides, the endo-chitosanases are more frequently used, therefore, it is the focus of this review (In the following section, all of the chitosanases refer to endo-chitosanase). Since 1973, the chitosanase was discovered by the Monaghan (Monaghan et al. 1973). He found that this enzyme is exclusively degrade chitosan, not chitin, indicating this is a specialized chitosanase. In 2004, the chitosanase was classified into EC3.2.1.132 by the International Commission of Enzymes. According to the amino acid sequence alignments, so far, the discovered chitosanases were found in the six glycoside hydrolase families 5, 7, 8, 46, 75 and 80. In the nature, the chitosanase could be discovered from bacteria, fungi and plants. And most of the well-characterized chitosanase were from the bacteria, such as *Bacillus* sp., *Serratia,* sp., *Janthinobacterium* sp., *Acinetobacter* sp., *Streptomyces* sp., *Microbacterium* sp. (Lyu et al. 2014). Most of the known chitosanases are from GH families 46, 75 and 80, which currently only contain chitosanases (Viens et al. 2015; Qin et al. 2018a). Below is the table (Table 3.1) that summarizes most of the characterized chitosanases based on the Cazy severs (http://www.cazy.org/). However, as the endo-type catalytic mode, endo-chitosanases often randomly cleave β-1,4-glycosidic bond inside of chitosan substrate, and release a mixture of uncontrollable DP chitooligosaccharides (Qin et al. 2018a). The biological activities of chitooligosaccharides are often studied using different random chitooligosaccharides mixtures (Li et al. 2016). It is essential to produce a well-defined chitooligosaccharide monomers or mixtures to study the structure-function relationship of chitooligosaccharides. Therefore, obtaining novel specific chitosanase and investigating controllable bioconversion process are highly beneficial for the production of specific DP chitooligosaccharides mixture or chitooligosaccharide monomers in industrial scale (Qin et al. 2018b).

Table 3.1 Summary of the characterized chitosanases (January 2019)

GH family	Number of the characterized chitosanases	Resource	GenBank
GH5	1	*Streptomyces griseus* HUT6037	BAC65342.1
GH7	1	*Aspergillus aculeatus* F-50	BAA25183.1
GH8	23	*Bacillus cereus* ATCC 14579	AAP09638.1
		Bacillus cereus H-1	AAO49750.1
		Bacillus cereus KNUC51	AAQ19678.1
		Bacillus cereus KNUC55	AAQ75086.1
		Bacillus circulans WL-12	CAA37062.1
		Bacillus sp. K17	PDB ID:1V5D
		Bacillus sp. KCTC 0377BP	AAK07481.1
		Bacillus sp. No.7-M	BAB19277.1
		Bacillus thuringiensis JAM-GG01	BAJ05248.1
		Bacillus thuringiensis serovar alesti	ABO61887.1
		Bacillus thuringiensis serovar canadensis	ABO61894.1
		Bacillus thuringiensis serovar darmstadiensis	ABO61891.1
		Bacillus thuringiensis serovar israelensis	ABO61892.1
		Bacillus thuringiensis serovar morrisoni	ABO61890.1
		Bacillus thuringiensis serovar san diego	ABO61888.1
		Bacillus thuringiensis serovar sotto	ABO61893.1
		Bacillus thuringiensis serovar thompsoni	ABO61889.1
		Bacillus thuringiensis serovar tochigiensis	ABO61895.1
		Lysobacter sp. IB-9374	BAE86850.1
		Paenibacillus cookii SS-24	BAL46897.1
		Paenibacillu fukuinensis D2/IK-5	BAB64835.1
		Paenibacillus sp. 1794	AFP58892.1
		Paenibacillus sp. X4	AGV55649.1
GH46	26	*Amycolatopsis* sp. CsO-2	BAA94840.1
		Bacillus amyloliquefaciens UTK	BAB19276.1
		Bacillus circulans MH-K1	BAA01474.2
		Bacillus coagulans CK108	AAL40906.1
		Bacillus sp. DAU101	ABC66094.1
		Bacillus sp. KFB-CO4	AAF24188.1
		Bacillus subtilis CH2	ACX55932.1

(continued)

Table 3.1 (continued)

GH family	Number of the characterized chitosanases	Resource	GenBank
		Bacillus subtilis subsp. subtilis str. 168	CAB14630.1
		Bacillus subtilis V26	WP_003237265.1
		Burkholderia gladioli CHB101	BAA82154.1
		Janthinobacterium sp. 4239	ADB89905.1
		Kitasatospora setae KM-6054	BAJ27342.1
			BAJ29876.1
			BAJ29883.1
		Microbacterium sp. OU01	ABM91442.1
		Nocardioides sp. N106	AAA63405.1
		Paenibacillus ehimensis EAG1	BAA23489.1
		Paenibacillus sp. BH-2005	ABC17783.1
		Pseudomonas sp. A-01	BAC06189.1
		Streptomyces coelicolor A3(2)	CAB61194.1
		Streptomyces lividans TK24	ACV74390.1
		Streptomyces sp. N174	AAA19865.1
		Streptomyces sp. SirexAA-E	AEN13266.1
		Uncultured bacterium	AJA30433.1
		Paramecium bursaria Chlorella virus 1	AAC96660.1
			BAA20342.1
		Paramecium bursaria Chlorella virus CVK2	
GH75	16	*Streptomyces avermitilis* MA-4680	BAC69561.1
		Aspergillus clavatus W-2	AJG44374.1
		Aspergillus fumigatus Af293	EAL84291.1
		Aspergillus fumigatus NCPF7367	CAE54966.1
		Aspergillus fumigatus Y2K	AAO41660.1
		Aspergillus oryzae IAM2660	BAD08218.2
			BAA92250.1
		Aspergillus oryzae RIB40	BAE63840.1
		Aspergillus sp. CJ22-326	ABZ88800.1
		Beauveria bassiana	AAG17940.1
		Chaetomium globosum CBS 148.51	EAQ87617.1
		Fusarium solani 0114	ABX57824.1
		Fusarium solani PHASEOLI SUF386	BAA12799.1
		Gongronella sp. JG-2005	ABY77913.1
		Penicillium chrysogenum AS51D	ADG96019.1
		Penicillium sp. D-1	AFG33049.1

(continued)

Table 3.1 (continued)

GH family	Number of the characterized chitosanases	Resource	GenBank
GH80	4	beta *proteobacterium* KNU3	AAP58254.1
		Mitsuaria chitosanitabida 3001	BAA32084.2
		Mitsuaria sp. 141	AEA29710.1
		Sphingobacterium multivorum	BAA82599.1

3.4.2.1 Three Dimensional Structural Analyses for Chitosanases

To understand the underlying catalytic mechanism and substrate binding specificity, several crystal structure of chitosanase were determined (PDB: 1CHK, 1V5D, 5B4S), including the complex structure with substrates (PDB: 4OLT, 4QWP). These determined structures provide valuable fundamental information for the applications chitosanases in the chitooligosaccharide generation.

Native Chitosanase Crystal Structure Determination

The X-ray crystal structure of chitosanase N174 is the first determined chitosanase structure (PDB: 1CHK) (Marcotte et al. 1996). Chitosanase N174 is one member of GH46 family, containing two globular domains connected by a bent backbone helix. The domains form a pronounced cleft, which involves the two key catalytic residues. The structure of chitosanase N174 is similar to that of bacteriophage T4 lysozyme in spite of essentially no sequence homology (Fig. 3.2a). So far, the crystal structures from three different GH families were determined (GH8, GH46, GH80) (Fig. 3.2b–d), The accumulating chitosanase structures indicate that the chitosanases belonging to GH46 and GH80 exhibit topological similarity with chitosanase N174, which suggests that these proteins originate from a lysozyme-type scaffold. However, chitosanase ChoK is constructed based on the double-α6/α6-barrel structure, which is significant different with other reported chitosanase structures (Adachi et al. 2004). This difference suggests that chitosanases with different tertiary structures may evolved from different structural ancestors. Also, the multiple sequence alignment derived from the well-characterized chitosanases demonstrated that these chitosanases did not share high sequence similarity (Fig. 3.3). For example, the chitosanase N174 from *Streptomyces* N174 (PDB: 1CHK), which belongs to GH46, only shared less than 15% sequence identity compared to the chitosanase McChoA from *Mitsuaria chitosanitabida* with PDB accession number: 5B4S.

Catalytic Mechanism

Two types of glycoside hydrolysis mechanisms have been described previously: 'retaining' and 'inverting' mechanisms (Wang et al. 1994). Based on the biochemical analysis and structural characterization, most of the well-characterized chitosanase from GH46 follows the "inverting" mechanism, including the N174 and OU01 (Lyu et al. 2014). Two significant features that distinguishes these two mechanisms are depicted based our determined first chitosanase OU01 complex

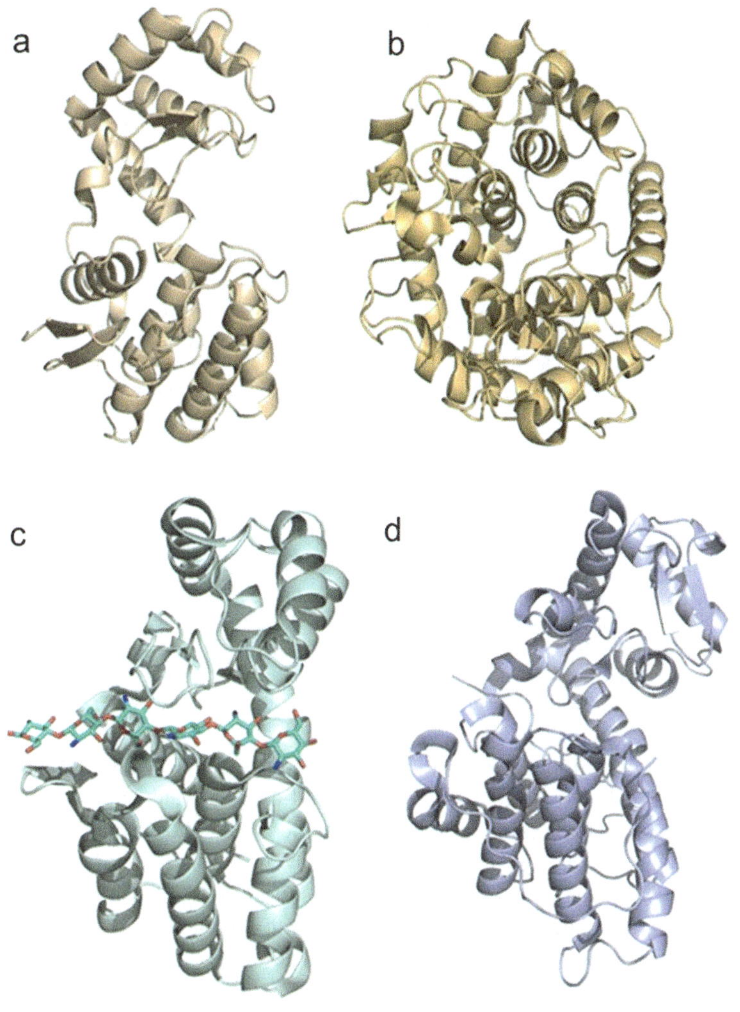

Fig. 3.2 The overall structures for several determined chitosanases. (**a**) Chitosanase N174 from *Streptomyces* N174 (PDB: 1CHK); (**b**) chitosanase ChoK from *Bacillus* sp. K17 (PDB: 1V5D); (**c**) chitosanase OU01 from *Microbacterium* sp. OU01 (PDB: 4OLT) in complex with its substrate ($(GlcN)_6$); (**d**) chitosanase McChoA from *Mitsuaria chitosanitabida* (PDB: 5B4S)

structure: (1). the distance between two catalytic residues. For the endo-chitosanases that follow the "inverting" mechanism, the distance between the two key is much longer than that in the enzymes that follow the "retaining "mechanism. For example, the distance between the two catalytic residues is 13.8 Å for apo chitosanase N174). And in our chitosanase OU01 complex structure, the distance between the backbone carbonyl of Asp43 and E25A is 14.3 Å. In contrast, the distance between the two key residues is ~5 Å for the retaining mechanism; (2). A key water molecule is present but functions differently between these two catalytic mechanisms. To

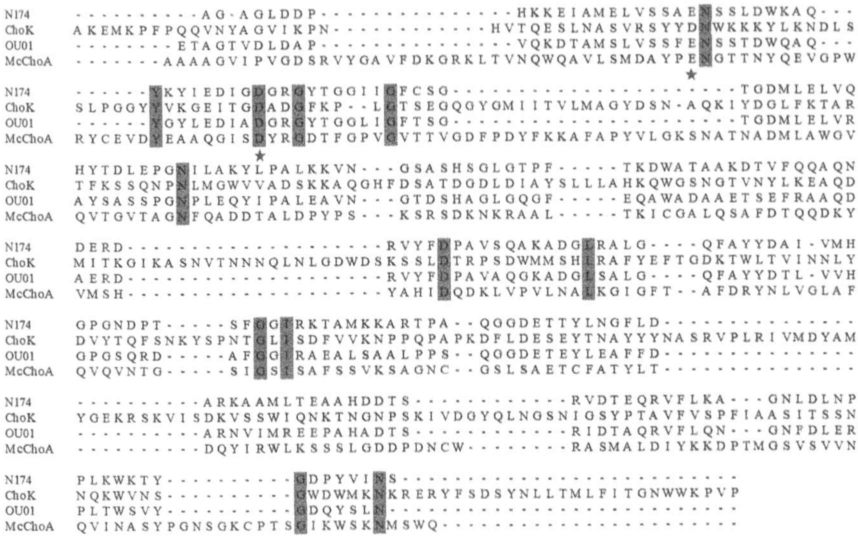

Fig. 3.3 Amino acid sequence alignment of four well-characterized chitosanases. The crystal structures of these four chitosanases from different GH families were determined. They include GH46 chitosanase N174 from *Streptomyces* N174 (PDB: 1CHK) and chitosanase OU01 from *Microbacterium* sp. (PDB: 4OLT), GH8 chitosanase ChoK from *Bacillus* sp. K17 (PDB: 1V5D); GH80 chitosanase McChoA from *Mitsuaria chitosanitabida* (PDB: 5B4S). Identical amino acids are shaded in gray. Two catalytic residues are marked with closed star

retain the anomeric atom configuration, the reaction occurs in a double displacement process for the "retaining" mechanism, and the key water molecule acts as a general base to attack the intermediate to complete the reaction from the same side. However, in the "inverting" mechanism, the key water molecule is activated by Asp43 and is supposed to attack the C^{-1} atom in the −1 position sugar from the opposite side. Elucidation of the detailed catalytic mechanism facilitates our in-depth understanding of the chitosan depolymerization mechanisms.

3.4.2.2 Summary of the Extensively Studied Chitosanases

To explore the substrate binding mechanism for chitosanase, we solved the first structure of chitosanase OU01 in complex with its substrate $(GlcN)_6$ (Lyu et al. 2014). The $(GlcN)_6$ is located at the cleft formed between the upper and the lower lobes and formed interactions with chitosanase OU01. The six sugar units of $(GlcN)_6$ are numbered −3 to +3 from the non-reducing end to the reducing end and the cleavage reaction occurs between −1 and +1 positions. The detailed interactions formed between $(GlcN)_6$ and chitosanase OU01 was analyzed. In combination with mutagenesis analysis, we determined the essential residues for substrate binding. Although the complex structure provided a clear oligosaccharide binding profile, the mechanism for the polymeric substrate binding is limited. Because chitosan is a cationic polysaccharide, the negative-charged residues around the substrate binding cleft are suggested to function in polymeric substrate binding. Mutagenesis analysis showed that the substitution of Asp40 can abolish enzymatic activity towards polymeric substrate, whereas retain almost 70% enzymatic activity towards oligomeric substrate. Therefore, we conclude Asp40 functions as a polymeric substrate binding or recognition subsite localized out of the binding cleft (Lyu et al. 2015). Moreover, the enzymatic activity of the other important residues determined by the complex structure were analyzed using two different substrate (i.e. polymeric substrate and oligomeric substrate). Through the combination of structural and biochemical analysis, we proposed a three-step polymeric substrate recognition mechanism for chitosanase OU01. The polymeric substrate is initially recognized by the enzyme through the crucial residues around −2 and −1 subsites. After the initial interaction, residues interact with sugars in the +1, +2, +3, and −3 subsites to further stabilize the substrate-enzyme. Then, the Asp40 helps to recognize the polymeric substrate located out of the main binding cleft (Lyu et al. 2015). These above comprehensive structural and functional analyses significantly facilitate the applications of chitosanase in generation of well-defined chito-oligosaccharides.

The Cleavage Mode Analysis for Chitosanase

Generally, enzymes that degrade carbohydrate polymers act in two different modes: "processive" and "non-processive" modes. For the enzymes acting in a processive mode, they remain associated with the substrate until every cleavable linkage in the chain has been hydrolyzed or until a given average number of attacks has been performed. However, the enzyme-substrate complex dissociates after each reaction in the reactions involving the enzymes acting in a non-processive mode (Lyu et al. 2015).

Previous studies demonstrated some chitinases act in a processive mode. Also, two important structural features possessed by these chitinases are summarized: (1). the presence of a well-defined tunnel-like substrate binding cleft; (2). and the presence of non-polar aromatic residues in the chitin binding cleft that allows non-specific binding of substrate (Sorbotten et al. 2005). Obviously, this two structural features have not been observed in the solved structures of chitosanase. And products analysis as a function of the time of degradation revealed a continuum of oligomers generated by chitosanses from GH46 or GH75, which demonstrated that these chitosanses act in a non-processive mode. In addition, the hydrolysis process contains an initial rapid phase and followed by a second slower phase, namely chitosanases exhibit more efficient hydrolysis towards the polymeric substrate than the oligomeric substrate. These results are consistent with the proposed polymeric substrate-binding mechanism, where more specific interactions form between chitosanase OU01 and polymeric substrate, compared to the oligomeric substrate. Further our determined complex structures provide the structural basis of non-processivity (PDB: 4OLT). In the apo structure, the substrate-binding cleft located between two lobes was observed in an open form. The substrate binding causes conformational changes and then results in closing the substrate binding cleft. Because all the residues involved in substrate recognition and catalysis are located in the interface between the two lobes, the closing of cleft allows the formation of proper orientation between substrate and chitosanase and then the occurrence of catalysis. After cleavage, the chitosanase turns to an open-form along with product disassociation. Therefore, the chitosanase must transition from an open-closed-open state during catalytic cycle, which is the structural basis for enzyme-substrate complex dissociates after each reaction (Lyu et al. 2015).

Cleavage Specificity Analysis

In degradation of partially acetylated chitosan, chitosanases exhibit different cleavage site specificities. According to the earliest and most popular classification, chitosanases can be divided into three subclasses to illustrate their cleavage sites specificities. All of them are able to cleave the GlcN-GlcN linker. Class I members such as chitosanase N174 is able to cleave the GlcNAc-GlcN linker as well. In contrast, class III chitosanase MH-K1 only cleaves the GlcN-GlcNAc linker. However, class II members, including *Bacillus* sp. K17 chitosanase, are only able to cleave the GlcN-GlcN linker (Saito et al. 1999). Obviously, this classification is based on specificity for sugars in the −1 and +1 sites and the subclass of one chitosanase can be determined by analyzing its final product. However, our structural and biochemical data suggest this classification is oversimplified. Product analysis indicated chitosanase OU01 is able to hydrolyze the polymeric chitosan into the final disaccharides (including DD and DA, D represents GlcN, A represents GlcNAc) and trisaccharides (including DDA and DAA). Generally, several cycles of hydrolysis are required to generate these final products. Therefore, all of the sugars will have potentially been selected by virtue of the interactions that they make in the −2/−1 sites. Our complex structure clearly showed that the interactions formed between the sugar in the −2 site and the residues can stabilize the positive charge from the C^2 amino group, which makes it hard to introduce an acetyl group in this site. Also, molecular modelling suggests that the acetyl group would make binding of GlcNAc at −2 site energetically unfavourable and GlcNAc can be readily accommodated in the −3, −1 and +1 positions with GlcN in the −2 subsite. Thus, chitosanase OU01 shows substrate specificity for GlcN in the −2 site and we suggest that chitosanase OU01 should be classified as a GlcN-GlcN/GlcNAc- GlcN/GlcNAc [(−2)-(−1)-(+1)] type chitosanase. We had reported these findings in 2014, however, we have not performed more detailed investigation to reassess the former classification. In 2017, Tobias et al. published their work on reassessment of chitosanase substrate specificities and classification. Weikert et al. assessed the substrate specificities by quantitatively sequencing the oligosaccharides. They demonstrated chtiosanase N174 is a GlcN-GlcN/GlcNAc- GlcN/GlcNAc-GlcN/GlcNAc [(−2)-(−1)-(+1)-(+2)] type chitosanase, which confirmed our conclusion on chitosanase OU01. Although our structure-based method for assessing substrate specificities had been ignored by Weikert et al., we believe the combination of structural analysis and their novel quantitative sequencing technique will promote the assessment of chitosanase substrate specificities (Weikert et al. 2017).

The Chitooligosaccharides Product Analyses Generated by the Chitosanases

The research group from Norwegian Biopolymer Laboratory (NOBIPOL) has done much work on the degradation of chitosan with widely varying the fraction of acetylated units (F_A) using chitosanases from different GH families. A GH46 chitosanase from *Streptomyces coelicolor* A3, named ScCsn46A, could extensively degrade the chitosans with varying F_A (F_A = 0.008, 0.32, 0.46, and 0.64) in a non-processive action mode, although to a decreasing degree of scission as F_A increases (Heggset et al. 2010). The most dominant product generated by ScCsn46A degrading the fully deacetylated chitosan (F_A = 0.008) is the dimer, and the product also contains considerable amounts of monomer. The degradation of chitosan (F_A = 0.32) by ScCsn46A reveals multiphasic kinetics. At the initial degradation, ScCsn46A exhibits rapid degradation rate. And the reducing ends of the products obtained in the initial phase are almost exclusively deacetylated. With the degradation reaction proceeds, a slower second kinetic phase dominates followed by a third, even slower, kinetic phase. And new acetylated reducing ends appear in this slow kinetic phase. These observations indicate ScCsn46A exhibits different of rates of degradation for different sequences in the chitosan and intermediate products (Fig. 3.4). Such difference on rates of degradation is caused by different substites specificities. Through products analysis, the ScCsn46A has been demonstrated possessing a relatively low substie specificity toward acetylated/deacetylated units but with a preference for deacetylated units. Therefore, ScCsn46A may have advantage in preparing fully deacetylated oligomers, and controlling the degree of scission may contribute to generate an oligomer with a predicted length.

This research group also studies the degradation mode of a GH75 chitosanase (ScCsn75A) using similar method (Heggset et al. 2010). Although ScCsn75A degrades chitosan with varying F_A (F_A = 0.008, 0.11, 0.31, 0.52, and 0.63) in a non-processive action mode similar to ScCsn46A, the catalytic properties of ScCsn75A differ from the properties of ScCsn46A. For example, the dominating tetramer obtained during the initial phase of degradation is DDDA for ScCsn75A but DDDD for ScCsn46A. Product analysis indicated ScCsn75A preferred binding with acetylated unit (A) in subsite −1, and deacetylated units (D) in subsite −2, while absolute preferred binding with deacetylated unites in subsite +1. These observations suggested that ScCsn75A shows higher subsite specifity than ScCsn46A. Consequently, the chitooligosaccharide mixtures generated by ScCsn75A are less complex than that generated by ScCsn46A.

These studies demonstrate that both the chain length and the chemical composition (including sequence) of chitooligosaccharides change as a function of the degree of scission, which is determined by the affinities of chitosanses for different sugar units on the substrate. Considering the fact that the bioactivities of the chitooligosaccharides are relate to chains length and composition/sequence, more chitosanses with varying subsite specificities are required to produce chitooligosaccharides with distinct sequence.

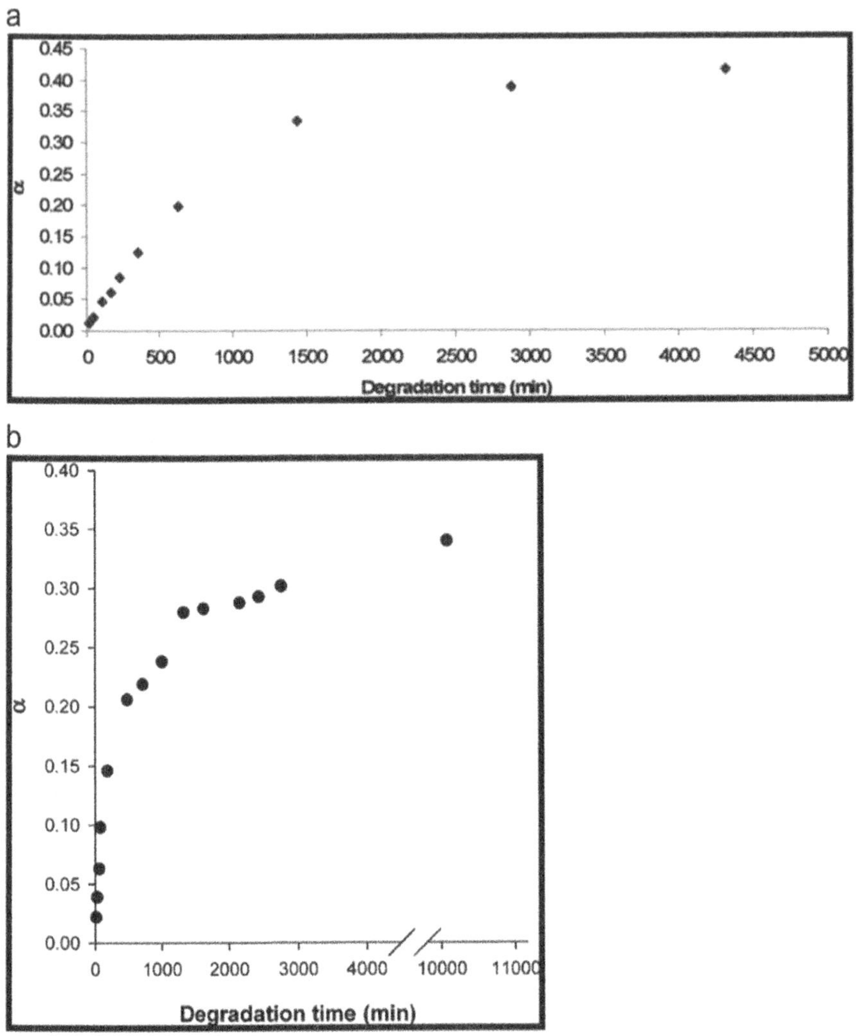

Fig. 3.4 Time course for degradation of chitosan ScCsn46A with different F_A (**a** $F_A = 0.008$; **b** $F_A = 0.32$). (Heggset et al. 2010)

3.4.2.3 Protein Engineering for Chitosanase and Applications of Chitosanase in the Chitosan Oligomer Generation

Chitooligomers derived from chitosan have a variety of bioactivities, such as antimicrobial, antioxidant, inhibition of tumor growth and stimulation of calcium deposition in bones. Moreover, the properties of the chitooligomers are correlated well with their degree of polymerization (DP) and acetylation (DA). Although many physical or chemical approaches were proposed to degrade chitosan into chitooligomers, the enzymatic-based hydrolysis of chitosan is often preferred for its advantage in environmental protection and products control. Therefore, the usage of chitosanase to generate a chitooligomers with a well-defined composition is critical for the wide application of chitosan. Obviously, a suitable chitosanase is the determinant for generation of chitooligomers with high quality. Thus, the accumulation of more chitosanases and in-depth knowledge of the mechanisms of substrate binding or specificities are required. Owing to the development of high-throughput sequencing, more sequences encoding chitosanases are available, which provides useful information for exploring the conserved/variable interaction patterns of chitosanases. For example, Liu et al. revealed the roles of key active-sites residues involved in substrate binding using structural bioinformatics and statistical analysis (Liu et al. 2015), which offers a guide for possible conversion among chitinolytic enzymes, monofunctional and multifunctional enzymes, and may also be applied to the rational design of chitosanases. More recently, a chito-oligosaccharides product GO2KA1 generated by the chitosan depolymerization using chitosanase (from *Bacillus circulans*) was found to be as the dietary supplements to help maintain the blood sugar level (Kim et al. 2014). The GO2KA1 product was composed of 45.1% dimers, 24.8% trimers, 12.8% tetramers.

The determined chitosanases structures provide structural basis for substrate binding. Therefore, protein-engineering of chitosanases is feasible basing on the increasing solved chitosanases structures. Recently, the subsite specificities of a chitosanase from *Bacillus* sp. MN, named CSN (short for BspCsnMN), are engineered (Regel et al. 2018). CSN has an absolute specificity for GlcN (D) at subsite −2 and −1, while it can accept the dyads GlcNAc-GlcN (A-D) and GlcN-GlcN (D-D) at subsite +1 and +2. Although the three-dimensional of CSN had not been solved, the structural model of CSN can be obtained by homology modelling using the crystal structure of *Bacillus* sp. K17 chitosanase as a template. The substrate cleft of CSN is highly electronegative, which is the structural basis for binding the electropositive amino group of GlcN, especially for subsite −2. To decrease the affinity to GlcN at subsite −2, the acidic amino acid E309 was substituted with the basic amino acid arginine. Meanwhile, a hydrophilic asparagine at position 308 was replaced by a hydrophobic valine to increase the affinity to the hydrophobic acetyl group of GlcNAc. Analysis of the subsite specificities of the engineered mutants demonstrated that the specificity of subsite −2 was altered as intended. The mutations increased the acceptance of GlcNAc at subsite −2, and the subsite specificity of subsite +1 was changed as well.

Early in 2008, the chitosanase from *Bacillus circulans* MH-K1 (Csn) using an endo-type action mode was converted into an exo-chitosanase by structural simulation and protein engineering as well (Yao et al. 2008). Structural comparison indicated that the structure of endo-1-4-glucanase (exo-type) is similar to that of cellobiohydrolase (endo-type), except that two obviously additional loops from opposite walls of the substrate-binding cleft was observed in endo-1-4-glucanase but not in cellobiohydrolase. This difference was considered as the determinant for exo-type action mode (Fig. 3.5). Accordingly, two peptides were designed and inserted into Csn to prepare a chimeric chitosanase (Chim-Csn). And products analysis demonstrated the Chim-Csn was converted into an exo-type chitosanase successfully. This study provides us more possibilities or methods to rational design a

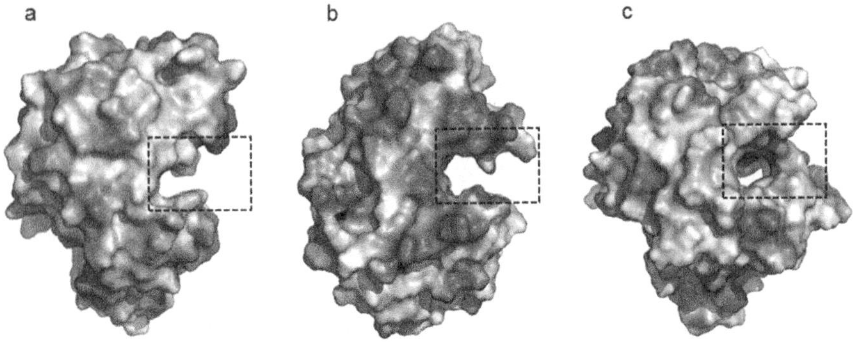

Fig. 3.5 Comparison of three-dimensional structures of endo- and exo-type glycoside hydrolases. The substrate binding clefts were boxed. (**a**) Endo-1,4-glucanase E2 of *Thermobifida fusca* YX (PDB: 2BOD); (**b**) endo-chitosanase of *B. circulans* MH-K1 (PDB: 1QGI); (**c**) cellobiohydrolase, i.e. exo-glucanase of *Humicola insolens* (PDB: 1BVW)

chitosanase. With the accumulation of more glucoside hydrolase structures, we believe increasingly approaches will be proposed to rational design chitosanases with promising functions.

3.5 Chitooligosaccharide Generation Based on the Chemoenzymatic Approach

In addition, some researchers focused on the synthesis of chitooligomer via a chemoenzymatic approach to obtain a relative higher degree of polymerization. For example, based on a chitinase mutant with lower hydrolytic activity, Yoshida used an one-pot chemoenzymatic approach through the 1,2-oxazoline derivatives of $(GlcNAc)_5$ and $(GlcNAc)_2$ to successfully achieve the $(GlcNAc)_7$ (Yoshida et al. 2012). Pallinti found that a GH18 chitinase SpChiD from *Serratia proteamaculans* 568 was an endo-acting processive enzyme, which was able to exhibit a hyper transglycosylation (TG) activity when using the trimer and hexamer as the substrates, therefore, the SpChiD has potential applications for the chitooligosaccharide generation (Purushotham and Podile 2012). This enzyme SpChiD was further investigated by Madhuprakash to explore its application for the chitooligosaccharide generation. It was found that SpChiD could be used for the defined DA (degree of acetylation) and PA (pattern of acetylation) chitooligosaccharide generation (Madhuprakash et al. 2015). Additionally, it was reported that a chitosanase from *Bacillus cereus* was used to produce the high degree polymerized chitooligosaccharide. In the aqueous condition, it was found that this chitosanase will be able produce octamer from a mixture of dimer to heptamer, indicating the transglycosylation activity. Moreover, it was discovered that in the reversed microreactors formed by AOT (sodium bis-2-(ethylhexyl) sulfosuccinate), the ratio of high degree polymerized chitooligosaccharide significantly enhanced (Hsiao et al. 2008).

3.6 Future Directions of Chitooligosaccharides Preparation

Compare to the canonical chemical methods, the enzymatic-based method is more environmental friendly. To make these enzymes are suitable for the large-scale chitosan oligosaccharides generation, more work need to be performed to enhance the yield of chitosanase in the heterologous over-expression system. More importantly, the fundamental catalytic and substrate binding mechanisms should be elucidated, especially toward the new discovered chitosanases. This will facilitate the protein engineering design of novel chitosanase for single chitosan oligomer generation.

Although enzymatic-based method is best known and widely used for large-scale production of chito/chitin oligosaccharides. The source of chitin materials is restricted as the environmental pollution of manufacturing process and limited marine chitin resources. The preparation of chito/chitin oligosaccharides by enzymatic/biological synthesis has attracted much attention in these years. Some glycosyl transferases, transglycosidases/transglycosylases and glycosynthases showed outstanding potential for the synthesis of chito/chitin oligosaccharides by monosaccharides. The enzymatic/biological synthesis of chito/chitin oligosaccharides by monosaccharides will describe in detail in the next chapter.

3.7 Monomer (Monosaccharide) Generation: Chemical Hydrolysis, Fermentation and Enzymatic Methods

3.7.1 Chemical Hydrolysis

The monomer of chitooligosaccharides is glucosamine (GlcN), which also has a variety of bioactivities, including functions in plant organogenesis and invertebrate embryogenesis. Glucosamine and its acetylated derivative (N-acetylglucosamine (GlcNAc)), both can be derived from chitin/chitosan, have been widely applied in food, economic feed-stocks, cosmetics, and pharmaceutical industries, especially in arthritis treatment (Chen et al. 2010; Nakamura 2011). Currently, the global sales of GlcN and GlcNAc products can reach 2 billion dollars per year. Accordingly, high efficiency processes for producing glucosamine are urgently needed in industry. Absolutely, chitin is a suitable biosource for the production of glucosamine, because of its abundant yield (about 100 billion tons of chitin are produced in nature every year). The conversion of chitin into glucosamine including chitin deacetylation and hydrolysis.

For the longest time, most of the commercial glucosamine is manufactured by chemical hydrolyzing chitin from the shells of shellfish, such as shrimp, lobsters, and crabs. Instead of chitin as feed-stock, some glucosamine products are manufactured by fungus *Aspergillus niger* through fermenting corn to meet the demands of vegetarians and others with objections to shellfish. Generally, hydrochloric acid (HCl) was used to generate glucosamine from crab shell chitin. The following is a typical chemical hydrolysis process: Decomposition of chitin (4 kg) was carried out by stirring with concentrated hydrochloric acid at 45 °C and water and soda ash for neutralization were added. The solution was decolored by adding 800 g of activated carbon and the salt filtered by electro dialyzer. The solution (10%) containing N-acetyl-D-glucosamine (73%) and N-acetyl chitooligosaccharide (27%) was obtained. Separation of N-acetyl-D-glucosamine was carried out by a RO film with a rate of salt blocking of 10% at 20 °C, repeatedly. The permeated liquid was collected, purified using an ion exchange resin column and crystallized. N-acetyl-D-glucosamine of 99% purity and a melting point of 201 °C, was obtained. N-acetyl-D-glucosamine was subsequently deacetylated with hydrochloric acid to prepare D-glucosamine. With the diverse market demand for GlcN products increasing, more high-quality GlcN products are required. Although amounts of GlcN products are manufactured by chemical hydrolysis approach per year, this method is always limited by variable raw material supply. In addition, the level of contaminants in the products cannot be ignored, because pretreatment of the raw material from shrimp or crab shell is complicated.

3.7.2 Enzymatic Methods

Enzymatic hydrolysis is an environmentally friendly method for the preparation of sugar monomer. Researchers have extensively studied the related enzyme systems for chitin metabolism and monomer preparation. The degradation pathway of chitin in bacteria and eukaryotes has been extensively studied, and the chitin catabolic pathways are summarized as shown in Fig. 3.6. Chitin is a natural carbohydrate composed of N-acetyl-D-glucosamine by β-1,4-glycosidic bond. The enzymes involved in a chitinolytic system are summarized as follows: endo-chitinases (EC 3.2.1.1.4), exo-chitinases (EC 3.2.1.14), chitobiases (EC 3.2.1.30), and β-N-acetyl-hexosaminidases (EC 3.2.1.52). Theses enzymes work cooperatively to convert chitin into N-acetyl-D-glucosamine. Endo-chitinases hydrolyze chitin into oligomers, such as chitotetraoses and chitotrioses. And N-acetyl-chitobioses are predominant in the end-products of hydrolysis of chitin by endo-chitinases. In contrast, exo-chitinases release N-acetyl-chitobiose without production of GlcNAc or oligomers. β-Nacetyl-hexosaminidases can cleave N-acetyl-chitobiose, N-acetyl-chitotriose and N-acetyl-chitotetraose into GlcNAc monomers.

Furthermore, in some thermophilic microorganism, $(GlcNAc)_2$ is specifically deacetylated to GlcN-GlcNAc by deacetylase and then was cleaved into GlcN and GlcNAc by β-N-acetyl-hexosaminidases. The resulting GlcNAc monomers can further be deacetylated into GlcN by deacetylase. In the glucosamine industrial preparation, a key reaction is deacetylated of GlcNAc to GlcN, which is catalyzed by chemical regents. Every year, the production of GlcN by chemical hydrolysis is over 29,000 tons scale. During chemical hydrolysis, harsh reaction conditions are required and additional separation for side products are also required. Some researchers identified that deacetylases can exhibit high selectively deacetylation activity towards GlcNAc under a mild reaction condition. The high selectively of deacetylase for GlcNAc over other biomolecules makes it suitable for industrial

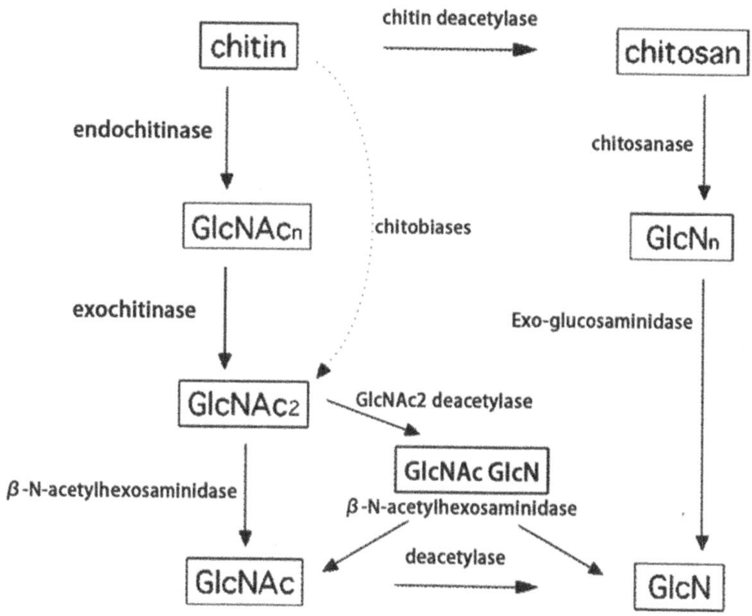

Fig. 3.6 The metabolic pathway of chitin degradation

GlcN production. In combination with chitinase and deacetylase, one-pot production of glucosamine from crude solid chitin is feasible.

Exo-β-D-glucosaminidase (exo-chitosane, EC 3.2.1.165) cleaves chitosan or chitosan oligomers to release D-glucosamine from the non-reducing terminal successively. Thus exo-β-D-glucosaminidase based hydrolysis is a potential D-glucosamine preparation method. Exo-β-D-glucosaminidase exhibits higher cleavage rate on chitotetraose and chitopentaose. And it can cleave GlcN-GlcNAc but not GlcNAc-GlcNAc.

Previous studies indicated enzymatic hydrolysis for glucosamine production is a promising approach (Lv et al. 2017). However, this approach is more complex than chemical hydrolysis, because both enzyme production and glucosamine production should be considered. Also, enzymatic processes still need extensive optimization to solve questions on low yields and incomplete conversion of chitin into its monomers. To apply the enzymatic approach to the production of glucosamine, more studies on maintaining enzyme activity/stability and managing enzyme cost should be carried out.

3.7.3 Fermentation

The GlcN and GlcNAc products manufactured using natural chitin as source are unfriendly to vegetarians and those objective to shellfish, and may carry the potential risk of allergic reactions. GlcN and GlcNAc can also be produced by microbial fermentation using recombinant microorganism such as *Escherichia coli* and *Bacillus subtilis*. Fermentation method as an environmentally friendly process is attracting increasing attention. Metabolic engineering is a potential approach to increase the production of a certain substance in cell by optimizing intracellular inheritance and regulating metabolic processes.

These processes are chemical networks that utilize a series of biochemical reactions and enzymes that allow cells to convert raw materials into molecules necessary for the cell's survival. Microbial synthesis offers an attractive alternative for sustainable production of desired biopolymers. The development of metabolic engineering is to be able to use microorganism to generate GlcN and GlcNAc on an industrial scale.

The strains used for production of GlcN and GlcNAc include fungi, viz., *Aspergillus* sp. BCRC 31742 (Sitanggang et al. 2010), recombinant *Escherichia coli* (Deng et al. 2005; Chen et al. 2012) and recombinant *Bacillus subtili* (Gu et al. 2017). For production of GlcN and GlcNAc by filamentous fungal fermentation, GlcN (or GlcNAc) production from fungal cell walls requires acid hydrolysis and long fermentation periods. Compared with traditional extraction and hydrolysis methods, the low productivity and high cost of this method weaken its economic competitiveness. Besides filamentous fungal fermentation method, GlcN and GlcNAc can also be produced using engineered bacteria, such as *Escherichia coli* and *Bacillus subtili*. For example, researchers have constructed a recombinant *Bacillus subtilis* strain to produce GlcNAc by blocking the input of extracellular GlcNAc and the catabolism of intracellular GlcNAc, while simultaneously eliminating the formation of acidic by-product lactate. Compared with *E. coli*, the gram-positive bacterium *B. subtilis* is considered as a generally recognized as safe (GRAS) organism, without exotoxin and endotoxin, no phage infection problems problem during the industrial fermentation. For this reason, *B. subtilis* system is highly favored over the use of *E. coli* system to produce GlcNAc and GlcN for food and medicinal products. In addition, *B. subtilis* is one of the most well-known Gram-positive microorganisms, its genome has been sequenced. And there are series of genetic manipulation tools available for the metabolic engineering studies of *B. subtilis*. And the regulatory mechanisms and metabolic pathways of GlcNAc/GlcN in *B. subtilis* have been extensively studied (Fig. 3.7). Thus, *B. subtilis* system offers several advantages as a possible chassis microorganism for the industrial production of GlcN and GlcNAc.

3.8 Conclusions

Now chitooligosaccharides can be prepared successfully by physical, chemical and enzyme-based methods, of which enzyme-based method is the most promising method used for chitooligosaccharides preparation. Because chitooligosaccharides exhibit remarkable biological activities, more well-defined single chitooligosaccharides with high value are required in the future. However, the preparation of well-defined single chitooligosaccharides in large-scale is still limited. We believe

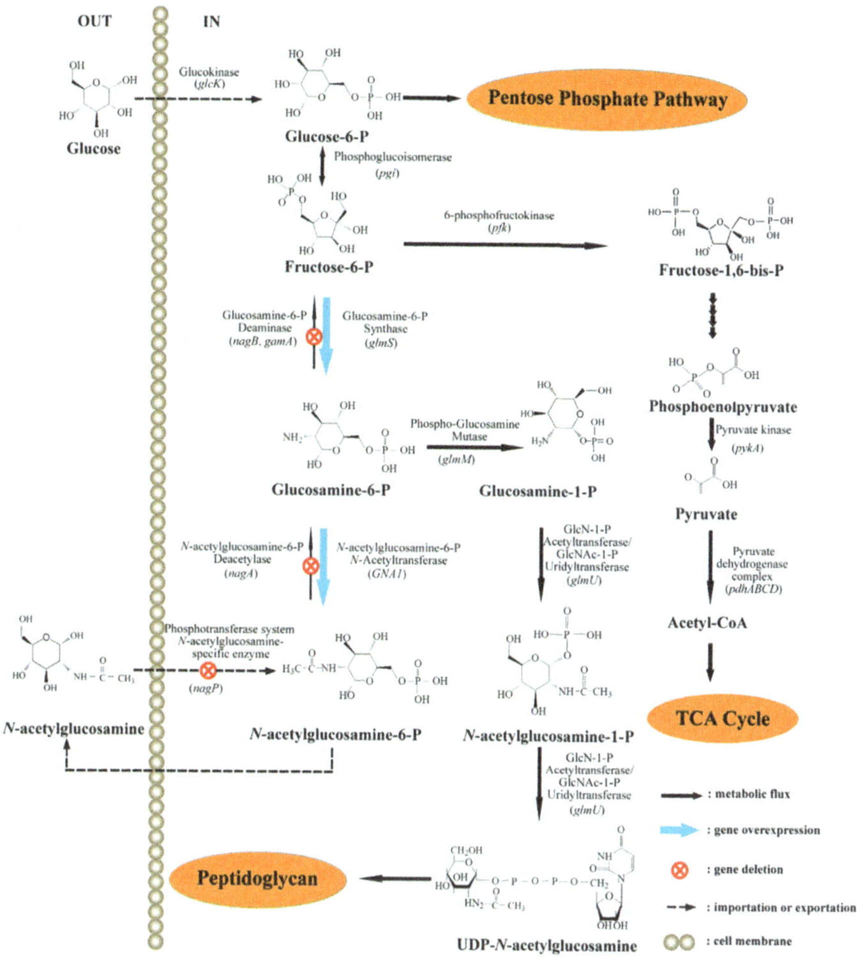

Fig. 3.7 Metabolic pathway of GlcNAc and GlcN in *B. subtilis* (Liu et al. 2013)

enzyme-based method is the key avenue for the generation of well-defined chitooligosaccharides. With the development of genomics of bacterias, more and more chitosanases with different properties will be discovered and characterized, which will accelerate the enzyme-based well-defined single chitooligosaccharides generation.

References

Adachi W, Sakihama Y, Shimizu S, Sunami T, Fukazawa T, Suzuki M, Yatsunami R, Nakamura S, Takenaka A (2004) Crystal structure of family GH-8 chitosanase with subclass II specificity from Bacillus sp K17. J Mol Biol 343(3):785–795

Bornscheuer UT, Huisman GW, Kazlauskas RJ, Lutz S, Moore JC, Robins K (2012) Engineering the third wave of biocatalysis. Nature 485(7397):185–194 van Bueren AL, Ghinet MG, Gregg K, Fleury A, Brzezinski R, Boraston AB (2009) The structural basis of substrate recognition in an exo-beta-D-glucosaminidase involved in chitosan hydrolysis. J Mol Biol 385(1):131–139

Chao C, Chen Y, Cheng C, Li Y (2013) Catalytic function of a newly purified exo-beta-D-glucosaminidase from the entomopathogenic fungus Paecilomyces lilacinus. Carbohydr Polym 93(2):615–621

Chen J-K, Shen C-R, Liu C-L (2010) N-Acetylglucosamine: production and applications. Mar Drugs 8(9):2493–2516

Chen X, Liu L, Li J, Du G, Chen J (2012) Improved glucosamine and N-acetylglucosamine production by an engineered Escherichia coli via step-wise regulation of dissolved oxygen level. Bioresour Technol 110:534–538

Deng MD, Severson DK, Grund AD, Wassink SL, Burlingame RP, Berry A, Running JA, Kunesh CA, Song L, Jerrell TA, Rosson RA (2005) Metabolic engineering of Escherichia coli for industrial production of glucosamine and N-acetylglucosamine. Metab Eng 7(3):201–214

Gohi BFCA, Zeng H, Pan AD (2016) Optimization and characterization of chitosan enzymolysis by pepsin. Bioengineering-Basel 3(3):17

Gu Y, Deng J, Liu Y, Li J, Shin H, Du G, Chen J., Liu L (2017) Rewiring the glucose transportation and central metabolic pathways for overproduction of N-acetylglucosamine in Bacillus subtilis. Biotechnol J 12(170002010SI)

Heggset EB, Dybvik AI, Hoell IA, Norberg AL, Sorlie M, Eijsink VGH, Varum KM (2010) Degradation of chitosans with a family 46 chitosanase from Streptomyces coelicolor A3(2). Biomacromolecules 11(9):2487–2497

Hsiao Y, Lin Y, Su C, Chiang B (2008) High degree polymerized chitooligosaccharides synthesis by chitosanase in the bulk aqueous system and reversed micellar microreactors. Process Biochem 43(1):76–82

Kim HJ, Ahn HY, Kwak JH, Shin DY, Kwon Y, Oh C, Lee JH (2014) The effects of chitosan oligosaccharide (GO2KA1) supplementation on glucose control in subjects with prediabetes. Food Funct 5(10):2662–2669

Kumar A, Varadara MC, Gowda LR, Tharanathan RN (2005) Characterization of chito-oligosaccharides prepared by chitosanolysis with the aid of papain and Pronase, and their bactericidal action against Bacillus cereus and Escherichia coli. Biochem J 391(2):167–175

Li K, Xing R, Liu S, Li P (2016) Advances in preparation, analysis and biological activities of single chitooligosaccharides. Carbohydr Polym 139:178–190

Liu Y, Liu L, Shin H, Chen RR, Li J, Du G, Chen J (2013) Pathway engineering of Bacillus subtilis for microbial production of N-acetylglucosamine. Metab Eng 19:107–115

Liu S, Shao S, Li L, Cheng Z, Tian L, Gao P, Wang L (2015) Substrate-binding specificity of chitinase and chitosanase as revealed by active-site architecture analysis. Carbohydr Res 418:50–56

Lv YM, Laborda P, Huang K, Cai ZP, Wang M, Lu AM, Doherty C, Liu L, Flitsch SL, Voglmeir J (2017) Highly efficient and selective biocatalytic production of glucosamine from chitin. Green Chem 19(2):527–535

Lyu Q, Wang S, Xu W, Han B, Liu W, Junes DNM, Liu W (2014) Structural insights into the substrate-binding mechanism for a novel chitosanase. Biochem J 461(2):335–345

Lyu Q, Shi Y, Wang S, Yang Y, Han B, Liu W, Jones DNM, Liu W (2015) Structural and biochemical insights into the degradation mechanism of chitosan by chitosanase OU01. BBA-Gen Subjects 1850(9):1953–1961

Madhuprakash J, El Gueddari NE, Moerschbacher BM, Podile AR (2015) Production of bioactive chitosan oligosaccharides using the hypertransglycosylating chitinase-D from Serratia proteamaculans. Bioresour Technol 198:503–509

Mao SR, Shuai XT, Unger F, Simon M, Bi DZ, Kissel T (2004) The depolymerization of chitosan: effects on physicochemical and biological properties. Int J Pharm 281(1–2):45–54

Marcotte EM, Monzingo AF, Ernst SR, Brzezinski R, Robertus JD (1996) X-ray structure of an anti-fungal chitosanase from streptomyces N174. Nat Struct Biol 3(2):155–162

Mine S, Watanabe M, Kamachi S, Abe Y, Ueda T (2017) The structure of an archaeal beta-glucosaminidase provides insight into glycoside hydrolase evolution. J Biol Chem 292(12):4996–5006

Monaghan RL, Eveleigh DE, Tewari RP, Reese ET (1973) Chitosanase, a novel enzyme. Nat New Biol 245(142):78–80

Nakamura H (2011) Application of glucosamine on human disease—Osteoarthritis. Carbohydr Polym 84(2):835–839

Nordtveit RJ, Varum KM, Smidsrod O (1994) Degradation of fully water-soluble, partially N-acetylated chitosans with lysozyme. Carbohydr Polym 23(4):253–260

Pan A, Zeng H, Foua GB, Alain C, Li Y (2016) Enzymolysis of chitosan by papain and its kinetics. Carbohydr Polym 135:199–206

Popa-Nita S, Lucas J, Ladaviere C, David L, Domard A (2009) Mechanisms involved during the ultrasonically induced depolymerization of chitosan: characterization and control. Biomacromolecules 10(5):1203–1211

Purushotham P, Podile AR (2012) Synthesis of long-chain chitooligosaccharides by a hypertransglycosylating processive endochitinase of Serratia proteamaculans 568. J Bacteriol 194(16):4260–4271

Qin Z, Chen Q, Lin S, Luo S, Qiu Y, Zhao L (2018a) Expression and characterization of a novel cold-adapted chitosanase suitable for chitooligosaccharides controllable preparation. Food Chem 253:139–147

Qin Z, Luo S, Li Y, Chen Q, Qiu Y, Zhao L, Jiang L, Zhou J (2018b) Biochemical properties of a novel chitosanase from Bacillus amyloliquefaciens and its use in membrane reactor. LWT-Food Sci Technol 97:9–16

Regel EK, Weikert T, Niehues A, Moerschbacher BM, Singh R (2018) Protein-engineering of chitosanase from Bacillus sp MN to alter its substrate specificity. Biotechnol Bioeng 115(4):863–873

Saito J, Kita A, Higuchi Y, Nagata Y, Ando A, Miki K (1999) Crystal structure of chitosanase from Bacillus circulans MH-K1 at 1.6-angstrom resolution and its substrate recognition mechanism. J Biol Chem 274(43):30818–30825

Sitanggang AB, Wu H, Wang SS, Ho Y (2010) Effect of pellet size and stimulating factor on the glucosamine production using Aspergillus sp BCRC 31742. Bioresour Technol 101(10):3595–3601

Sorbotten A, Horn SJ, Eijsink V, Varum KM (2005) Degradation of chitosans with chitinase B from Serratia marcescens – production of chito-oligosaccharides and insight into enzyme processivity. FEBS J 272(2):538–549

Su P, Wang S, Shi Y, Yang Y (2013) Application of cellulase-polyamidoamine dendrimer-modified silica for microwave-assisted chitosan enzymolysis. Process Biochem 48(4):614–619

Sun T, Zhou D, Xie J, Mao F (2007) Preparation of chitosan oligomers and their antioxidant activity. Eur Food Res Technol 225(3–4):451–456

Tegl G, Oehlknecht C, Vielnascher R, Rollett A, Hofinger-Horvath A, Kosma P, Guebitz GM (2016) Cellobiohydrolases produce different oligosaccharides from chitosan. Biomacromolecules 17(6):2284–2292

Tian F, Liu Y, Hu K, Zhao BY (2003) The depolymerization mechanism of chitosan by hydrogen peroxide. J Mater Sci 38(23):4709–4712

Tishchenko G, Simunek J, Brus J, Netopilik M, Pekarek M, Walterova Z, Koppova I, Lenfeld J (2011) Low-molecular-weight chitosans: preparation and characterization. Carbohydr Polym 86(2):1077–1081

Trombotto S, Ladaviere C, Delolme F, Domard A (2008) Chemical preparation and structural characterization of a homogeneous series of chitin/chitosan oligomers. Biomacromolecules 9(7):1731–1738

Viens P, Lacombe-Harvey M, Brzezinski R (2015) Chitosanases from family 46 of glycoside hydrolases: from proteins to phenotypes. Mar Drugs 13(11):6566–6587

Wang Q, Graham RW, Trimbur D, Warren RAJ, Withers SG (1994) Changing enzymatic reaction mechanisms by mutagenesis: conversion of a retaining glucosidase to an inverting enzyme. J Am Chem Soc 116(25):11594–11595

Weikert T, Niehues A, Cord-Landwehr S, Hellmann MJ, Moerschbacher BM (2017) Reassessment of chitosanase substrate specificities and classification. Nat Commun 8:1698

Wu S, Pan S, Wang H, Wu J (2013) Preparation of chitooligosaccharides from cicada slough and their antibacterial activity. Int J Biol Macromol 62:348–351

Xia W, Liu P, Liu J (2008) Advance in chitosan hydrolysis by non-specific cellulases. Bioresour Technol 99(15):6751–6762

Xing RE, Liu S, Yu HH, Guo ZY, Wang PB, Li CP, Li Z, Li PC (2005) Salt-assisted acid hydrolysis of chitosan to oligomers under microwave irradiation. Carbohydr Res 340(13):2150–2153

Xing R, Liu Y, Li K, Yu H, Liu S, Yang Y, Chen X, Li P (2017) Monomer composition of chitooligosaccharides obtained by different degradation methods and their effects on immunomodulatory activities. Carbohydr Polym 157:1288–1297

Yao Y, Shrestha KL, Wu Y, Tasi H, Chen C, Yang J, Ando A, Cheng C, Li Y (2008) Structural simulation and protein engineering to convert an endo-chitosanase to an exo-chitosanase. Protein Eng Des Sel 21(9):561–566

Yoshida N, Tanaka T, Noguchi M, Kobayashi A, Ishikura K, Ikenuma T, Seno H, Watanabe T, Kohri M, Shoda S (2012) One-pot chemoenzymatic route to chitoheptaose via specific transglycosylation of chitopentaose-oxazoline on chitinase-template. Chem Lett 41(7):689–690

Zhang H, Lu Y, Wang Y, Zhang X, Wang T (2018) D-glucosamine production from chitosan hydrolyzation over a glucose-derived solid acid catalyst. RSC Adv 8(10):5608–5613

Preparation of Chitin Oligosaccharides and Its Monomer

4

Shaoqing Yang, Zhengqiang Jiang, Yihao Liu,
and Shuai Ma

Abstract

Functional oligosaccharides have drawing increasing attention in recent years due to their important biological activities. To date, more than 20 kinds of functional oligosaccharides have been commercially produced and widely used in various fields. However, industrial production of *N*-acetyl chitooligosaccharides is still not available due to that the direct degradation of chitin is difficult and conversion rate is low. Production of *N*-acetyl oligosaccharide has some advantages when compared with other functional oligosaccharides, such as maltooligoaccharide, isomaltooligosaccharide, fructooligosaccharide, mannooligosaccharide etc. For example, the cost of raw materials for *N*-acetyl oligosaccharide production is nearly negligible as chitin is one of the most abundant regenerated biomass only next to cellulose. In addition, the conversion of *N*-acetyl oligosaccharide from chitin can not only partially reduce the burden of environmental pollution caused by the disposal of waste chitin materials, but also offer a kind of functional product helpful for human health. Hence, some studies have been performed for efficient and environmental friendly production of *N*-acetyl oligosaccharide from chitin materials. In the present chapter, the major conversion methods (including chemical, physical and enzymatic methods) as well as several novel strategies are introduced in details.

Keywords

N-acetyl chitooligosaccharides · Pretreatment · Chitinolytic enzymes · Enzymatic hydrolysis · Synthesis

---

S. Yang (✉) · Z. Jiang · Y. Liu · S. Ma
College of Food Science and Nutritional Engineering, China Agricultural University, Beijing, China
e-mail: ysq@cau.edu.cn

L. Zhao (ed.), *Oligosaccharides of Chitin and Chitosan*,
https://doi.org/10.1007/978-981-13-9402-7_4

4.1 Functions of Chitin Oligosaccharides

Chitin is a β-1,4-linked linear insoluble polymer of *N*-acetyl-D-glucosamine (GlcNAc), which represents the second most abundant polysaccharide after cellulose (Monge et al. 2018). More than 100 billion tons of chitin materials have been produced annually worldwide without utilization, the discarding of which in nature not only results in environmental problems, but also a waste of biomass (Wei et al. 2017; Wakita et al. 2017). Chitin oligosaccharides (*N*-acetyl COSs) and GlcNAc are the partially degraded or completely degraded products of chitin by chemical, physical or enzymatic methods. *N*-acetyl COSs, especially with degree of polymerization (DP) of 2–6, have attracted considerable attention in recent years due to their water-solubility and temperature and pH stability, as well as excellent biological activities, such as antimicrobial, anti-tumor and hypoglycemic activities (Bhattacharya et al. 2007; Wu et al. 2017). In addition to these biological activities, *N*-acetyl COSs are also considered as effective plant elicitors (Yin et al. 2016) and potent angiogenic inhibitor (Wang et al. 2007), involving in the promotion of bifidus proliferation (Chen et al. 2002), the characteristics of which could be widely used in agricultural, food and pharmaceutical industrials (Bhattacharya et al. 2007; Wendland et al. 2009; Suresh and Kumar 2012). Thus, production of *N*-acetyl COSs is a promising way for the disposal and recycling of enormous waste chitin materials.

4.2 Chemical Methods for Preparation of Chitin Oligosaccharides

N-acetyl COSs have historically been commercially produced via a process on the basis of concentrate acid hydrolysis of crude chitin materials (Jung and Park 2014). Commonly, the degradation of chitin by chemical method can be performed by strong acid hydrolysis, such as hydrochloric acid. The acid concentration and operation temperature should be carefully selected and strictly controlled. The two parameters must be high enough to degrade chitin sufficiently, but not too high since the *N*-acetyl COSs products could be further degraded. Suitable reaction conditions include 15–36% (w/v) hydrochloric acid and about 40–80 °C (Chen et al. 2010). Briefly, crude chitin was ground, sieved and then incubated in 12 M Conc. hydrochloric acid. The resulted solution is neutralized with sodium hydroxide, filtered, desalted and then qualitatively analyzed by thin layer chromatography (TLC) method. The hydrolysis products are freeze dried and the final product contains *N*-acetyl COSs. Meanwhile, the required acid could be replaced by phosphoric acid or hydrofluoric acid. It has been reported that chitin was suspended with 37% (w/v) hydrochloric acid, stirred for 2 h at 4 °C, cooled to 0 °C and the suspension was then adjusted to pH 7 with 50% sodium hydroxide. The suspension mixture was centrifuged and the supernatant was filtered using a glass filter and concentrated on a rotary evaporator. The obtained sample was further fractionated on a Bio-gel P2 acrylamide column and quantitatively analyzed by high performance liquid chromatography (HPLC). The fractions contain a serious of *N*-acetyl COSs derivatives, including GlcNAc, $(GlcNAc)_2$, $(GlcNAc)_3$, $(GlcNAc)_4$, $(GlcNAc)_5$ and $(GlcNAc)_6$.

GlcNAc can also be prepared by an alternative process using GlcN as the substrate. Briefly, chitin is dissolved in concentrated hydrochloric acid and then heated in boiling water for 3 h. The drastic conditions can not only convert chitin to its monomer completely, but also could remove the acetyl group of GlcNAc at the same time. So in the process, an additional *N*-acetylation reaction must be performed for GlcNAc production. Briefly, GlcN is dissolved in methanol solution (10%, w/v) and *N*-acetylated in acetic anhydride in the presence of a kind of ion exchange resin (Dowex1 in carbonate form). An overall GlcNAc yield of 43% was obtained after filtration and crystallization. Zhan et al. described another improved *N*-acetylation procedure, in which GlcN was *N*-acetylated using pyridine as a solvent in the presence of acetic anhydride and tributylamine (Chen et al. 2010). The highest GlcNAc yield of 70% was obtained with purity high up to 99% after a series of purification steps.

Although chemical methods for *N*-acetyl COSs and GlcNAc production is considered to be an economic and sample route, the products are not deemed natural materials due to their chemical modifications (Sashiwa et al. 2003). Moreover, trace amounts of *O*-acetylated and di-acetylated products have also been detected by liquid chromatograph-mass spectrometer (LC-MS) except *N*-acetylated products. Furthermore, gas chromatography-mass spectrometer (GC-MS) analysis revealed that solvent and tributylamine are also present in the products (Wang and Chang 1997). Thus, these residual substances may attribute to bitter tastes of chemically produced *N*-acetyl COSs, which limited their commercial application in industries, especially in food industry (Fu et al. 2016; Yang et al. 2016a).

4.3 Physical Methods for Preparation of Chitin Oligosaccharides

In recent years, several alternative methods which can generate high energy microenvironments such as ultrasound, microwaves and ball-milling are considered to be complementary and promising tools for the efficient conversion of biomass (Tabasso et al. 2015). The localized temperatures and pressures developed in mechanochemical processes offers an effective method to induce reactivity in organic materials including chitin and cellulose. The approach overcomes the need for solvent and can efficiently destroy the crystalline structure of polysaccharides by weakening the network bonded by intermolecular hydrogen. It has been reported that the reductions in crystallinity are essential for chitin conversion.

Mechanochemical treatments have already exhibited great potential for environmental friendly sustainable processing of chitin within biomass valorization. X-ray diffraction analysis revealed that crystallinity of chitin can be reduced by 50% in 2 h by using a ball mill in a controlled way (Fig. 4.1). Furthermore, the spectra analysis revealed that glycosidic linkage content decreased and *N*-acetyl groups retained.

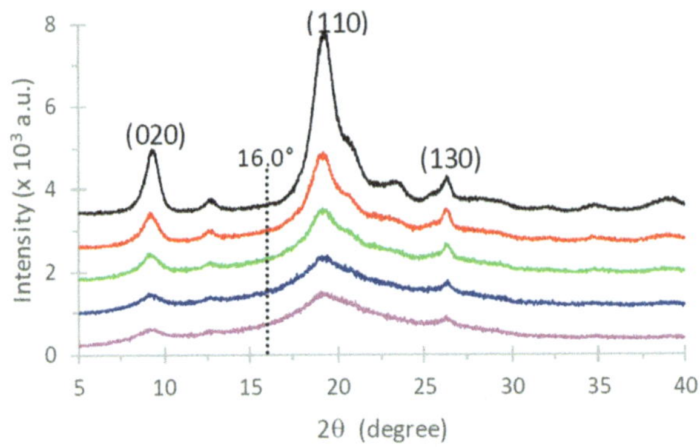

Fig. 4.1 XRD analysis of untreated (black signal) and milled α-chitin by 2 × 0.5″ balls for 30 (red), 60 (green), 90 (blue) and 120 (pink) min, respectively (Margoutidis et al. 2018)

The addition of kaolinite or natural clay in the ball mill results in a significant increase in the solubility of the milled materials (more than 75% water soluble products with kaolinite, while 35% without kaolinite). The products in this process are identified as *N*-acetyl COSs with DP 1–5 by a new quantitative matrix-assisted laser desorption-ionization (MALDI-TOF) mass spectrometric method (Margoutidis et al. 2018). These data are in accordance with a colorimetric assay of reducing ends and size-exclusion chromatography (SEC). GlcNAc and $(GlcNAc)_2$ yields of 5.1% and 3.9% were obtained within 6 h, respectively, which is comparable to that of glucose and cellobiose from cellulose degraded by the same method.

4.4 Enzymatic Methods for Preparation of Chitin Oligosaccharides

N-acetyl COSs can be produced by chemical and physical methods. However, concentrated acids used in the chemical procedure can result in a series of associated environmental problems (Wei et al. 2017). Besides, the yield of *N*-acetyl COSs by physical methods are still too low to be scaled up, except for the huge amount of energy consumption. Consequently, the mild enzymatic approach for the production of *N*-acetyl COSs is highly preferred owing to product consistency, environmental protection and sustainable development (Patil and Jadhav 2014). Suitable hydrolytic enzymes are the key factors for enzymatic conversion of chitin. Till now, several kinds of enzymes have been used in the hydrolysis of chitin materials for *N*-acetyl COSs production, such as chitinolytic enzymes and lipases.

Chitinolytic enzymes are the major enzymes which have been widely used for *N*-acetyl COSs production from chitin materials. On the basis of their action modes, chitinolytic enzymes have been classified into two categories: (1) chitinases (EC 3.2.1.14) that randomly cleave the glycosidic bonds in internal sites of chitin; (2) β-*N*-acetylhexosaminidases (EC 3.2.1.52) that successively catalyze the removal of GlcNAc residues from the non-reducing end of the chitin chain backbone (Adrangi and Faramarzi 2013). To date, a large amount of chitinolytic enzymes have been identified, characterized and applied with the aim of improving the hydrolysis efficiency of chitin for *N*-acetyl COSs and GlcNAc production. Moreover, several commercial non-chitinase hydrolytic enzymes have also been reported to be capable of hydrolyzing chitin for *N*-acetyl COSs and GlcNAc production, such as lipase, lysozyme, pectinase, papain, glucanase and tannase (Jung and Park 2014).

4.4.1 The Pretreatments of Chitin

In order to achieve an efficient bioconversion, pretreatments constitute a key preliminary step to affect the supramolecular structure of biopolymers. It consists of an incubation step, by analogy to the biochemical pathway on lignocellulosic biomass. Pretreatments have been widely described for cellulosic biomass, as it is necessary prior to enzymatic hydrolysis due to that chitin has strong crystallinity and insolubility in an aquatic environment. For instance, chitin is normally treated with strong acids, such as phosphoric acid (for swollen chitin) or hydrochloric acid (for colloidal chitin) to break down its crystal structure thus increasing the accessibility of the substrate to the degrading enzymes (Jung and Park 2014). A large panel of pretreatments have been developed and applied in chitin degradation process for enzymatic production of *N*-acetyl COSs and GlcNAc, such as mechanochemical grinding, strong acid treatment, sub- or super-critical fluid treatment, steam explosion and ultrasonication etc (Osada et al. 2012). In recent years, a series of innovative and environmental friendly pretreatments have been applied to chitin conversion for *N*-acetyl COSs and GlcNAc production, such as treatment with high pressure homogenization (Wei et al. 2017), ionic liquids (Husson et al. 2017) and bacterial fermentation (Zhang et al. 2018).

High pressure homogenization has been proved to be an efficient pretreatment method for chitin conversion. Wei et al. found that high pressure homogenization could change the crayfish shell to exhibit a fluffy netted structure, and 3.9 g/L GlcNAc yield (25 g/L of crayfish shell) in a batch enzymatic reaction was achieved by chitinase hydrolysis when the crayfish shell was pretreated by five treatment cycles (with pressure 400 bar) (Fig. 4.2). The results suggested that high pressure homogenization may be an efficient method for enzymatic production of GlcNAc from crayfish shell directly.

Husson et al. revealed that the enzyme accessibility to the chitin substrate increased when chitin was pretreated by a room temperature ionic liquid (1-ethyl-3-methyl imidazolium, [C_2mim][OAc], RTIL). Ionic liquid pretreatment provided an efficient GlcNAc production of 185 mg/g chitin and a $(GlcNAc)_2$ production of 668 mg/g chitin, converted by the chitinases from *Trichoderma viride* and *Streptomyces griseus*, respectively. Alternatively, a promising simultaneous route was developed. On the basis of an one-pot enzymatic hydrolysis of chitin in [C_2mim][OAc]-aqueous medium, it can selectively generate GlcNAc or $(GlcNAc)_2$ by decreasing the required [C_2mim][OAc] amount and the number of steps (Fig. 4.3). Finally, the synergistic action of the two chitinases from *T. viride* and *S. griseus* was

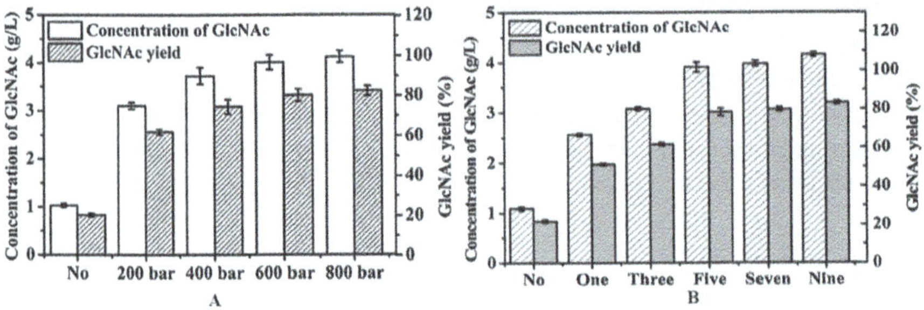

Fig. 4.2 Effects of high pressure treatment on the GlcNAc conversion efficiency from crayfish shells. (**a**) Effect of different pressures applied in five cycle; (**b**) effect of cycle numbers treated at 400 bar (Wei et al. 2017)

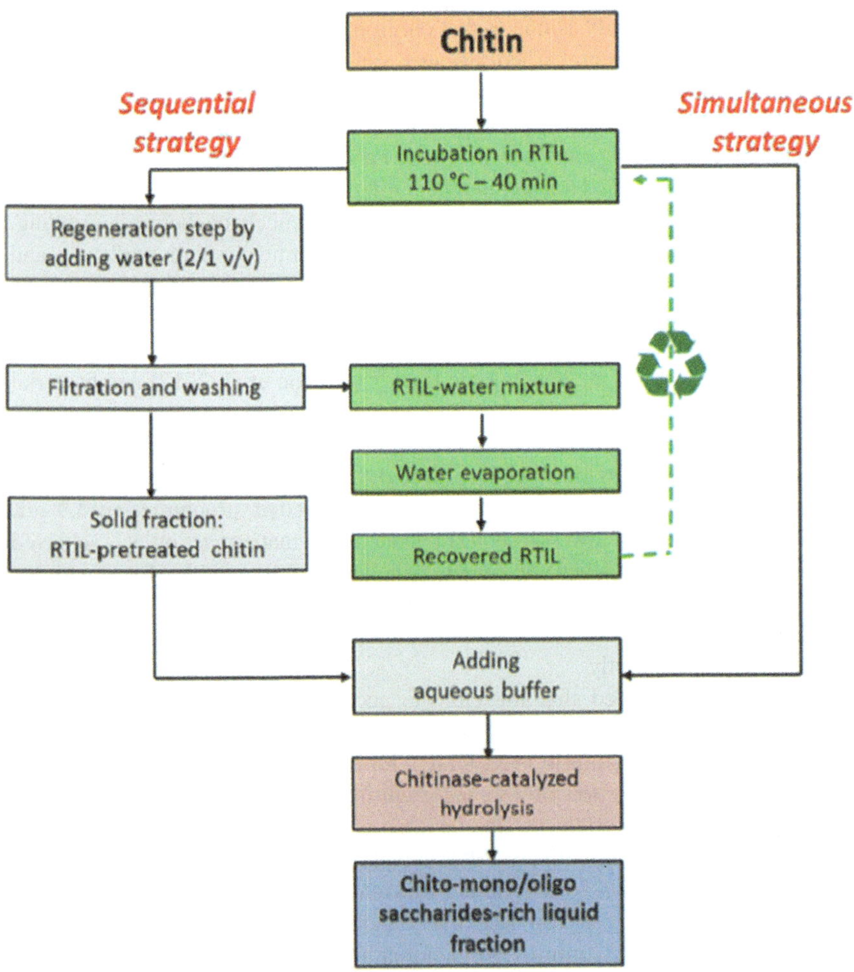

Fig. 4.3 Skeleton of sequential and simultaneous strategies for the bio-catalytic conversion of chitin (Husson et al. 2017)

confirmed to be relevant to the significant increase of GlcNAc production (760 mg/g chitin).

In addition, bacterial fermentation can also be helpful to the *N*-acetyl COSs and GlcNAc production. Zhang et al. reported that bacterial fermentation (*Chitinolyticbacter meiyuanensis* SYBC-H1) pretreatment made chitin powder to exhibit a visible fleeciness structure. Scanning electron microscopy (SEM) analysis further revealed that the surface of the chitin showed a fiber-like structure with diameter range of 10–200 nm. By using the pretreatment method, followed by an enzymatic reaction (6 h), the highest GlcNAc yield of 96% (19.2 g/L) was achieved from 20 g/L chitin. The new approach provided an efficient, mild and environmentally friendly method for reducing the crystallinity of chitin. The method can enhance the GlcNAc conversion rate without the use of acids, oxidants or other chemical reagents.

4.4.2 The Enzymes Used for Preparation of Chitin Oligosaccharides: Chitinases and β-*N*-Acetylglucosaminidases

4.4.2.1 Classification, Structure and Catalytic Mechanism

The reported glycoside hydrolases (GH) have been collected and classified into different GH families on the base of their sequence homology. A continuously updated list of these GH families is available at CAZy database (http://www.cazy.org). Most of the chitinases fall into GH families 18, 19, 23 and 48 (Ueda et al. 2009; Fujita et al. 2006), while most of the β-*N*-acetylhexosaminidases are classified into GH families 3, 20, 84 and 116 (Yang et al. 2014; Qin et al. 2015; Ferrara et al. 2014). The catalytic domains of members in different GH families all fold into a common three-dimensional structure (Fig. 4.4). Members from GH families 20 and 84 all have similar $(\beta/\alpha)_8$ barrel domains (Sumida et al. 2011), while enzymes from GH families 19 and 23 adopt an α+β structure (Udaya Prakash et al. 2010). Differently,

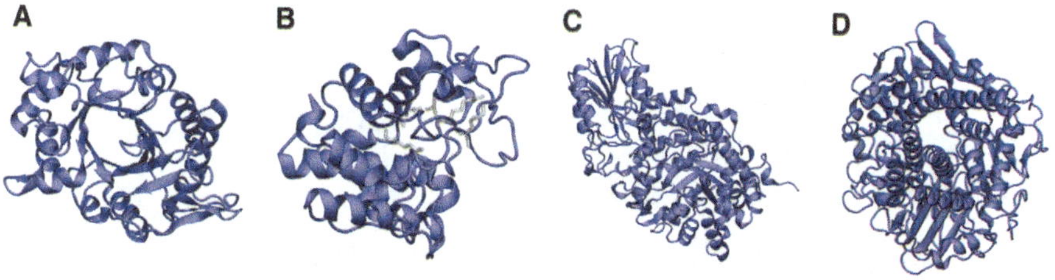

Fig. 4.4 Three-dimensional structure of several representative chitinases from different GH families. (**a**) A GH family 18 chitinase from *Aspergillus fumigatus* displaying a $(\beta/\alpha)_8$ barrel (PDB ID: 2XVP). (**b**) A GH family 19 chitinase from *Carica papaya* exhibiting α+β structure (PDB ID: 3CQL). (**c**) A GH family 3 glucohydrolase from *Hordeum vulgare* having a bipartite domain consisting of a $(\beta/\alpha)_8$ barrel (lower right) and an $(\alpha/\beta)_6$ sandwich (upper left) (PDB ID: 1IEQ). (**d**) A GH family 48 cellobiohydrolase from *Clostridium thermocellum* exhibiting an $(\alpha/\alpha)_6$ barrel (PDB ID: 1L1Y) (Adrangi and Faramarzi 2013)

the catalytic domains of GH family 3 enzymes display bipartite structure consisting of a $(\beta/\alpha)_8$ barrel followed by an $(\alpha/\beta)_6$ sandwich. However, some GH family 3 enzymes lack of $(\alpha/\beta)_6$ sandwich. GH family 48 enzymes commonly exhibit an $(\alpha/\alpha)_6$ barrel structure characterized by six central α-helices surrounded by six external α-helices (Adrangi and Faramarzi 2013).

Generally, most of chitinases are modular proteins. They contain auxiliary domains such as the carbohydrate-binding module (CBM), except a catalytic domain (Guillen et al. 2010). The CBM is helpful in enhancing the enzyme activity towards insoluble substrates by anchoring the enzyme to the substrate. Besides, it can also destroy the crystalline structure of the substrate thus leads to the formation of free chain ends in substrate. Similar to the catalytic modules, CBMs have also been classified into different homologous protein families, and they can be accessed at the CAZy database (http://www.cazy.org). The second feature that helpful for the chitinases to overcome their low accessibility to insoluble substrates is that some chitinolytic enzymes have a deep and narrow substrate-binding cleft lined with aromatic residues (Fig. 4.5) (Zakariassen et al. 2010). Such chitinases act in a processive manner. It means that once the enzyme bonds to a substrate chain, they make the substrate chain pass over their catalytic cleft and cut at several hydrolytic sites, but do not release the substrates after each cleavage. The aromatic residues play an important role in offering the necessary environment for the flexible binding and movement of the substrate via the active site. Though processive reaction is an efficient strategy for the insoluble substrates hydrolysis, it usually lead to the activity decrease of the enzyme towards soluble or more accessible polymeric substrates.

Generally, chitinolytic enzymes catalyze the de-polymerization of their substrates through two pathways *viz.* single- and double-displacement mechanisms. There are two distinct catalytic groups involved in both of the catalysis pathways. One is a carboxyl group which acts as a proton donor and is usually provided by a

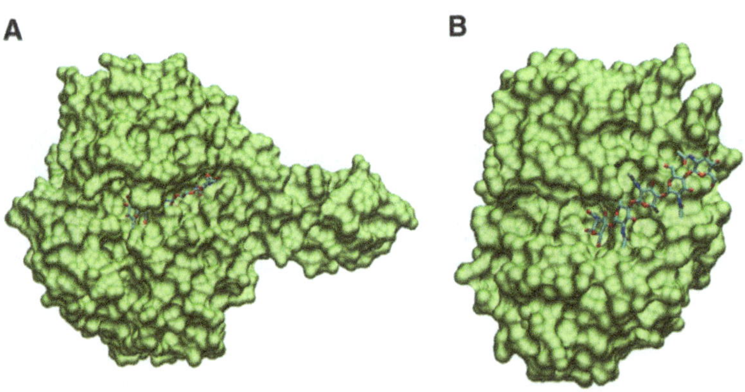

Fig. 4.5 Three-dimentional structure of substrate-binding cleft of processive and non-processive chitinases. (**a**) A chitinase from *Serratia marcescens* with processive manner (PDB ID: 1E6N). (**b**) A chitinase from *Hevea brasiliensis* with non-processive manner (PDB ID: 1KQY). Substrates are shown in stick representation (Adrangi and Faramarzi 2013)

conserved glutamate residue at the enzyme's active site. In some special cases, glutamate residue may be substituted by aspartate residue, such as some chitinases from GH family 84. The other catalytic group acts as either a base (in single-displacement mechanism) or a nucleophile (in double-displacement mechanism). The latter group may be a carboxyl moiety provided by a conserved aspartate or glutamate residue, or may be an *N*-acetyl group of the sugar located in the −1 subsite of the enzyme (Aam et al. 2010). Subsites of the substrate are numbered from −n to +n, where negative signs mean the non-reducing end of the chain, while positive signs mean reducing ends, and the cleavage occurred between the −1 and +1 subsite. The single-displacement mechanism is also known as the inverting mechanism, as it leads to the inversion of the anomeric configuration of the hydrolyzed GlcNAc residue. The double-displacement mechanism also refers to the retaining mechanism, with the anomeric configuration retained.

4.4.2.2 Exploration and Properties of Chitinases and β-*N*-Acetylglucosaminidases

Discovery and explore original chitinolytic enzymes have drowned growing attention in recent years. Chitinolytic enzymes are distribute in a wide range of organisms including bacteria, fungi, plants, insects and animals for different functions, such as morphogenesis, nutrition and defense against chitin containing pathogens. Many of the above mentioned organisms have several chitinolytic enzyme encoding genes. For example, filamentous fungi generally have 10–20 different chitinolytic enzyme genes, while mycoparasitic species have high up to 30 or even more such genes. These chitinolytic enzymes degrade chitin in a synergetic or successive manner. Some higher organisms have also been reported to contain a large amount of chitinolytic encoding genes such as arabidopsis. However, some of these genes do not code active enzymes. For example, some organisms including plants, invertebrates and animals, express so-called chitinase-like lectins encoding genes without chitinolytic activity. This may be due to the substitutions in their key catalytic residues. Though these proteins retain the ability to bind chitin, they lacked of enzyme activity.

The molecular masses of chitinases vary widely in the range of 20–120 kDa. Generally, microbial chitinases have a molecular mass in the range of 20–110 kDa, which is similar to that of chitinases from plant (25–40 kDa), and smaller than that of the chitinases from insect (40–85 kDa). Depending on the sources of microorganisms isolated, chitinases are found to be active over wide ranges of pH and temperature (Yang et al. 2016b). For example, endochitinases from *Streptomyces violaceusniger* and *Streptomyces thermoviolaceus* are most active at 28 °C and 80 °C, respectively. Also, the enzyme from *S. thermoviolaceus* was had an optimal pH in the range of 8.0–10.0, while the chitinase from *Stenotrophomonas maltophilia* C3 was most active in the pH range of 4.5–5. The chitinases from different sources also exhibited a broad range of isoelectric points (*p*I) from 3.0 to 10.0.

The molecular masses of microbial β-*N*-acetylglucosaminidases are in the range of 40–90 kDa. Similar to chitinases, β-*N*-acetylglucosaminidases from different microorganisms are also found to have wide ranges of optimal temperatures

(30–70 °C) and pH (3.5–10.0) (Zhang et al. 2018). For instance, β-*N*-acetylglucosaminidases from *Streptomyces* sp. NK5 and *Trichoderma reesei* have optimum pH values of 10.0 and 4.0, respectively, while the optimum temperatures of 60 and 70 °C, respectively.

4.4.2.3 Future Prospective of Chitinases and β-*N*-Acetylglucosaminidases

Many attempts have been made to enhance the catalytic properties through directed evolution, site directed mutagenesis and domain fusion. Through these studies, more information about the importance of domains, specific loops, and residues in chitinases and β-*N*-acetylglucosaminidases have been elucidated and can be harnessed for useful chitinase and β-*N*-acetylglucosaminidase application. Further exploration of domains and their respective functions will be of advantage in designing chitinases and β-*N*-acetylglucosaminidases with enhanced properties. Studies on the interaction between chitinases and different forms of chitin can also help to better understand the varied antifungal properties of chitinases.

Predetermination and comparison of behaviors of various chitinases by molecular dynamics may reveal novel mechanisms of interactions. Consequently, more design and mutation strategies can be employed to develop chitinases with enhanced antifungal, hydrolytic or transglosylating properties. Finally, the fast-growing field of bioinformatics, the more and more computational tools and the increasing databases which describe and predict structure and function of proteins can help to streamline engineering of chitinases in a more rational manner to yield chitinases with improved activity and stability specific for the various biotechnological applications.

4.4.3 Specific Enzymatic Methods

4.4.3.1 Preparation of GlcNAc

GlcNAc is commonly produced by the complete hydrolysis of chitin through the coaction of chitinases and *N*-acetyl-β-hexosaminidase. To date, some chitinolytic enzymes have been reported to be capable of hydrolyzing chitin to release GlcNAc. For example, the chitinase from *Burkholderia cepacia* TU09 was used to hydrolyze β-chitin and α-chitin to produce GlcNAc, and the highest yields of 85% and 80% were obtained, after 1 and 7 days, respectively. The chitinase from *Bacillus licheniformis* SK-1 was applied for GlcNAc and $(GlcNAc)_2$ productions from β-chitin, and yields of 75% and 20% were obtained, respectively within 6 days. The *Bacillus licheniformis* SK-1 chitinase hydrolyzed α-chitin to produce GlcNAc with a conversion yield of 41% (Pichyangkura et al. 2002). For GlcNAc production, β-chitin from squid pen is better than α-chitin from crab or shrimp shells as the substrate in most cases. It has been reported that *N*-acetyl COSs and GlcNAc were produced from colloidal α-chitin by a crude enzyme from *Paenibacillus illinoisensis*. In the above procedure, the GlcNAc production rate increased continuously with the increase of incubation time, whereas that of *N*-acetyl COSs decreased, and the highest GlcNAc production of 62.2% (1.71 mg mL^{-1}) was obtained after 24 h of incubation. The yields of $(GlcNAc)_2$, $(GlcNAc)_3$, $(GlcNAc)_7$ and $(GlcNAc)_8$ were 4.9%, 1.2%, 4.1% and 9.6%, respectively, at the same time (Jung et al. 2007).

Some non-chitinase enzymes have also been used to hydrolyze chitin for GlcNAc production. For example, crude cellulase culture from *Trichoderma viride* and *Acremonium cellulolyticus* were used to hydrolyze α-chitin (from crab shell) and β-chitin (from squid pen) for GlcNAc production, separately, and the yield of GlcNAc was enhanced by combination of the two types of cellulases when compared that by sole enzyme (Sashiwa et al. 2003). Crystalline chitin from bee and α-chitin from crab shell were hydrolyzed by the crude enzyme culture from *Trichoderma reesei*, containing both cellulose and β-glucanase activities, and the highest GlcNAc yield of 86% was obtained after 10 days of incubation. The mixed enzymes of cellulose and lipase (9:1) from *Aspergillus niger* was applied in the hydrolysis of β-chitin. The GlcNAc yield increased with the increase of substrate concentration, while keeping the enzyme concentration constant, and the highest GlcNAc yield of 61% from β-chitin (10 mg mL^{-1}) was achieved after 4 days of incubation. Flake type of chitin can also be used as the substrate for the N-acetyl COSs production, as well as swollen chitin, colloidal chitin and powder chitin. It has been reported that the crude enzyme extract from *Aeromonas hydrophila* hydrolyzed crab shell α-chitin flake to yield 66–77% of GlcNAc after 10 days of incubation (Sashiwa et al. 2002).

N-acetyl COSs can be used as substrates for (GlcNAc) production by exo-type chitinases. For example, $(GlcNAc)_2$ was completely converted to its monomer GlcNAc by an exochitinase (ChiA71) from *Bacillus thuringiensis* with extension of incubation time (Thamthiankul et al. 2001). It was also completely hydrolyzed by the *Stenotrophomonas maltophilia N*-acetyl-β-hexosaminidase StmHex (Katta et al. 2013). In addition, GlcNAc was also prepared from $(GlcNAc)_4$ by an exochitinase *Aeromonas hydrophila*. Though this procedure is not suitable for commercial production of monomers as oligomers are more difficult to prepare and much more expensive than GlcNAc, it is suitable for the elucidation of the enzyme properties (Lan et al. 2004).

Several attempts have been performed for the selective productions of GlcNAc and $(GlcNAc)_2$ (Kuk et al. 2005a). For instance, the crude enzyme culture from *Aeromonas* sp. GJ-18 contained both *N*-acetyl-β-hexosaminidase and *N*-acetyl chitobiase. The *N*-acetyl-β-hexosaminidase was inactive at above 50 °C, while *N*-acetyl chitobiase was stable at this temperature. So the crude enzyme effectively hydrolyzed α-chitin to release GlcNAc at the incubation temperatures below 45 °C, while $(GlcNAc)_2$ was the major product when the incubation temperatures were higher than 50 °C. Thus, GlcNAc and $(GlcNAc)_2$ were selectively prepared from α-chitin by changing incubation temperatures. The highest yields of GlcNAc (74%) and $(GlcNAc)_2$ (35%) were obtained at 45 °C and 55 °C, respectively, both after 5 days of incubation (Kuk et al. 2005b).

4.4.3.2 Preparation of Chitin Oligosaccharides

$(GlcNAc)_2$ was also obtained as the major hydrolytic product from chitin materials by controlling the ratio of *N*-acetyl-β-hexosaminidase to *N*-acetyl chitobiase activities of the culture broth of *Aeromonas* sp. After 7 days of incubation, $(GlcNAc)_2$ yields of 78.9% and 56.6% were obtained from swollen α-chitin and powdered β-chitin, respectively, by crude enzyme which had been incubated at 50 °C for *N*-acetyl-β-hexosaminidase inactivation (Kuk et al. 2006). Generally, most of $(GlcNAc)_2$ products were released from colloidal chitin as it is water soluble and easy to be hydrolyzed when compared to chitin. So far, some attempts on bio-production of $(GlcNAc)_2$ from colloidal chitin through enzymatic hydrolysis have reported. A chitinase from a marine bacterium *Vibrio anguillarum* was used to hydrolyze colloidal chitin for $(GlcNAc)_2$ production, and the highest $(GlcNAc)_2$ yield of 40.3% was achieved (Takiguchi and Shimahara 1988). Besides, a commercial bovine pepsin was used to degrade shrimp α-chitin for $(GlcNAc)_2$ production, and the final yields of the products GlcNAc, $(GlcNAc)_2$ and $(GlcNAc)_3$ were 19%, 75% and 9.5%, respectively (Ilankovan et al. 2006). A chitinase from a hyperthermophilic archaeon *Thermococcus kodakaraensis* KOD1 also hydrolyzed colloidal chitin to yield $(GlcNAc)_2$ as the predominant end product (Tanaka et al. 2003). Bulk-produced crude enzyme can be used to partially hydrolyze marine biopolymer chitin to produce $(GlcNAc)_2$. A pepsin with high chitinolytic activity hydrolyzed pepsin treated chitin at the conditions of pH 5.4 and 44 °C yielded 71.5% $(GlcNAc)_2$ after 24 h of incubation. A novel recombinant chitinase (PbChi70) from marine bacterium *Paenicibacillus barengoltzii* was expressed and purified to homogeneity. PbChi70 hydrolyzed colloidal chitin to produce 21.6 mg mL^{-1} $(GlcNAc)_2$ with the highest yield of 89.5% (Fig. 4.6). $(GlcNAc)_2$ was further separated to a purity of 99% with a final yield of 61% (Yang et al. 2016b). In addition, $(GlcNAc)_2$ can also be prepared by a successive microbial fermentation, firstly using *Lactobacillus paracasei* fermentation for demineralization, then *Serratia marcescens*

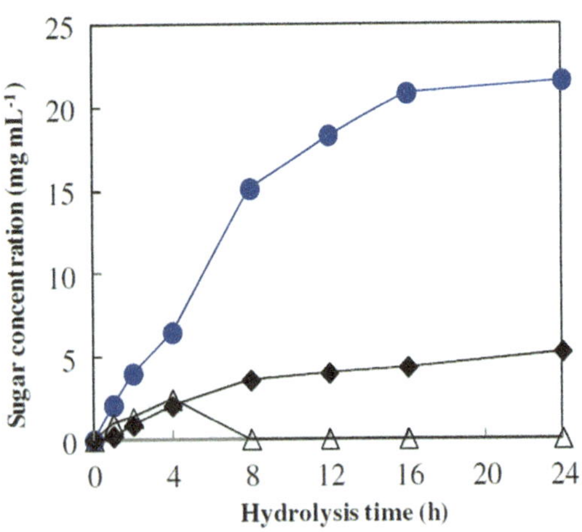

Fig. 4.6 Effect of incubation time on the production of $(GlcNAc)_2$ from colloidal chitin (Yang et al. 2016b) by PbChi70 from *P. barengoltzii*. Symbols: (◆) GlcNAc; (●) $(GlcNAc)_2$; (△) $(GlcNAc)_3$

fermentation for deproteinization, and finally *Aeromonas* sp. fermentation for the hydrolysis of the remained chitin.

Due to the catalytic mechanisms, most of the chitinolytic enzymes hydrolyze chitin to generate mainly GlcNAc and $(GlcNAc)_2$. Some chitinolytic enzymes also display transglycosylation ability except for the hydrolytic activity. So they have the ability to transfer the oligosaccharide moieties released to a suitable acceptora to form new glycosidic bonds. The transglycosylation activity enable these chitinolytic enzymes exhibits great potential in the synthesis of size- and stereo-specific *N*-acetyl COSs, or their correspondent derivatives. The transglycosylation reaction of glycolytic enzymes including chitosanase, chitinase and other glycosidases can also be used to synthesize high DP *N*-acetyl COSs. For example, *Trichoderma reesei* chitinase hydrolyzed $(GlcNAc)_4$ to yield $(GlcNAc)_2$ (55.7%) and $(GlcNAc)_6$ (39.6%) via a transglycosylation reaction. Akiyama et al. successfully synthesized *N*-acetyl COSs with DP 4–12 by a lysozyme-catalyzed transglycosylation reaction using *N,N′,N″*-triacetylchitotriose and *N,N′,N″*-tri(monochloro)acetylchitotriose as the substrates. The *N*-monochloroacetyl groups in the synthesized products were further removed by a base-catalyzed reaction. Hattori et al. successfully synthesize *N*-acetyl COSs with high DPs (6–15) through a lysozyme-mediated transglycosylation using $(GlcNAc)_3$ as starting substance in an aqueous system (containing 30% $(NH_4)_2SO_4$) (Jung and Park 2014). Partial production of *N*-acetyl COSs and GlcNAc by microbial enzymes can be found in Table 4.1.

Because of the catalytic mechanisms of most chitinolytic enzymes, chitin oligosaccharides mainly include $(GlcNAc)_2$ and GlcNAc, which can be easily purified. However, chitin oligosaccharides produced by enzymes mainly consist of $(GlcNAc)_2$ and GlcNAc lead to low diversity of oligosaccharides, which limited its commercial application. Although chitin oligosaccharides with high DPs can be produced by some chitinases with transglycosylation activity, the yields still remain low. Thus, to overcome these inefficiencies, screening chitinases with high transglycosylation activity or looking for other efficient synthetic strategies for oligosaccharides production from chitin is still of great value.

4.5 Purification and Characterization of Chitin Oligosaccharides

Chemical or enzymatic hydrolysis of chitin usually leads to a mixture of oligomers. So far, several separation and purification techniques for *N*-acetyl COSs production have been developed and widely used, such as gel filtration (Sørbotten et al. 2005), ultrafiltration (Lopatin et al. 2009), ion-exchange (Haebel et al. 2007), HPLC and TLC. Generally, preparative separation of *N*-acetyl COSs commonly use size exclusion chromatography (SEC) methods on the basis of their molecular sizes. A SEC system equipped with Superdex TM30 (GE Healthcare) columns (in series) can be used to separate *N*-acetyl COSs with similar DP values in the range of 2–20 efficiently.

Table 4.1 Partial enzymatic production of GlcNAc and *N*-acetyl COSs (Pichyangkura et al. 2002; Sashiwa et al. 2002, 2003; Thamthiankul et al. 2001; Lan et al. 2004; Kuk et al. 2005a; Kuk et al. 2006; Tanaka et al. 2003)

Chitin source	Enzyme	Enzyme source	Mol. Wt.	Condition	Product and yield	Analysis method
Swollen chitin	Crude enzyme	*Aeromonas* sp. GJ-18	–	40 °C, 9 days	GlcNAc 94.9%	HPLC (NH2P-50 4E)
Swollen chitin	Crude enzyme	*Aeromonas* sp. GJ-18	–	45 °C, 5 days	GlcNAc 74%, $(GlcNAc)_2$ 4.8%	HPLC (NH2P-50 4E)
				55 °C, 5 days	GlcNAc 3.9%, $(GlcNAc)_2$ 34.7%	
α-chitin, β-chitin	Crude enzyme	*A. hydrophila* H-2330	–	17 °C, 10 days	GlcNAc 64–77%	HPLC (NH2P-50)
Chitin	Chitinase	*A. hydrophila* H-2330	–	37 °C, 12 h	GlcNAc, $(GlcNAc)_2$~$(GlcNAc)_5$	TLC
Colloidal chitin	Exochitinase	*Bacillus thuringiensis* subsp. *pakistani*	66/60/47/32 kDa	37 °C, 24 h	GlcNAc	TLC
Swollen chitin	Chitinase	*Paenibacillus illinoisensis* KJA-424	38/54/63 kDa	37 °C, 24 h	GlcNAc 62.2%	HPLC
α-Chitin	Cellulase	*Trichoderma viride*	–	37 °C, 3 days	GlcNAc 16%	HPLC/NMR
		Acremonium cellulolyticus			GlcNAc 22%	
Chitin	Chitinase	*Thermococcus kodakaraensis* KOD1	90 kDa	70 °C, 3 h	$(GlcNAc)_2$	TLC
Chitin	Exochitinase	*Vibrio anguillarum* E-383a	–	–	$(GlcNAc)_2$ 40.3%	HPLC

In addition, there are some easier methods for *N*-acetyl COSs separation. For example, $(GlcNAc)_2$ can also be purified by an active charcoal column. Namely, the hydrolysate was firstly loaded onto an active charcoal column, and the bound sugars were eluted ethanol solution with a gradient of 0–40% (v/v) at a flow rate of 1 mL min^{-1}. The fractions were collected and checked for purity by TLC and HPLC, respectively (Yang et al. 2016b). Katano et al. presented a simple and rapid method for separation and purification of high DP *N*-acetyl COSs. Briefly, 10 mg of commercially available *N*-acetyl COSs sample consisting of $(GlcNAc)_n$ (n = 1–7) is mixed with 100 μL of the 1 M hydrochloric acid. All the $(GlcNAc)_n$ compounds are dissolved in the aqueous solution, and then mixed with 900 μL ethyl alcohol. The precipitate was collected by centrifugation and qualitatively analyzed. The precipitate consisted mainly $(GlcNAc)_n$ ($n \geqq 5$), indicating that the short-chain $(GlcNAc)_n$ compounds (with $n \leqq 3$) are quite different from the longer ones in solubility. By repeating the procedure, long-chain $(GlcNAc)_n$ sample ($n \geqq 5$) with high purity could be prepared. The long chain $(GlcNAc)_n$ compounds would be useful in the study of plant pathology, as well as chitin chemistry, since the long-chain $(GlcNAc)_n$ compounds are known to be good elicitors (Katano et al. 2017).

Although advances for separation of *N*-acetyl COSs have been achieved in last decade, the methods for separation and analysis of native *N*-acetyl COSs in low concentrations is still scare, and most of them are semi-quantitative, requiring derivatization, or having a relatively high detection limit or requiring specialized equipment. High performance anion exchange chromatography (HPAEC) combined with pulsed amperometric detection (PAD) is a useful tool for the sensitive detection, quantitation and separation of a wide range of underivatized carbohydrates. This method allows quantification and separation of a broad range of non-labeled carbohydrates. Munster et al. described and validated an efficient and rapid method for the separation and sensitive detection of non-labeled *N*-acetyl COSs with DP 1–6. In the procedure, 25 mM sodium hydroxide supplemented with 22.2 mM sodium acetate was used as mobile phase, and $(GlcNAc)_{1-6}$ could be separated well within 6 min. Namely, the CarboPac PA-1 column is operated at 25 °C, solution A, B and C were 0.1 M sodium hydroxide, 0.1 M sodium hydroxide with 0.6 M sodium acetate, and ultrapure water, respectively (van Munster et al. 2015).

The column was pretreated in five steps: (1) washing with solution A at a flow rate of 1 ml min^{-1} for 10 min; (2) washing with 100% solution B for 10 min; (3) washing with 3.7% B and 96.3% A for 10 min; (4) washing with 3.7% B, 21.3% A and 75% C for 20 min; (5) washing with solution C for 10 min to equilibrate the column prior to sample injection. Samples were separated by wash the column with solution C. Briefly, 25 μL $(GlcNAc)_{1-6}$ standards or samples containing internal standard (IS) (10 μM L-fucose) were injected using the auto sampler equipped with a sample loop. Subsequently, 225 mM sodium hydroxide was flowed at a speed of 2 ml min^{-1} and mixed in the mixing coil for few seconds. Sugar detection was done in 16 min after injection using a standard quadruple potential waveform, and the concentrations of *N*-acetyl COSs in the samples were calculated using a calibration curve showing the relation between the analyte amounts and the ratios of the peak areas of standards prepared with known amounts of *N*-acetyl COSs.

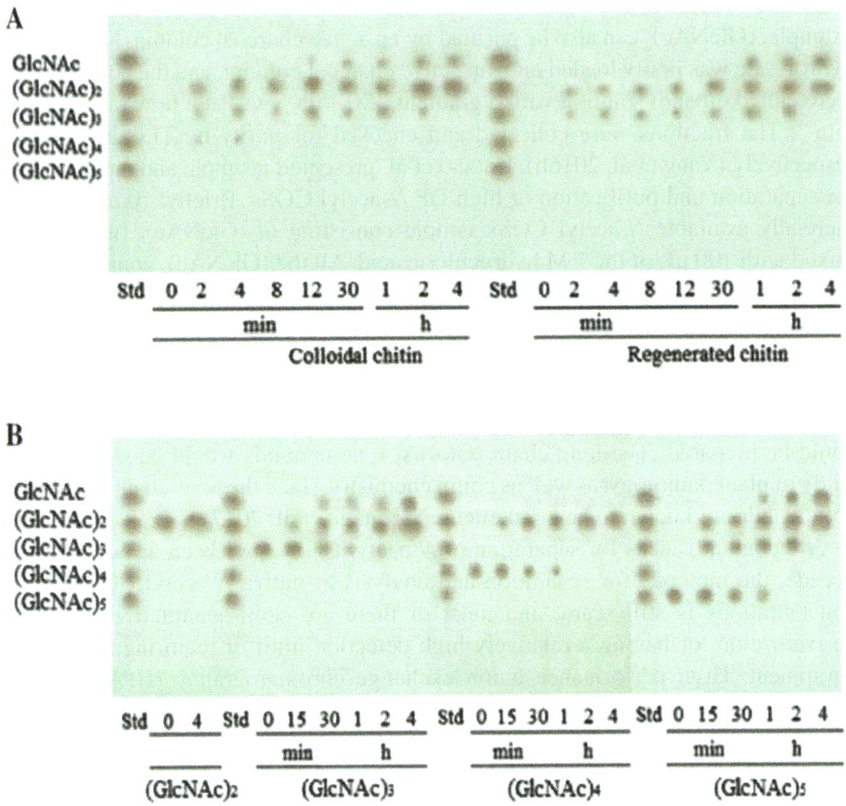

Fig. 4.7 TLC analysis (Yang et al. 2016b) of hydrolysis products of colloidal chitin (**a**) and *N*-acetyl COSs (DP 2–5) (**b**) by the chitinase from *P. barengoltzii*

After separation of *N*-acetyl COSs, nuclear magnetic resonance (NMR) can be a traditional and efficient method for further structural identification. Meanwhile, the methods mentioned above, such as TLC (Fig. 4.7), HPLC (Fig. 4.8a) and HPAEC-PAD (Fig. 4.9) could also be simple and efficient methods for preliminary characterization of *N*-acetyl COSs. For molecular weight analysis of *N*-acetyl COSs, MALDI-TOF mass spectrometry is a classical method (Fig. 4.8c). Briefly, the *N*-acetyl COSs sample is dissolved in the ultrapure water, and then mixed with an equal volume of 2, 5-dihydroxybenzoic acid (DHB) solution, and finally 0.75 μL of the mixture is spotted and dried at room temperature. All mass spectrometric data are collected on a spectrometer at 337 nm pulsed nitrogen laser (Zhang et al. 2017).

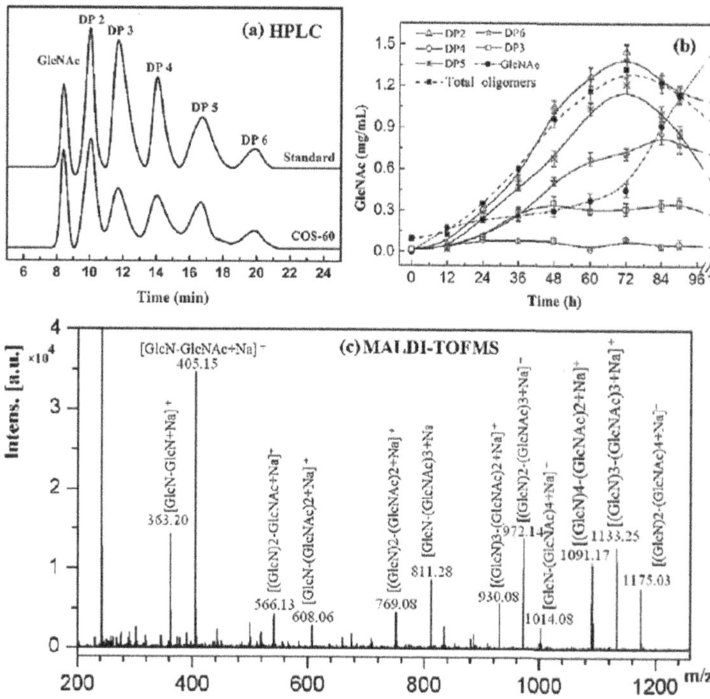

Fig. 4.8 Detection and identification of hydrolyzed products of chitin by chitinolytic enzymes. (**a**) HPLC chromatogram of *N*-acetyl COSs from HLP by chitinolytic enzymes; (**b**) Effect of hydrolysis time on the concentration of *N*-acetyl COSs from HLP by chitinolytic enzymes. (**c**) MALDI-TOF MS of *N*-acetyl COSs from HLP by chitinolytic enzymes (Zhang et al. 2017)

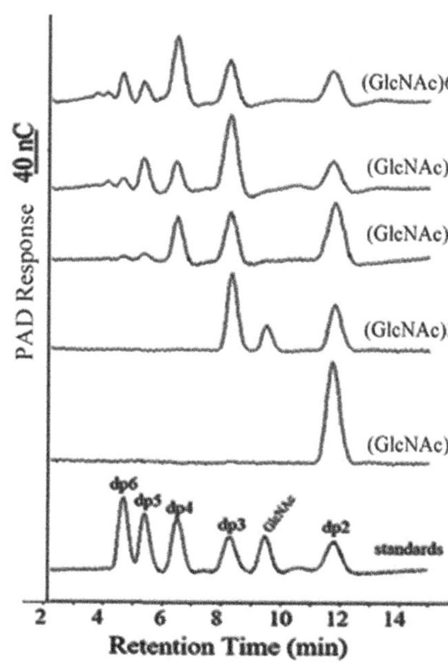

Fig. 4.9 HPAEC-PAD analysis of the hydrolysis products from *N*-acetyl COSs by a chitinase ChiEn1 (van Munster et al. 2015)

4.6 Enzymatic Synthesis of Chitin Oligosaccharide

4.6.1 Preparation of Chitin Oligosaccharide by Reverse Hydrolgesysis and Transglycosylation Reaction

Chitin oligosaccharide produced by chemical synthesis requires multiple protection and deprotection steps, and is not considered as bioactive material due to low yield rate and chance of toxicity. The preparation of chitin oligosaccharide by biosynthesis has attracted much attention. Two kinds of enzymes, glycosyltransferase and glycoside hydrolase, are widely used in the biosynthesis of chitin oligosaccharide. In comparison with glycosyltransferases, some glycoside hydrolases, which have the reverse hydrolysis or transglycosylation activity, have the advantage of broad substrate specificity, accessibility and much cheaper substrates. In nature, glycoside hydrolases generally undergo glycosidic hydrolysis, the leaving group in the reaction is oligosaccharide, and the nucleophile (glycosyl acceptor) is water. However, monosaccharides can also be used as nucleophiles to replace water, which provides the possibility that the reaction of glycosidase can be converted from hydrolysis to synthesis. Glycoside hydrolases have two modes for oligosaccharide synthesis: reverse hydrolysis and transglycosylation.

4.6.1.1 Reverse Hydrolysis Reaction

N-acetylglucosamine and β-*N*-acetylglucosaminidase are used to synthesize chitin oligosaccharide in reverse hydrolysis. Under the influence of leaving group, more reverse hydrolysis reactions use modified substrates. *p*NP-GlcNAc, which is easy to leave as the leaving group, is often used as the donor for the reverse hydrolysis reaction of β-*N*-acetylhexosaminidase. Only few literatures directly use *N*-acetylglucosamine as substrate. Rauvolfova et al. synthesized GlcNAc (β-1,3) GlcNAc, GlcNAc (β-1,4) GlcNAc and GlcNAc (β-1,6) GlcNAc with the β-*N*-acetylhexosaminidase from *Penicillium funiculosum* as the substrate, and the yield was 3.8%, 1.7% and 10.0%, respectively in 2004 (Rauvolfová et al. 2004). The reverse hydrolysis reaction is thermodynamically controlled and the final yield is limited by equilibrium. The reverse hydrolysis is mainly affected by the following aspects: (I) Concentration of substrates. By adding a high concentration of *N*-acetylglucosamine, the reaction can be facilitated in the direction of synthesis and also show a preference on the isomer formation. When the concentration of GlcNAc increased from 0.5 to 1.0 M, the yields of product GlcNAc (β-1,3) GlcNAc, and GlcNAc (β-1,6) GlcNAc also tripled. A continued increase in substrates may backfire by inhibiting enzyme activity. (II) Concentration of enzyme. Appropriately increasing the amount of β-*N*-acetylglucosaminidase can increase the transformation yield within a defined range, but excessive enzyme will promote the hydrolysis reaction. (III) Water activity. Each glycosidase has a critical water activity. When the water activity is below the critical level, the activity of the enzyme will decrease, but a high yield can be achieved. With high water activity, the reaction speed is fast, but the yield decreases. In the above study, the addition of salts ($(NH_4)_2SO_4$ or $MgSO_4$) resulted in increasing yield and improving the regioselectivity of β-*N*-acetylhexosaminidase. (IV) Concentration of product. Because the reaction is restricted by equilibrium, the reaction can also be promoted in the direction of synthesis through selective adsorption of products, thus improving the total yield. (V) The source of enzyme. So far, β-*N*-acetylglucosaminidases from fungal sources and glycoside hydrolase family 20 have been reported for synthesis of amino oligosaccharides, but that derived from bacteria have not been reported (Slámová et al. 2015).

4.6.1.2 Transglycosylation Reaction

The transglycosylation (TG) reaction is the reaction from new glycosidic bonds between the donor and acceptor sugar molecules. Glycosyl-enzyme intermediate will be formed in the reaction of the retaining glucoside hydrolase. If water attack the glycosyl-enzyme intermediate, the hydrolysis reaction, or attack by a sugar molecular, transglycosylation reaction occurs. Longer chitin oligosaccharide can be produced using chitinases and chitin oligosaccharides as substrates by transglycosylation reaction. According to the reports, most transglycosylation reactions are based on chitin oligosaccharide with DP 3–6. A chitinase-D from *Serratia marcescens* showed transglycosylation acitivity on $(GlcNAc)_{3-6}$ and produced longer chain COSs (Aronson et al. 2006). Endochitinase of *Flavobacterium johnsoniae* exhibited transglycosylation activity with $(GlcNAc)_5$ and $(GlcNAc)_6$ as substrate (Vaikuntapu et al. 2018). Only Asaki Uehara et al. has reported the chitinase from *Stenotrophomonas maltophilia* could synthesize 40% trisaccharide within 3 h with 10% $(GlcNAc)_2$ as substrate (Uehara et al. 2018). Easily to find, the yield of the product synthesized by the transglycosylation reaction is larger than reverse hydrolysis reaction, and the reaction time is shorter than that of reverse water (generally more than 4 days). In addition, the substrate specificity of chitinases in transglycosylation reaction is different.

Transglycosylation reactions are controlled by kinetics, and the amount of the donor usually needs to reach 2–4 times that of the receptor. The yield of transglycosylation is determined by the balance between the donor synthesis rate and the hydrolysis rate, and is controlled by the product hydrolysis (Sinha et al. 2016). In transglycosylation reaction process, when the reactants are exhausted and the concentration of products reaches the maximum. Chitinases are one of the important factors affecting the transglycosylation reaction, which are mainly from glycoside hydrolases family 18 and 19. The sources of chitinase with transglycosylation activity are extensive, e.g., *Serratia marcescens*, *Vibrio carchariae*, *Stenotrophomonas maltophilia*, *Cycas revolute*, *Enterobacter cloacae* and so on (Bhuvanachandra et al. 2018; Mallakuntla et al. 2017; Suginta et al. 2010; Vaikuntapu et al. 2016; Umemoto et al. 2016).

The site-specific mutation of key amino acids of chitinases is a promising approach to enhance the transglycosylation activity. Aromatic residues at glycoside binding sites are considered to be the key to the transglycosylation activity of GH18 chitinases (Martinez et al. 2012). Transglycosylation activity can be enhanced in chitinases mutants by removing or introducing side chains of tryptophan (Trp). Madhuprakash et al. confirmed that the indole rings of two adjacent Trp residues create a major hindrance for the product DP4 movement towards the catalytic center by studying the molecular dynamics simulations of mutants G119W (Madhuprakash et al. 2014). Aronson et al. studied conserved amino acid mutations in the binding grooves of chitinase derived from *Serratia marcescens* (Aronson et al. 2006). The mutant W167A performed transglycosylation with 45% $(GlcNAc)_3$ as substrate more efficiently than the wild type (8%). The above strategies serve the same purpose, allowing the substrate to remain in the active site for a longer time. Other studies have found that the nucleophile of chitinases is the key to the transglycosylation reaction. The glycosides near the glycosidic cleavage site was mutated into glutamic. As a result, the mutation of D246E in *Af*ChiB resulted the decrease of its chitinolytic activity and increase of its transglycosylation activity, while the mutation of W246E maintained its hydrolytic activity and loss its transglycosylation activity at all (Lü et al. 2009). These results suggest the strategy of knocking out nucleophiles Asp can improve the activity of transglycosylation.

4.6.2 Preparation of Chitin Oligosaccharide by Glycosidic Synthase and Glycosyl Transferases

4.6.2.1 Glycosidic Synthase

Glycosidases can not only catalyze the hydrolysis of glycosidic linkages, but also have the ability to form new glycosidic linkages. Transglycosylation is a common mode of reaction, which is usually controlled by kinetics catalyzed by a regenerative exoglycosidase using an activated glycoside. However, hydrolysis of the substrate and the transglycosylation product reduces the yield, typically from 20% to 40%. The classical enzymatic synthesis of glycosides has turned to specialized applications in the past decade, which has led to the need to the development of new methods such as substrate and enzymatic engineering (Honda and Kitaoka 2006).

A glycosidic synthase is a nucleophilic residue mutant of a glycoside hydrolase. The site-directed mutagenesis of a nucleophilic attack group of a glycoside hydrolase to an amino acid that does not have a nucleophilic function can destroy the first step of the hydrolysis reaction, which abolishes the natural hydrolysis activity. At the same time, using a-D-fluorinated sugar with self-capacitance ability as a glycosyl donor, the transglycoside reaction can be realized at the active center of the enzyme, which converting the glycoside hydrolase into a glycoside synthase, and effectively catalyzes the synthesis of glycosidic bonds. According to the different glycosidic synthesis mechanism, the existing glycoside synthase can be divided into two types: retained type and inverted type (Gloster et al. 2008). Retained glycoside synthase is a type of glycoside synthase engineered based on the catalytic mechanism of retained glycoside hydrolase. At present, most glycoside synthase enzymes are mutated by retained glycoside hydrolase, but inverted glycoside hydrolysis enzyme modification is also an important source of glycoside synthase (Cobucci-Ponzano et al. 2012).

Mucor hiemalis endo-β-*N*-acetylglucosaminidase (Endo-M) is a GH family 85 enzyme (Umekawa et al. 2008). Umekawa found that it not only catalyzes the hydrolysis but also catalyzes the transglycosylation reaction, transferring the released sugar chains to new glycosidic bonds formed on receptors other than water. The mutation of Asn 175, acts like a glycoside synthase, was found to "knock out" the hydrolytic activity, possessing the mutant transglycosylation ability using highly active glucoazolamide as donor substrate. This is the first report on a sugar synthase from endo-β-*N*-acetylglucosaminase, which is carried out by a substrate-assisted mechanism. Martinez found that the mutations on the key catalytic amino acids of chitinases from two GH families can enable them ability to act as sugar synthase enzymes to bind oxazoline-activated donors to chitin oligosaccharide receptors obtained from fermentation (Martinez et al. 2012). Chitinase A1 from *Bacillus circulans* (BcChiA1) and chitinase 42 from *Trichoderma harzanium* (ThChit42) were mutated on the three conserved carboxylic acids individually. The results suggested that all the amino acid residues (the general acid/base glutamate, and the two aspartates defined as the putative stabilizer and its assistant) play roles in the catalysis. These mutants can also be considered as chitin transglycosylases as they have the ability to catalyze the natural chitin oligosaccharide transglycosylation reaction, but exhibited no hydrolytic activity any more. Slámová successfully designed a GH family 20 β-*N*-acetylhexosaminidase from a fungus, and obtained three mutants of the corresponding transglycosidase by single amino acid mutation at the active site of the enzyme (Slámová et al. 2015). The specific procedure is to change the tyrosine 470 that interacts with the heterocyclic reaction intermediate to asparagine phenylalanine and histidine. These β-*N*-acetylhexosaminidase mutants were successfully used in the synthesis of two chitooligosaccharides with different aglycone moieties and no significant product hydrolysis, and the GOS products with DP 1–4 were purified and biochemically characterized. The synthesis method is simple and easy to be extended. Many typical β-*N*-acetyl-aminohexosaminease substrates have been tested as glycosyl donors, however, only *p*NP-GlcNAc was found to be suitable for higher oligomeric oligomers preparation (Bojarová et al. 2011). There is an exception that when chitotriose, a natural hexosaminidase, was used as the substrate, the higher oligomers were formed. This represents the first report on an exoglycosidase derived from transglycosidase using a substrate-assisted catalytic mechanism. In addition, different from the previous study using mutant chitinase and endo-β-*N*-acetylglucosaminidase, the product of the transglycosylation reaction was purified and characterized and further identified by NMR spectra.

At present, the research center of glycosidic synthase is in the principle verification stage, and there are some problems to be solved, such as poor stability of sugar donor, low vitality, and difficulty in industrial application. However, it has become one of the important tools for scientists to modify the target molecules of glycosylation, in order to increase water solubility and achieve structural and biological diversity. The glycosidic synthetase for industrial application should have the advantages of good enzyme stability, high catalytic efficiency, broad substrate spectrum and stable and easy availability of sugar donors. However, the development process of each industrial enzyme needs to be continuously improved and improved. Therefore, the acquisition of ideal glycosidic synthase still depends on the breakthrough development of related science and technology and the continuous efforts of scientific researchers.

4.6.2.2 Chemical Enzymatic Synthesis (Based on Glycosyl Transferases)

Glycosyltransferase is usually limited by the source, and difficult to isolate, purify and express heterogeneously. The synthesis catalyzed by this enzyme is highly specific and highly selective to both donor and receptor. Generally, nucleotide sugar (e.g. UDP-glucoside) is used as the donor for glycosylation reaction, which is expensive, difficult to prepare and cannot be obtained in large quantities. Thus, novel strategy, named chemical enzymatic method, based on glycosyl transferases was studied for synthesizing oligosaccharides in recent years. It is a new type of highly efficient synthesis technology based on enzyme-catalyzed synthesis and supplemented by chemical methods. The enzymatic synthesis has high selectivity, high specificity, no need for protection base operation, and high synthesis yield, thus greatly reducing the time and money investment. In addition, the method has a wet reaction condition, which reduces the risk of unsafeness and environmental pollution. Chemical enzymatic synthesis combines the advantages of chemical and enzymatic methods, and the two complement each other, and it is expected to become a new ideal method for synthesizing oligosaccharides.

An example of successful research was chemical enzymatic synthesis of heparin oligosaccharides. Researchers simulated the synthesis pathway of heparin in vivo, identify key enzymes for heparin synthesis, and construct engineering bacteria that can produce related enzymes. Efficient synthesis of heparin oligosaccharides is carried out under controlled conditions in vitro. Firstly synthesize a glycosyl donor and a sulfate-based donor, and then synthesize a di-hexasaccharide skeleton from a commercial p-nitrophenyl-β-D-glucuronide, and then chemically modify to synthesize different degrees of modification. Heparin oligosaccharides, due to the similarity of the backbone of chitin oligosaccharides and heparin oligosaccharides, it is suspected that chitin oligosaccharides can be synthesized using the following method:

GlcNAc (N-acetylglucosamine) or GlcNTFA (*N*-trifluoroacetylglucosamine), ATP (adenosine triphosphate) and UTP (uridine triphosphate) were used as raw materials to synthesize glycosyl donor UDP-GlcNTFA (Uridine diphosphate-*N*-trifluoroacetylglucosamine)/UDP-GlcNAc (Uridine diphosphate-*N*-acetylglucosamine) under the action of three recombinant enzymes NahK (*N*-acetyl hexosamine 1-position kinase), GlmU (1-phospho-N-acetylglucosamine uridine transaminase) and PPA (Pyrophosphorylase). The commercial nitrobenzene-β-D-glucuronide (GlcA-*p*NP) was used as a starting material to synthesize a two to six sugar skeleton under the catalysis of glycosyltransferases KfiA and PmHS2.

The chemical enzymatic method is highly efficient, specific and simple, and the reaction steps are simple. After a simple purification process, a gram-grade glycosyl donor can be prepared, and the purity of the product is as high as 98%, which provides high quality and sufficient for the synthesis of oligosaccharides. The yield of the oligosaccharide skeleton can be guaranteed to be about 99%. This method is worthy of being a new research direction for chitin oligosaccharides synthesis.

4.7 Perspectives

Though the production and economic coefficient of *N*-acetyl COS prepared by bio-production with enzymes or microbial fermentation is obviously lower than that of by traditional chemical methods, the bio-production is considered as a more promising route as the procedure is environmental friendly and the composition of product is easy to be controlled. From the commercial and industrial points of view, preparation of crude enzymes rather than pure enzymes, including chitosanase, chitinase, cellulase, lysozyme, protease, pepsin, and lipase are preferred for this purpose for oligosaccharides production. It appears that synthesis by the transglycosylation activity of chitinolytic enzymes is another promising way for *N*-acetyl COSs and their derivatives, especially for those *N*-acetyl COSs with special or desired molecules.

There is a big challenge in the bio-production of *N*-acetyl COSs from the raw chitin materials of different sources directly, including crab and shrimp shells and microbial cell walls. However, the production of oligosaccharides from chitin and crab shells are available through a mechanochemical grinding pretreatment followed by enzymatic hydrolysis. The existence of the crystal structure in chitin and crab shells significantly reduced the accessibility of the chitinolytic enzymes to the substrates, thus resulted in low enzymatic conversion yield. Hence, breaking down of the crystal structure is crucial in improving the enzymatic degradation efficiency of chitin materials. Solubilization is the preferred choice for the destruction of chitin crystallinity. However, in the route the substrate must be hydrolyzed by the certain enzymes with special properties, as the reactions should be performed in some organic or heterogeneous solvent systems. Hence, identification of novel enzymes that possess good stability in extreme environments is of great value for *N*-acetyl COSs bio-production.

Optimization of reactor system is another strategy to overcome the low efficiency of enzymatic bio-production. Some pretreatments have been development and combined with enzymatic methods, including boll milling, steam explosion, high pressure homogenization, ionic liquids, bacterial fermentation etc. These methods have been confirmed to be useful ways for improving chitin degradation efficiency. For instance, a reactor system with a dual ultrafiltration membrane has been described to be able to overcome the problems of reuse of enzyme in batch reactors as well as the poor affinity of the enzymes to substrate in column reactor systems.

Although enzymatic method is widely used for production of chito/chitin oligosaccharides. The chitin polysaccharide materials are restricted as the environmental pollution of manufacturing process and limited marine chitin resources. Thus, some novel biosynthetic pathway was investigated in recent years. Enzymatic synthesis of chito/chitin oligosaccharides using monosaccharides shows great potential in the future. The tool enzymes for the enzymatic synthesis of chito/chitin oligosaccharides are glycosyl transferases, transglycosidases/transglycosylases and glycosynthases.

References

Aam BB, Heggset EB, Norberg AL et al (2010) Production of chitooligosaccharides and their potential applications in medicine. Mar Drugs 8:1482–1517

Adrangi S, Faramarzi MA (2013) From bacteria to human: a journey into the world of chitinases. Biotechnol Adv 31:1786–1795

Aronson JN, Halloran BA, Alexeyev MF et al (2006) Mutation of a conserved tryptophan in the chitin-binding cleft of *Serratia marcescens* chitinase a enhances transglycosylation. Biosci Biotechnol Biochem 70:243–251

Bhattacharya D, Nagpure A, Gupta RK (2007) Bacterial chitinases: properties and potential. Crit Rev Biotechnol 27:21–28

Bhuvanachandra B, Madhuprakash J, Podile AR (2018) Active-site mutations improved the transglycosylation activity of, *Stenotrophomonas maltophilia*, chitinase A. Biochim Biophys Acta Protein Proteomics 1866:407–414

Bojarová P, Slámová K, Křenek K et al (2011) Charged hexosaminides as new substrates for β-*N*-acetylhexosaminidase-catalyzed synthesis of immunomodulatory disaccharides. Adv Synth Catal 353:2409–2420

Chen HC, Chang CC, Mau WJ et al (2002) Evaluation of *N*-acetyl chitooligosaccharides as the main carbon sources for the growth of intestinal bacteria. FEMS Microbiol Lett 209:53–56

Chen JK, Shen CR, Liu CL (2010) *N*-acetylglucosamine: production and applications. Mar Drugs 8:2493–2516

Cobucci-Ponzano B, Perugino G, Strazzulli A et al (2012) Thermophilic glycosynthases for oligosaccharides synthesis. Methods Enzymol 510:273–300

Ferrara MC, Cobucci-Ponzano B, Carpentieri A et al (2014) The identification and molecular characterization of the first archaeal bifunctional exo-β-glucosidase/*N*-acetyl-β-glucosaminidase demonstrate that family GH116 is made of three functionally distinct subfamilies. Biochim Biophys Acta 1840:367–377

Fu X, Yan QJ, Wang J et al (2016) Purification and biochemical characterization of novel acidic chitinase from *Paenicibacillus barengoltzii*. Int J Biol Macromol 91:973–979

Fujita K, Shimomura K, Yamamoto K et al (2006) A chitinase structurally related to the glycoside hydrolase family 48 is indispensable for the hormonally induced diapause termination in a beetle. Biochem Biophys Res Commun 345:502–507

Gloster TM, Turkenburg JP, Potts JR et al (2008) Divergence of catalytic mechanism within a glycosidase family provides insight into evolution of carbohydrate metabolism by human gut flora. Chem Biol 15:1058–1067

Guillen D, Sanchez S, Rodriguez-Sanoja R (2010) Carbohydrate-binding domains: multiplicity of biological roles. Appl Microbiol Biotechnol 85:1241–1249

Haebel S, Bahrke S, Peter MG (2007) Quantitative sequencing of complex mixtures of heterochitooligosaccharides by MALDI-linear ion trap mass spectrometry. Anal Chem 79:5557–5566

Honda Y, Kitaoka M (2006) The first glycosynthase derived from an inverting glycoside hydrolase. J Biol Chem 281:1426–1431

Husson E, Hadad C, Huet G et al (2017) Effect of room temperature ionic liquids on selective biocatalytic hydrolysis of chitin via sequential or simultaneous strategies. Green Chem 19:4122–4131

Ilankovan P, Hein S, Ng CH et al (2006) Production of *N*-acetyl chitobiose from various chitin substrates using commercial enzymes. Carbohydr Polym 63:245–250

Jung WJ, Park RD (2014) Bioproduction of chitooligosaccharides: present and perspectives. Mar Drugs 12:5328–5356

Jung WJ, Jo GY, Kuk JH et al (2007) Production of chitin from red crab shell waste by successive fermentation with *Lactobacillus paracasei* KCTC-3074 and *Serratia marcescens* FS-3. Carbohydr Polym 68:746–750

Katano H, Noba S, Sato K et al (2017) Solubility-based separation and purification of long-chain chitin oligosaccharides with an organic-water mixed solvent. Anal Sci 33:639–642

Katta S, Ankati S, Podile AR (2013) Chitooligosaccharides are converted to *N*-acetylglucosamine by *N*-acetyl-β-hexosaminidase from *Stenotrophomonas maltophilia*. FEMS Microbiol Lett 348:19–25

Kuk JH, Jung WJ, Jo GH et al (2005a) Selective preparation of *N*-acetyl-D-glucosamine and *N,N'*-diacetylchitobiose from chitin using a crude enzyme preparation from *Aeromonas* sp. Biotechnol Lett 27:7–11

Kuk JH, Jung WJ, Jo GH et al (2005b) Production of *N*-acetyl-β-D-glucosamine from chitin by *Aeromonas* sp. GJ-18 crude enzyme. Appl Microbiol Biotechnol 68:384–389

Kuk JH, Jung WJ, Jo GH et al (2006) Production of *N,N'*-diacetylchitobiose from chitin using temperature-sensitive chitinolytic enzyme preparations of *Aeromonas* sp. GJ-18. World J Microbiol Biotechnol 22:135–139

Lan X, Ozawa N, Nishiwaki N et al (2004) Purification, cloning, and sequence analysis of β-*N*-acetylglucosaminidase from the chitinolytic bacterium *Aeromonas hydrophila* strain SUWA-9. Biosci Biotechnol Biochem 68:1082–1090

Lopatin SA, Derbeneva MS, Kulikov SN et al (2009) Fractionation of chitosan by ultrafiltration. J Anal Chem 64:648–651

Lü Y, Yang H, Hu H et al (2009) Mutation of Trp137 to glutamate completely removes transglycosyl activity associated with the *Aspergillus fumigatus* AfChiB1. Glycoconj J 26:525

Madhuprakash J, Tanneeru K, Karlapudi B et al (2014) Mutagenesis and molecular dynamics simulations revealed the chitooligosaccharide entry and exit points for chitinase D from *Serratia proteamaculans*. Biochim Biophys Acta 1840:2685–2694

Mallakuntla MK, Vaikuntapu PR, Bhuvanachandra B et al (2017) Transglycosylation by a chitinase from *Enterobacter cloacae* subsp. cloacae generates longer chitin oligosaccharides. Sci Rep 7(1):5113

Margoutidis G, Parsons V, Bottaro CS et al (2018) Mechanochemical amorphization of α-chitin and conversion into oligomers of *N*-acetyl-D-glucosamine. ACS Sustain Chem Eng 6:1662–1669

Martinez EA, Boer H, Koivula A et al (2012) Engineering chitinases for the synthesis of chitin oligosaccharides: catalytic amino acid mutations convert the GH-18 family glycoside hydrolases into transglycosylases. J Mol Catal B Enzym 74:89–96

Monge EC, Tuveng TR, Vaaje-Kolstad G et al (2018) Systems analysis of the family glycoside hydrolase family 18 enzymes from *Cellvibrio japonicus* characterizes essential chitin degradation functions. J Biol Chem 293:3849–3859

van Munster JM, Sanders P, Geralt A et al (2015) Kinetic characterization of *Aspergillus niger* chitinase CfcI using a HPAEC-PAD method for native chitin oligosaccharides. Carbohydr Res 407:73–78

Osada M, Miura C, Nakagawa YS et al (2012) Effect of sub- and supercritical water pretreatment on enzymatic degradation of chitin. Carbohydr Polym 88:308–312

Patil NS, Jadhav JP (2014) Enzymatic production of *N*-acetyl-D-glucosamine by solid state fermentation of chitinase by *Penicillium ochrochloron* MTCC 517 using agricultural residues. Int Biodeterior Biodegrad 91:9–17

Pichyangkura R, Kudan S, Kuttiyawong K et al (2002) Quantitative production of 2-acetamido-2-deoxy-D-glucose from crystalline chitin by bacterial chitinase. Carbohydr Res 337:557–559

Qin Z, Xiao YB, Yang XB et al (2015) A unique GCN5-related glucosamine *N*-acetyltransferase region exist in the fungal multidomain glycoside hydrolase family 3 β-*N*-acetylglucosaminidase. Sci Rep 5:18292

Rauvolfová J, Weignerová L, Kuzma M et al (2004) Enzymatic synthesis of *N*-acetylglucosaminobioses by reverse hydrolysis: characterisation and application of the library of fungal β-*N*-acetylhexosaminidases. J Mol Catal B Enzym 29:259–264

Sashiwa H, Fujishima S, Yamano N et al (2002) Production of *N*-acetyl-D-glucosamine from α-chitin by crude enzymes from *Aeromonas hydrophila* H-2330. Carbohydr Res 337:761–763

Sashiwa H, Fujishima S, Yamano N et al (2003) Enzymatic production of *N*-acetyl-D-glucosamine from chitin. Degradation study of *N*-acetylchitooligosaccharide and the effect of mixing of crude enzymes. Carbohydr Polym 51:391–395

Sinha S, Chand S, Tripathi P (2016) Enzymatic production of glucosamine and chitooligosaccharides using newly isolated exo-β-D-glucosaminidase having transglycosylation activity. Biotechnology 6:13

Slámová K, Krejzová J, Marhol P et al (2015) Synthesis of derivatized chitooligomers using transglycosidases engineered from the fungal GH20 β-*N*-Acetylhexosaminidase. Adv Synth Catal 357:1941–1950

Sørbotten A, Horn SJ, Eijsink VG et al (2005) Degradation of chitosans with chitinase B from *Serratia marcescens*. Production of chito-oligosaccharides and insight into enzyme processivity. FEBS J 272:538–549

Suginta W, Vongsuwan A, Songsiriritthigul C et al (2010) Enzymatic properties of wild-type and active site mutants of chitinase A from *Vibrio carchariae*, as revealed by HPLC-MS. FEBS J 272:3376–3386

Sumida T, Fujimoto K, Ito M (2011) Molecular cloning and catalytic mechanism of a novel glycosphingolipid-degrading beta-*N*-acetylgalactosaminidase from *Paenibacillus* sp. TS12. J Biol Chem 286:14065–14072

Suresh PV, Kumar PKA (2012) Enhanced degradation of α-chitin materials prepared from shrimp processing byproduct and production of *N*-acetyl-D-glucosamine by thermoactive chitinases from soil mesophilic fungi. Biodegradation 23:597–607

Tabasso S, Carnaroglio D, Calcio GE et al (2015) Microwave, ultrasound and ball mill procedures for bio-waste valorisation. Green Chem 17:684–693

Takiguchi Y, Shimahara K (1988) *N,N'*-Diacetylchitobiose production from chitin by *Vibrio anguillarum* strain E-383a. Lett Appl Microbiol 6:129–131

Tanaka T, Fukui T, Atomi H et al (2003) Characterization of an exo-β-D-glucosaminidase involved in a novel chitinolytic pathway from the hyperthermophilic archaeon *Thermococcus kodakaraensis* KOD1. J Bacteriol 185:5175–5181

Thamthiankul S, Suan-Ngay S, Tantimavanich S et al (2001) Chitinase from *Bacillus thuringiensis* subsp *pakistani*. Appl Microbiol Biotechnol 56:395–401

Udaya Prakash NA, Jayanthi M, Sabarinathan R et al (2010) Evolution, homology conservation, and identification of unique sequence signatures in GH19 family chitinases. J Mol Evol 70:466–478

Ueda M, Ohata K, Konishi T et al (2009) A novel goose-type lysozyme gene with chitinolytic activity from the moderately thermophilic bacterium *Ralstonia* sp. A-471: cloning, sequencing, and expression. Appl Microbiol Biotechnol 81:1077–1085

Uehara A, Takahashi N, Moriyama M et al (2018) Synthesis of chitin oligosaccharides using dried *Stenotrophomonas maltophilia* cells containing a transglycosylation reaction-catalyzing β-*N*-acetylhexosaminidase as a whole-cell catalyst. Appl Biochem Biotechnol 184:673–684

Umekawa M, Huang W, Li B et al (2008) Mutants of mucor hiemalis endo-β-*N*-acetylglucosaminidase show enhanced transglycosylation and glycosynthase-like activities. J Biol Chem 283:4469–4479

Umemoto N, Ohnuma T, Osawa T et al (2016) Modulation of the transglycosylation activity of plant family GH 18 chitinase by removing or introducing a tryptophan side chain. FEBS Lett 589:2327–2333

Vaikuntapu PR, Rambabu S, Madhuprakash J et al (2016) A new chitinase-D from a plant growth promoting *Serratia marcescens* GPS5 for enzymatic conversion of chitin. Bioresour Technol 220:200–207

Vaikuntapu PR, Mallakuntla MK, Das SN et al (2018) Applicability of endochitinase of *Flavobacterium johnsoniae* with transglycosylation activity in generating long-chain chitooligosaccharides. Int J Biol Macromol 117:62–71

Wakita S, Kimura M, Kato N et al (2017) Improved fluorescent labeling of chitin oligomers: Chitinolytic properties of acidic mammalian chitinase under somatic tissue pH conditions. Carbohydr Polym 164:145–153

Wang SL, Chang WT (1997) Purification and characterization of two bifunctional chitinases/lysozymes extracellularly produced by *pseudomonas aeruginosa* K-187 in a shrimp and crab shell powder medium. Appl Environ Microbiol 63:380–386

Wang Z, Zheng LH, Yang SL et al (2007) *N*-Acetylchitooligosaccharide is a potent angiogenic inhibitor both in vivo and in vitro. Biochem Biophys Res Commun 357:26–31

Wei G, Zhang AL, Chen K et al (2017) Enzymatic production of *N*-acetyl-D-glucosamine from crayfish shell wastes pretreated via high pressure homogenization. Carbohydr Polym 171:236–241

Wendland J, Schaub Y, Walther A (2009) *N*-acetylglucosamine utilization by *Saccharomyces cerevisiae* based on expression of *Candida albicans NAG* genes. Appl Environ Microbiol 75:5840–5845

Wu X, Wang J, Shi YQ et al (2017) *N*-Acetyl-chitobiose ameliorates metabolism dysfunction through Erk/p38 MAPK and histone H3 phosphorylation in type 2 diabetes mice. J Funct Foods 28:96–105

Yang SQ, Song S, Yang QJ et al (2014) Biochemical characterization of the first fungal glycoside hydrolyase family 3 β-*N*-acetylglucosaminidase from *Rhizomucor miehei*. J Agric Food Chem 62:5181–5190

Yang SQ, Fu X, Yan QJ et al (2016a) Biochemical characterization of a novel acidic exochitinase from *Rhizomucor miehei* with antifungal activity. J Agric Food Chem 64:461–469

Yang SQ, Fu X, Yang QJ et al (2016b) Cloning, expression, purification and application of a novel chitinase from a thermophilic marine bacterium *Paenibacillus barengoltzii*. Food Chem 192:1041–1048

Yin H, Du YG, Dong ZM (2016) Chitin oligosaccharide and chitosan oligosaccharide: two similar but different plant elicitors. Front Plant Sci 7:522

Zakariassen H, Klemetsen L, Sakuda S et al (2010) Effect of enzyme processivity on the efficacy of a competitive chitinase inhibitor. Carbohydr Polym 82:779–785

Zhang Y, Zhou XL, Ji LS et al (2017) Enzymatic single-step preparation and antioxidant activity of hetero-chitooligosaccharides using non-pretreated housefly larvaepowder. Carbohydr Polym 172:113–119

Zhang AL, Wei G, Mo XF et al (2018) Enzymatic hydrolysis of chitin pretreated by bacterial fermentation to obtain pure *N*-acetyl-D-glucosamine. Green Chem 20:2320–2327

Detection and Separation of Chito/Chitin Oligosaccharides

5

Zhen Qin, Yanying Hou, Waheed Ahamed, Yun Li, and Liming Zhao

Abstract

The bioactivities of chito/chitin oligosaccharides were closely related to their structures, degree of deacetylation (DD) and degree of polymerization (DP). In that case, separation and purification of chito/chitin oligosaccharides is highly beneficial for the further research and application of chito/chitin oligosaccharides, which is essential for the production of high value chito/chitin oligosaccharides in industrial scale. In addition, detection and analysis of chito/chitin oligosaccharides is helps to understand the structure-function relationship more clearly. The chito/chitin oligosaccharides with different degrees of polymerization have different molecular weights, different charge numbers, and different strengths from the separation medium. The separation of chito/chitin oligosaccharides is mainly achieved by combining membrane techniques such as ultrafiltration, nanofiltration and reverse osmosis together to obtain chito oligosaccharides with high purity and narrow interval distribution. For chito/chitin oligosaccharide purification, the most common used methods are absorption chromatography, affinity chromatography, size exclusion chromatography and ion exchange chromatography. For chito/chitin oligosaccharides analysis, TLC, HPLC and electrophoresis were used for qualitative and quantitative analysis of chito/chitin oligosaccharide. MS, IR, NMR and TGA were developed for the structural and sequence analysis according to the sample chemical structure. This chapter mainly reviews recent developments in separation and analysis of chito/chitin oligosaccharides. A variety of separation and analysis methods and examples are discussed.

Z. Qin · Y. Hou · W. Ahamed · Y. Li · L. Zhao (✉)
State Key Laboratory of Bioreactor Engineering, School of Biotechnology, East China University of Science and Technology, Shanghai, China
e-mail: zhaoliming@ecust.edu.cn

L. Zhao (ed.), *Oligosaccharides of Chitin and Chitosan*,
https://doi.org/10.1007/978-981-13-9402-7_5

Keywords
Detection · Separation · Purification · Chitooligosaccharide · Chitinoligosaccharide

5.1 Introduction

The bioactivities of chito/chitin oligosaccharides were closely related to their structures, degree of deacetylation (DD) and degree of polymerization (DP). In that case, detection and separation of chito/chitin oligosaccharides are highly beneficial for the better understanding of the structure and functions relationship of chito/chitin oligosaccharides and production of high value chito/chitin oligosaccharides in industrial scale. However, due to the complex structures of chito/chitin oligosaccharides and their similar physicochemical properties between different degrees of polymerization, separation and purification of chito/chitin oligosaccharides remains to be further developed.

The separation of chito/chitin oligosaccharides now is mainly focused on two aspects: using membrane technology and chromatographic technology. Membrane technology such as nanofiltration, reverse osmosis and electrodialysis possess huge advantages in separating chito/chitin oligosaccharides on large scales in narrow range distribution with high purities. However, it is unattainable to purify chito/chitin oligosaccharides monomers with different DP. While chromatographic technology is relatively easy to get the monomer in specific DP with high purity but it has many problems when using on large-scale application. To solve the problem, many kinds of chromatographic technologies have been developed such as SEC, absorption chromatography, affinity chromatography and IEC. All these technologies made enormous contribution to the development on chito/chitin oligosaccharides separation.

Different methods were usually combined together to get the specific quantity and structure of chito/chitin oligosaccharides. TLC, HPLC and electrophoresis were used for qualitative and quantitative analysis of chito/chitin oligosaccharides. However, in these methods, standard substance needs to be provided to compare the contents of COS. In that case, MS, IR, NMR and TGA were developed for the structural and sequence analysis according to the sample chemical structure.

This chapter mainly focuses on the recent developments in separation and analysis of chito/chitin oligosaccharides. A variety of separation and analysis methods are discussed.

5.2 Separation and Purification of Chito/Chitin Oligosaccharides

COS are degraded products of chitosan and are composed of 2–10 glucosamine units linked by β-1, 4-O-glucoside bond (Li et al. 2016) which possess a variety of physiological activities. Due to the complex structures of COS and their similar physicochemical properties between different DP, it is very difficult to separate them and get the purified products. At present, the reported methods for separation and preparation of COS monomers are mainly by membrane and chromatographic technologies.

5.2.1 Separation by Membrane Technology

Membrane separation technology is a high efficiency technique used for separation, concentration and purification of molecules. It has been developed rapidly in the past 40 years and widely used in industrial fields. Compared with other separation technology, membrane separation has many advantages, such as no phase changes in process, easy and continuous operation at room temperature, simple equipment, low energy consumption, high separation efficiency, no secondary pollution and so on. Therefore, membrane separation technology has become an important high-tech and sustainable development technology in the twenty-first century. The principle of membrane separation is the separation of different substances by the selective permeability of membranes based on external forces. At present, membrane separation process is mainly divided into microfiltration, UF, NF, RO, ED and pervaporation. A schematic diagram of several membrane separation processes is shown in Fig. 5.1. This technology is mainly used in the fields of food, medicine, industrial wastewater treatment, ultra-pure water preparation and biotechnology field. COS have many advantages such as good solubility, good absorption moisture, good

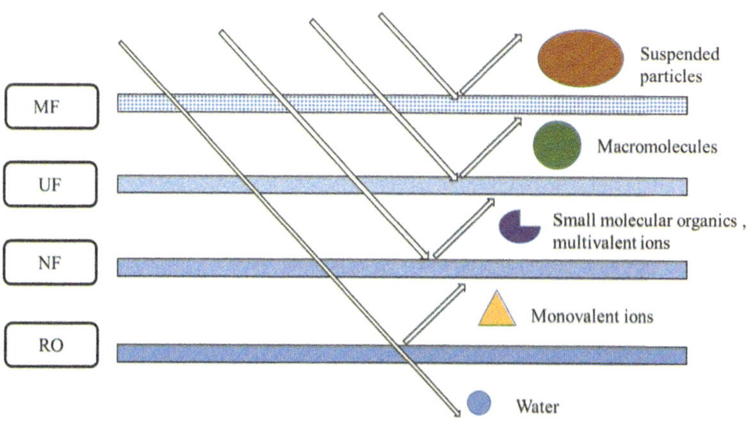

Fig. 5.1 Schematic diagram of membrane separation principal

biocompatibility, and have many functions like antibacterial, antitumor anti-cancer, anti-oxidation, lower cholesterol, regulating the body's immune function, etc. The biological activities of COS depend on their degree of polymerization. However, the degree of polymerization of COS mixture prepared directly was wide in range, which has influence to the study of COS. Technologies that could separate different DP of COS are meaningful for application and further functional research. Based on the molecular weight and charge characteristics of COS, the current separation and purification methods for COS mainly include membrane separation and chromatographic separation. Different types of membranes can be used for the separation of COS, such as UF membranes, NF membranes and RO membranes, UF can be used to remove macromolecular substances in the COS solution, especially the enzyme remaining in the chitosan solution. Small molecular substances such as salts and monosaccharides in the COS solution can be removed by NF or RO. Separation of COS by ED is a relatively novel approach. The principle of ED is the separation of COS on the base of electric charge under electric potential. In preparing of COS, UF, NF, RO and ED are the common separation methods (Table 5.1).

Table 5.1 Classification and characteristics of membrane separation techniques

Membrane separation technique	Principle	Driving force (kPa)	Intercept component	Application
Microfiltration	Sieving	20–100	0.1–20 um	Clarification, separation, removal of bacteria, and filtration
Ultrafiltration	Sieving	100–1000	5–100 nm	Concentration, grading, and purification of macromolecular solution
Nanofiltration	Dissolving diffusion, Donnan effect	500–1500	>1 nm	Separation, purification, and enrichment process of food, medicine, and biochemical industries
Reverse osmosis	Dissolving diffusion	1000–10,000	0.1–1 nm	Concentration of low molecular-weight components and removal of dissolved salts in aqueous solutions
Electrodialysis	Ion exchange	Electrochemical potential—penetration	Large ions and water	Removal of salt and deacidification of solutions containing neutral components
Pervaporation	Dissolving diffusion	Concentration difference	Insoluble or nonvolatile components	It is mainly used for volatile organic pollutants in the product separation and enrichment

5.2.1.1 Ultrafiltration

UF refers to the sieve separation process under the action of pressure force. Solutes with size of 5–100 nm, which are mainly macromolecule compounds, antibodies and viruses, were separated from the mixture. UF membranes with specific molecular weight cut-offs can be selected to effectively separate solutions with different molecular weight ranges. Under ideal conditions, the macromolecules with large molecular weight are retained in the solution by the UF membrane, on the contrary, the desired low molecular weight molecules pass through the membrane and collected in permeate. It was widely applied in the classification of oligosaccharides. In 1999, there were studies on obtaining oligosaccharides by enzymatic method and then maximizing the yield of products with a DP value of 3–10 by membrane separation. UF was successfully used to separate and prepare glucosamine oligosaccharides. Dong Huizhong (Dong et al. 2014) prepared the COS from DP 6 to DP 8 by a UF membrane with molecular weight cut-off 3000 Da and a molecular weight cut-off 500 Da NF membrane, the yield and purity of the product were 73.9% and 82.2%. Chen Jianguo (Chen 2006) prepared COS by fermentation then filtered out to remove the high molecular weight chitosan, chitosanase and other proteins by UF membrane (molecular weight cut-off, 10 kDa). Then the filtrate was filtered by UF membrane (molecular weight cut-off, 5 kDa and 2 kDa) to obtain COS solutions with different molecular weights, 5–10 kDa, 2–5 kDa and smaller than 2 kDa, respectively. COS were obtained by vacuum concentration and spray drying. Ngo D N (Ngo et al. 2009) used an UF membrane system to obtain two kinds of NA-COS with molecular weight 1–3 kDa (NA-COS 1–3 kDa) and below 1 kDa (NA-COS<1 kDa). Hu Rui (Hu 2012) used membrane to separate and purify the enzymatic hydrolysate of chitosan with average molecular weight of 500–1000 Da, the enzyme protein and unreacted chitosan were effectively rejected by the UF membrane, partial monosaccharide and monovalent ions are effectively rejected by NF membrane to achieve concentration of COS at a lower temperature by low energy consumption.

Recently, membrane-coupled enzymatic hydrolysis technology was widely used to prepare high DP oligosaccharides. In 2000, Jeon Y J (Jeon and Kim 2000) showed that the UF membrane reactor system can be used to produce COS by enzymatic hydrolysis. In addition, oligosaccharides with a relatively high DP (trimer to hexamer), were successfully obtained at a yield of above 80% by adjusting the permeation rate. Kuroiwa T (Kuroiwa et al. 2009) studied the production of COS using a continuous enzyme membrane bioreactor. DP 5 and DP 6 COS were obtained in the penetrants at a concentration of 2.3 g/L under optimized conditions. This concentration corresponds to a 46% value of the feed substrate concentration which is very high. The half-life of the productivity of the bioreactor system is estimated to be 50 days. The results showed that the proposed membrane bioreactor has great advantages in selectivity (yield of target product) and stable operation. The reactor should be useful for the actual production of high value COS. Sinha S (Sinha et al. 2014) used an enzyme membrane reactor equipped with an UF membrane of 3000 Da cut-off to fractionate COS produced by the purified chitosanase enzyme and an improved yield of chitosan pentamer (0.1 mg/mL) and hexamer (0.6 mg/mL) was achieved.

5.2.1.2 Nanofiltration and Reverse Osmosis

NF refers to the process of separating and removing small molecular substances with molecular weight of 300–500 Da from aqueous solution. NF is a new membrane separation technology, provides a selective medium between UF and reverse osmosis. The separation mechanism of NF is very complex; mainly contain dissolving diffusion and donnan effect. NF had applied to fractionate, concentrate and desalinate the sample solution. The loss of solute can be minimized. RO is the process of separating liquid mixture under the difference of pressure on the two sides of membrane, which can remove dissolved salts in aqueous solutions. Their application fields have developed from early desalination to the separation of some organic and inorganic substances in chemical, food, pharmaceutical and paper industries. Han Yongping (Han and Lin 2012) studied the purification of COS preparation liquid by NF. The optimum operating conditions for the NF were 35–40 °C, 1.0 MPa, 2.5 times diluted before NF, and 1.5 times concentrated for the intermittent constant volume percolation process. Under these conditions, monosaccharides, monovalent salt ions and most of the disaccharide from the preparation liquid were extracted, and the purity of the highly active COS was more than 92%, and the concentration of the liquid was nearly doubled. The chitin oligosaccharides, prepared by enzymatic hydrolysis, were classified by different molecular weights by UF and NF, and the salts and monosaccharides were filtered off, and then subjected to reverse osmosis concentration and spray drying to obtain chitin oligosaccharides with high purity and low polymerization degree. The purity of chitin oligosaccharides containing DP 2, DP 3 and DP 4 were 71.3%, 23.5% and 2.4% respectively (Qiao et al. 2010). Wei Xinlin prepared COS by enzymatic hydrolysis, enzymes were obtained from Aspergillus fumigatus BSF114. Pentamer (DP 5) and hexamer (DP 6) were isolated and purified from COS by the UF, NF, ethanol precipitation and the CM-Sephadex C-25 column (Wei et al. 2009).

5.2.1.3 Electrodialysis

When dialysis is carried out under the action of electric field, the phenomenon that the charged solute particles in the solution move through the membrane under electric driven force is called electrodialysis. The membrane used for ED is actually an ion-exchange membrane, divided into cationic exchange membrane (CEM) and anion exchange membrane (AEM) according to the charge. In the electrodialysis process, the positive membrane allows the anion to pass through and repel the cation, and the negative membrane allows the cation to pass through and repel the anion, which is the selective permeability of the ion exchange membrane. ED with UF membrane is a hybrid electro separation process combining conventional ED and UF membranes, which was recently tested for the separation of several single COS (DP 2–4) because of their cationic behavior in an acidic medium. The dimer showed the highest electromigration rate, followed by trimer, and then by tetramer. The effect of the molecular weight cut-off of the UF membrane, pH, EDUF cell

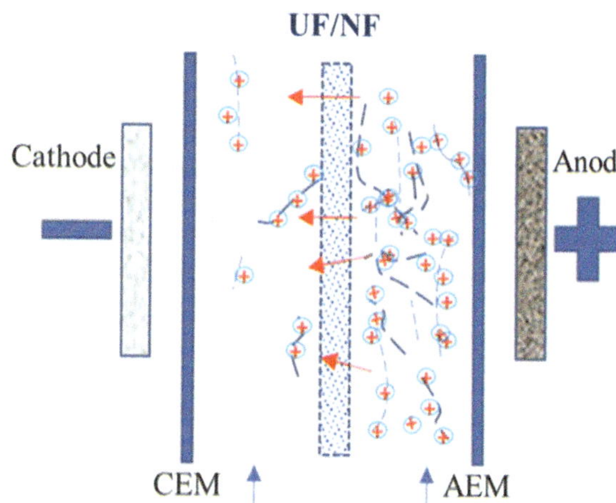

Fig. 5.2 Schematic diagram of ED process

configuration, solution flow velocity and electric field strength on the electromigration behavior of COS were studied. It was reported that the 10,000 Da MWCO UF membrane was the best membrane to separate of the COS tetramer (Aider et al. 2006a). COS electromigration decreased by increasing the pH (Aider et al. 2006b). COS migration and separation was observed only in the cell configuration of anode-AEM1-UFM-AEM2-cathode, solution flow velocity was not significant, while the electric field strength showed a significant impact on each COS electromigration rate, with 2.5 V/cm, it was possible to obtain a solution composed only of DP 2 and DP 3 (Aider et al. 2006a, b, 2008a, b, 2009) (Fig. 5.2).

Since the structure of different DP COS is similar and the difference of molecular weight is small, it is very difficult to prepare a single polymerization degree COS by membrane separation technology alone. Moreover, the concentration of COS enzymatic solution used in most membrane separation studies is currently low, which is unfavorable for the efficiency of separation. Therefore, it is necessary to carry out further fine separation and preparation by combining techniques such as chromatography.

5.2.2 Separation by Chromatographic Technology

Chromatographic separation technique is an effective method for separating the target composition from mixture, which has been widely applied in pharmaceuticals, functional food and fine chemicals. Different compositions distribute in the stationary phase and mobile phase repeatedly, and pure products are acquired by different distribution coefficient of the two phases. The biological activities of glucosamine oligosaccharides depend on their degree of polymerization. However, most of the glucosamine oligosaccharide products are mixture with a wide range of DP. The functional research of glucosamine oligosaccharides was limited by the lack of single DP product, especially high DP glucosamine oligosaccharides (DP $\geq$ 4). Size exclusion chromatography, absorption chromatography, affinity chromatography and ion exchange chromatography have been applied in the separation of glucosamine oligosaccharides to prepare single DP or certain DP range mixture.

5.2.2.1 Size Exclusion Chromatography

Size exclusion chromatography is also called gel-filtration or gel-permeation chromatography. Different compositions can be separated by SEC due to different sizes of the molecules. Matrix of SEC were constructed from different materials, such as cross-linked dextran and agarose, etc. The separating properties were determined by size and distribution of the pore in the chromatography matrix. There were no ligand interactions involved. Smaller solutes can diffuse into the pore, and larger solutes were excluded, as a result, large molecules were eluted firstly (Fig. 5.3). SEC was generally applied to separate biological molecules and to determine the molecular weight distributions of polymers.

At present, SEC has been successfully applied in the preparation of functional oligosaccharides with homogeneous DP. Yang applied an extra fine column (100 × 2.6 cm) of Bio-gel P-2 to separate inulin-oligosaccharides with more than 98% purities of different DP products (Yang et al. 2010). Glucosamine oligosaccharides were also successfully prepared by SEC method. Ken-ichi Amano and Eiji Ito prepared chitin oligosaccharides ($(GlcNAc)_n$, n = 1–4) by a column (1.0 × 48 cm) of Bio-Gel P-2, with 0.15 mol/L NaCl as eluent (AMANO and ITO 1978). Masato Izume applied Bio-Gel P-2 (2.65 × 165 cm) and Bio-Gel P-4 (2.65 × 175 cm) columns to separate the enzymatic hydrolysate mixture of 10% N-acetylated chitosan, and the yields of $(GlcNAc)_5$, $(GlcNAc)_6$ and $(GlcNAc)_7$ were 8.2%, 12.0% and 9.0% with more than 90% purities, respectively(Izume et al. 1992). Sei-ichi Aiba acquired hexamer, pentamer, tetramer, trimer, and dimer of $(GlcNAc)_n$ with high purities (more than 92%) by a Toyopearl HW-40S column (5 × 68 cm) (Aiba 1994). Compare

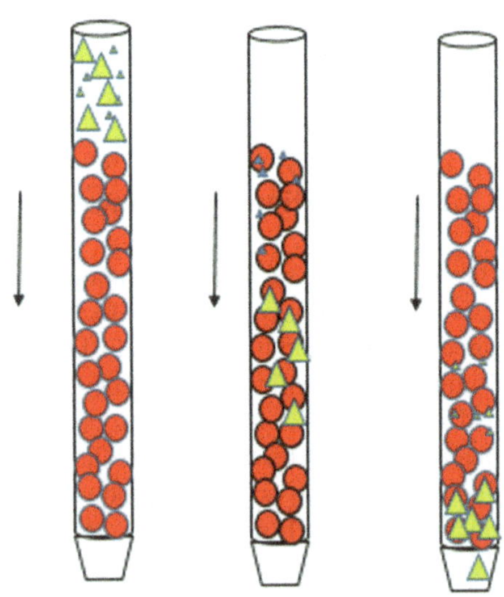

Fig. 5.3 Schematic diagram of size exclusion chromatography

with chitin oligosaccharides, COS mixture were more difficult to be separated by SEC method because of the difference of their structure. Gao Lixia and Li Chunxi obtained five different single DP COS by polyacrylamide gel Bio Gel P-6 and P-6 Fine column, and the purities of DP 3 and DP 4–7 were 95% and 90%, respectively (Gao et al. 2013).

5.2.2.2 Absorption Chromatography

Adsorption chromatography is often called the Liquid-solid chromatography (LSC). The process of separation was based on the interaction between solute and adsorbent. The adsorbent was active porous solid with large surface area, such as silica gel, alumina and activated carbon, etc. Active sites of the adsorbent could interact with the polar functional groups of the compound. Thus, adsorption chromatography was well suited for the separation of polar and nonpolar molecules.

Oligosaccharides could interact with adsorbent by the formation of hydrogen bond. And activated charcoal, as a common adsorbent, was widely used to prepare oligosaccharides, such as rehmannia oligosaccharides, glucan oligosaccharides, agar oligosaccharides and thallus laminariae oligosaccharide, etc. chito/chitin oligosaccharides could also interact with activated charcoal, but the result was not good. such as, Tomas Semenuk separated COS by charcoal column with Bio-Gel P-2 (Fine, 45–90 μm) and Bio-Gel P-4 (Fine, 45–90 μm) columns, and the results were shown in Table 5.2 (Semeňuk et al. 2001).

5.2.2.3 Affinity Chromatography

Affinity chromatography is a type of liquid chromatography, which based on the highly selective interaction between immobilized ligand and structural element of the target molecule. The target molecules could form reversible binding with immobilized ligand of affinity chromatography materials, and the mixture could be separated from each other by changing conditions of mobile phase. When the mixture flow with mobile phase, the component with affinity ligands are selectively adsorbed, and the rest of the impurities flow out firstly. The target products were obtained by using appropriate buffer to desorb. Based on the differences of the interactions between chelated copper (II) ions and various COS, metal affinity chromatography was developed to separate COS, and the purities of DP 2, DP 3 and DP 4 COS were 95%, 95% and 90%, respectively (Le Dévédec et al. 2008).

Table 5.2 Separation of COS on charcoal column

Etoh in eluent (v/v)	Composition of fraction	Volume (mL)
N/A (water)	Salts (>95%), GlcNAc (<5%)	500
4%	GlcNAc (>95%), chitobiose (<5%)	1000
8%	Chitobiose (>85%), chitotriose (<15%)	1000
10%	Chitotriose (>85%), chitotetraose	1000
13%	Chitotetraose (>85%), chitopentaose (<15%)	1000
30%	Chitopentaose (70%), (GlcNAc)n, a = 6–9	1500

5.2.2.4 Ion Exchange Chromatography

The stationary phases of IEC were charged groups, which could bind to ions with opposite charges by electrostatic interactions (Fig. 5.4). The interactions could be modulated by the change of ionic strength and pH. There were two different types of IEC, cation exchange chromatography (CEC) and anion exchange chromatography (AEC). The functional groups of anion exchange column were mainly $-NH_2$ and $-NH_3$, and the functional groups of cation exchange agent were mainly $-SO_3H$ and -COOH.

Ion-exchange chromatography has been widely applied to separate oligosaccharides due to its high resolution, large sample load and fast flow rate (Bultel et al. 2010; Chuang et al. 2001). COS is a kind of natural cation sugar with free amino groups. And different DP of COS could be acquired by IEC because of the difference numbers of amino groups. Studies showed that, CM Sepharose column were extensively applied in the preparation of single DP COS. XinlinWei obtained COS 5 (DP 4, 59.84%; DP 5, 40.16%) and COS 6 (DP 5, 6.89%; DP 6, 93.11%) products by CM-Sephadex C-25 column (Wei et al. 2009). Kecheng Li separated COS with DP more than 6 from the prepared fully deacetylated chitooligomers mixture by CM Sepharose Fast Flow column, and obtained five fractions, mainly contained glucosamine oligomers with DP 6–7 (41.31%, 50.22%), DP 7–8 (22.47%, 70.13%), DP 9–10 (53.06%, 27.99%), DP 10–12 (18.45%, 49.36%, 22.31%), and DP >12, respectively (Li et al. 2012b). Highly deacetylated COS mixture were also separated by CM Sephadex C25 column, and the purities of DP 2–6 and DP 7 COS were over 98% and 93%, respectively (Li et al. 2013b). COS with DP 2–12 was prepared by acid hydrolysis of chitosan and five fractions were separated from the prepared COS by CM Sephadex C-25 column. The fractions were desalted by activated charcoal extraction, and trimer (95.7%), tetramer and pentamer (90.9%), pentamer (85.6%), hexamer (89.5%) and heptamer to decamer (94.7%) were collected (Li et al. 2012a). Two partially acetylated chitotrioses (N-acetylchitotriose and N,N′-diacetylchitotriose) were also separated successfully by CM Sepharose fast flow column (Li et al. 2013a). Strong acidic ion-exchange resin also could be applied to prepare single DP COS. Xiong et al. prepared several COS with single DP by Dowex 50WX8-200 (strong acidic cation-exchange resin) eluted by a linear gradient solution of HCl, and the purities of DP 2–4, DP 5 and DP 6 were above 98%, 90% and 80%, respectively. The HCl was removed from the fractions by a charcoal

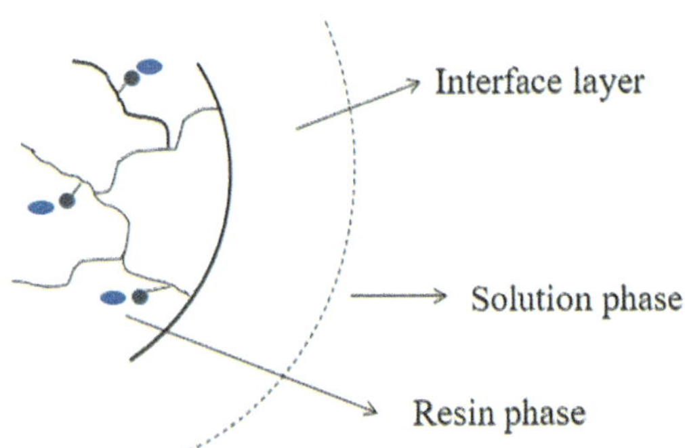

Fig. 5.4 Schematic diagram of ion diffusion process

column (Xiong et al. 2009). Qing Xu separated DP 2 and DP 3 COS by a homemade cation ion exchange resin (QY-H003) with hydrochloric acid solutions at different concentrations, the purities of separated chitobiose and chitotriose were 98.06% and 96.00%, respectively (Xu et al. 2017).

5.2.3 Separation by Combination of Membrane, Electric Field and Chromatography

Currently, COS produced by various processes are all mixtures of DP 2–10 and even some monosaccharides and lower molecular weight polysaccharides. However, the existing COS separation and purification technology can only be used to prepare COS with higher purity in small quantities, which is difficult to meet the needs of the application and that lead to the high price. In order to solve the above problems, assumptions taken to build up a device which can separate and purify COS in large quantities by combination of these technologies.

As shown in the Fig. 5.5, the MWCO of the UF/NF membrane gradually decreases from anode to cathode. The COS solution is added to the first sub electrophoresis region. Because COS are positively charged, they migrate to the negative electrode after electrification. When the conductivity is no longer changed, the migration of COS is completed. At this point, the baffle is placed, the power is turned off, and the COS solution is obtained from the valve. The obtained COS solution can be put back into the first electrophoresis region for secondary separation.

If the separation effect of the existing UF/NF membrane is not ideal, the next step is to carry out membrane modification – membrane chromatography. To make up for the above, membrane chromatography is a worth considering method. Membrane chromatography uses a membrane with a certain pore size as a medium and linking ligands, separating and purifying by the interaction between membrane ligands and target molecules such as proteins. Compared to membranes, membrane chromatography has the advantage of being able to separate bio macromolecules with similar properties. Compared to chromatography, membrane chromatography has advantages of lower operating pressure, much faster mass transfer process and short operating weeks phase, the higher separation efficiency.

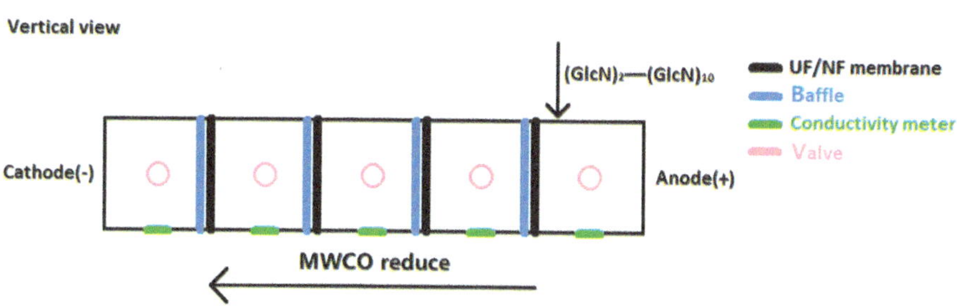

Fig. 5.5 Combination of membrane, electric field and chromatography together for COS separation

The bulk modification of the membrane can be divided into physical modification and chemical modification. Physical modification is the blending of membrane material with material of required functions to make the blend film and in the chemical modification, first functionalize the main chain of the membrane material and then making the film. Membrane surface modification is performed to modify the surface of the membrane by introducing some functional groups.

As the membranes referred in Fig. 5.5, the properties of the membrane can be improved by membrane modification and so that the separation efficiency could be improved too. The development of modern biological engineering technology has put forward higher and higher requirements for the separation and purification of macromolecules. As a new type of biochemical separation method, Membrane chromatography has not been widely used in industrial production, and lack of theoretical model and the support of a large number of experimental data. However, it's believed that with the development of the membrane technology, ligands coupling technology and automation control technology, membrane chromatographic separation technology in biological macromolecular separation and purification process will certainly play a more and more important role (Table 5.3).

5.3 Analysis of Chitooligosaccharides

The analysis of COS molecules based on contents, degree of polymerization, structure and sequence of molecules. The component analysis for contents and degree of polymerization are done by using thin layer chromatography and high performance

Table 5.3 Classification and characteristics of membrane chromatography techniques

Membrane chromatography technology	Membrane matrix	Ligands	Application
Affinity membrane chromatography (AMC)	Chitin, GFG, GMA-EDMA, PVA, Polydivinyl alcohol etc.	L-histidine, PAB, Chitin residue etc.	Insulin, Endotoxin, Immunoglobulin antibody etc.
Hydrophobic interaction membrane chromatography (HIMC)	Cellulose, quick disk, polyethylene etc.	Phenyl, hexamethylenediamine, dodecyl etc.	Pyrogen, bovine serum albumin, human tumor necrosis factor etc.
Ion-exchange membrane chromatography (IMC)	Cellulose, GMA, GMA-DEMA, modified cellulose, chitin etc.	DEAE, CM, EGDE, quaternary amino group, sulfonate etc.	Immunotoxin, oligonucleotide, prothrombin etc.
Multistage membrane chromatography (MMC)	Sartobind membrane, modified cellulose and Copolymer etc.	–	Urokinase, BSA, IgG, Formate dehydrogenase etc.

liquid chromatography. While the structural or sequence analysis that involve the determination of degree of acetylation or deacetylation, sequence of acetylation and molecular weight distribution, are analyzed by using advance methods such as mass spectrometry, infrared, nuclear magnetic resonance and thermal gravimetric analysis. According to the need and purpose of analysis, techniques are used for two type of analysis, qualitative and quantitative analysis and structural and sequence analysis.

5.3.1 Qualitative and Quantitative Analysis

5.3.1.1 Thin Layer Chromatography

The qualitative analysis of DP determination is usually performed by thin layer chromatography. TLC has its broad applications for qualitative determination or analysis of molecules from a solution. TLC required silica based stationary phase and solvent based mobile phase following by a dye solution and baking. The molecules of solution migrate with mobile phase on the stationary, mobile phase consists on different concentration of solvents used for COS determination. The mobile phase differently used by scientists and the ratio of solvents maintained. The mainly used ratios are given in Table 5.4.

5.3.1.2 High Performance Liquid Chromatography

High performance liquid chromatography is a widely used chromatography for non-volatile compounds on small stationary phases with low pressure and suited for the analysis of low and high molecular weight compounds such as peptides and proteins. HPLC is mainly combined with ultraviolet visible light wavelength spectroscopy or mass spectrometry detection. The small stationary phase column is the core of high-performance liquid chromatography, showing a strong selectivity, column efficiency and retention mechanism, so HPLC method becomes a common technique in various separation fields, and the qualitative analysis of chromatographic method is used to describe the well-defined COS and obtain quantitative data. For the first time, chito oligosaccharide was determined using Bondapak (3.9 × 300 mm)

Table 5.4 Mobile phase solvents

Sr#	Solvents	Ratios (v/v)	References
1	n-butanol, methanol, 25% ammonia solution-water	5:4:2:1	Purushotham et al. (2012) and Suma and Podile (2013)
2	28–30% ammonium water: water: isopropanol	2:4:14	Pechsrichuang et al. (2013, 2018)
3	n-butanol: methanol: 28% ammonia solution: H_2O	10:8:4:2	Songsiriritthigul et al. (2010)
4	*n*-propanol–water–ammonia water	7:3:1	Li et al. (2016)
5	Isopropanol–water–ammonia	4:2:3	Li et al. (2015)
6	Isopropanol: water: ammonium hydroxide	15:1:7.5	Qin et al. (2018a, b)

Column UV-210 nm detector (Ohtakara and Mitsutomi 1988), and the quantitative determination method of chitooligosaccharide was revealed. The HPLC was further classified on the basis of the material type, functioSnal group and separation mechanism of column like amino chromatography column, carbohydrate column, SEC, high performance anion exchange chromatography column (HPAEC), hydrophilic interaction liquid chromatography column (HILIC) and other stationary phase columns.

Amino Chromatography Column

Amino columns contain amino (NH_2) functional groups that involve in the retention and separation of desired molecules. These columns attached with detectors to analyze the desired COS molecules based on their DP as Shodex Asahipak NH_2P-50-an amino column-used with UV detector to reveal the determination of $(GlcNAc)_n$, n = 2–5 (Yoon 2005), with RI detector to determine DP 2 to DP 6 COS (Dong et al. 2015). Corona TM CAD detector was used to determine the COS molecules with DP 2–7 (Xiong et al. 2009). Tosoh TSK Amide 80 used to reveal DP 1–6 and DP 4–6 with UV and IR detectors, respectively (Ji et al. 2003; P et al. 2010). The ionic strength of elution solvents in the gel-filtration high performance liquid chromatography increased to suppress the electrostatic interaction of chito oligosaccharides. Thus Gel based amino columns (TSK-GEL NH_2-60, Tosoh) were also used for separation of DP 2 to DP 8 COS molecules with 60% acetonitrile eluent at of rate of 0.8 ml/min (Choi et al. 2002).

In our previous research, COS mixtures were analyzed by HPLC system (Shimadzu 20A, Shimadzu, Kyoto, Japan) equipped with evaporative light-scattering detector (ELSD). The COS were separated on high-performance sugar column (Shodex Asahipak NH2 P-50 4E, Shodex, Kyoto, Japan), eluted by acetonitrile and distilled water (70/30) mixture with a flow rate of 1.0 mL/min at 30 °C (Qin et al. 2018b). The result was showed in Fig. 5.6.

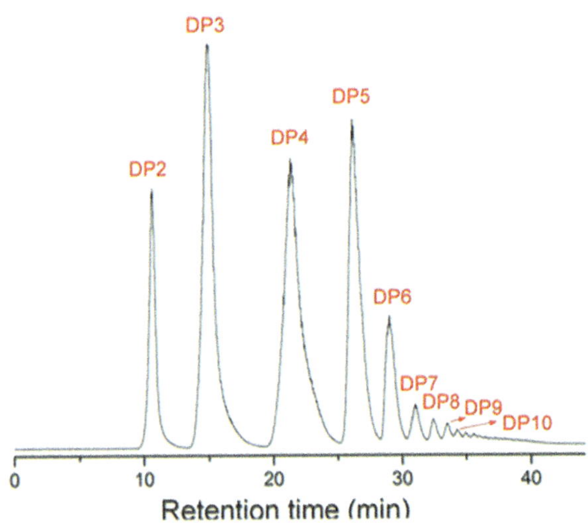

Fig. 5.6 HPLC analysis of COS mixture

Size Exclusion Chromatography

Size exclusion chromatography (SEC) technique used to separate molecules based on size/hydrodynamic volume, that do not involve the binding of molecules with resins and measured by how efficiently they penetrate the pores of the stationary phase. This technique known as gel permeation/filtration, its gel filtration when mobile phase is aqueous while its gel permeation when organic phase used. It is based on the discrimination of individual sample components by the pores of the packing material, large sample molecules cannot or can only partially penetrate the pores and elute from the column first, whereas smaller molecules can access all or a larger number of pores and elute later.

Einbu et al., separated oligosaccharides from reaction mixture using three Superdex™30 columns connected in series with refractive index detector. The columns were eluted with 0.15 M ammonium acetate at a flow rate of 0.8 mL/min. While investigation the relationship between detector response and molecules mass, they found the linear relationship in peak area and mass of molecules injected and calculate the molar fractions of the different oligomers from SEC (Einbu et al. 2007). In another study, Shodex OHpak SB-G precolumn connected in series with Shodex OHpak SB-802 and Shodex OHpak SB-802.5 columns (8 × 300 mm) investigated to determine the mass of molecules. One hundred millimolar sodium nitrate solution used as elution at flow rate of 0.5 mL/min with RI detector (Chambon et al. 2015).

Carbohydrate Column

Carbohydrate columns usually used for analysis and detection of mono and disaccharides. It also has been applied for detection of COS. High performance carbohydrate column used with MeCN–Water (7:3) mobile phase at a flow rate of 1 mL/min and detected by UV detector at 210 nm (UV) (Tokuyasu et al. 2000).

High Performance Anion Exchange Chromatography

Ion exchange chromatography worked on the base of opposite ions/charge in stationary phase, opposite ion/charge to target molecules make this strong, efficient and sensitive analytical method. HPAEC technique based on high-performance anion exchange chromatography used with pulsed amperometric detection (HPAEC-PAD). It was established for the simultaneous separation and determination of glucosamine $(GlcN)_1$ and COS ranging from DP 2–6 without prior derivatization. The detection limits of this technique investigated as 0.003–0.016 mg/L with 0.2–10 mg/L linear range. The analysis optimized and carried out using CarboPac-PA100 analytical column (4 × 250 mm) with 0.2 M aqueous sodium hydroxide-water mixture (10:90) elution at u 0.4 mL/min flow rate. This efficient, sensitive, and quick HPAEC-PAD method provided satisfactory separation and demonstrated as suitable for separating, identifying, and quantifying DP 1–6 within 25 min. High sensitivity, satisfactory linearity, precision and accuracy were achieved. This anion exchange based column showed high accuracy, efficiency and sensitivity so it's useful for routine analysis of COS (Cao et al. 2016). This method also carried out to estimate other sugars from the sample. Tokuyasu, Ono et al., used this method to determine the sugars with CarboPac™PA1 analytical column, eluting with 15 mM NaOH, connected with pulsed amperometric detector (Tokuyasu et al. 1999).

Hydrophilic Interaction Liquid Chromatography

Hydrophilic interaction liquid chromatography also an ion exchange chromatography that provides an alternative approach to effectively separate small polar compounds using polar stationary phases. Polar samples always show good solubility in the aqueous mobile phase used in HILIC, which overcomes the drawbacks of the poor solubility often encountered in NP-LC. Firstly, it was found that HILIC stationary phase possesses ion exchange chromatography retention mechanism and this concept was proposed by Zhu et al. (1991). HILIC/IEC is often used for the separation of polar and hydrophilic compounds. Jiang, M., *et al*, used this mechanism to separate COS from DP 2 to DP 6 and calculate their molecluar weight. They analysed the solution on X-Amide column using two mobile phases at different ratios phase: A was acetonitrile, and phase B was ammonium formate aqueous solution. Maintaining the temperature at 30 °C with 1.0 mL/min flow rate, analysis was carried out using evaporation light-scattering detector (ELSD) (Jiang et al. 2014). HILIC columns with different material and functional groups were studied for specific fractions of molecules. As click maltose column was developed in HILIC mode applied to separate oligosaccharides (Fu et al. 2010). COS contains different DPs was also separated by this column with ELSD.

Recently, a new HILIC stationary phase developed by Zhai X. et al., to separate chitooligosachharides from DP 2 to DP 6. They grafted 3-aminophenylboronic acid group as a functional compound on silica in their stationary phase. This compound give rise a HILIC, main force for the retention behavior was hydrogen bond, and showed great HILIC characteristics for COS separation. The method used refractive index detector (RID) and a mobile phase that prepared by mixing ACN and 200 mM of stock ammonium formate ($HCOONH_4$) (Zhai et al. 2018). Hence, HILIC showed excellent behavior with modified material for the determination of COS and their fractions based on their hydrophilic interactions.

5.3.1.3 Electrophoresis

Electrophoresis is a technique used to separate charged molecules under the application of electric potential. The charged molecules move under the electric potential through gel or capillary stationary stage. The electrophoresis is classified as capillary or gel electrophoresis based on the use of stationary method. Mostly the saccharides are neutral or negatively charged but COS has positive charged in acidic condition, this phenomenon of protonation distinguish from other saccharides, separation of COS using positive charge capillary can be possible. Hattori et al., successfully used these characteristics and study the separation of COS with positively coated capillary in acidic solution. They concluded that electrophoretic mobility of positively charged small oligomers depends upon the number of saccharide chains in acidic solution (Hattori et al. 2010).

Capillary electrophoresis (CE) is useful for COS separation because of its high separating resolving power, protonated charged on these molecules and small amount requirement of sample and solvent. Beaudoin et al. (2005) successfully quantify a sub-micromolar chitin and COS using CE with an aminopyrine–fluorophore conjugation procedure and laser. However, the method required derivatization of samples with relatively expensive materials and equipment. After all, Blanes et al. (2008) developed a direct method using capillary zone electrophoresis with contactless conductivity detection of COS. They used 10% (v/v) acetonitrile–water containing 10 mM NaOH electrolyte solution (pH > 12) to separate COS, alkalinity of electrolyte solution cause the successful separation of negatively charged COS and other saccharides (El Rassi 1999). Capillary electrophoresis with capacitively coupled contactless conductivity detection was successfully used in the literature, to quantify N-acetylglucosamine and five N-acetyl-COS (C_2–C_6) produced after reaction with a purified chitinase.

5.3.2 Structural and Sequence Analysis

COS molecules have bioactive properties and applied in biotechnological, therapeutic food, pharmaceutical and agricultural fields. The structure and function of COS are closely related. In order to further clarify the structure and activity relationship of COS, to reveal the biochemical processes and expand the application of COS in the field of biomedicine, it is necessary to further analyze the structural components and sequence of them. In this chapter, several methods of structural and sequence characterization of COS were introduced. These methods are including mass spectrometry, infrared spectrometry, nuclear magnetic resonance and thermal gravimetric analysis.

5.3.2.1 Mass Spectrometry

Mass spectrometry is a method of detecting mobile ions (charged atoms, molecules, or molecular fragments) by their gravimetric ratios using electric and magnetic fields. The core devices in mass spectrometry are the ion source mass analyzer and ion detector. The ion source can ionize the sample molecules under high vacuum conditions. Due to receiving high energy, ionized molecules shattered into smaller mass of multiple fragmentation ions and neutral particles. They obtain average kinetic energy under the action of accelerating electric field and enter the mass analyzer. The mass analyzer is a device that disconnects ions of different masses at the same time with different mass-to-charge ratio (m/e) and the separated ions then enter in the ion detector one by one. The detector collects and amplifies the ion signals, which are processed by computer and mapped into mass spectra. Various types of ion sources, mass analyzers, and ion detectors are applied in structural analysis.

Mass spectrometry is often used in combination with LC to determine the structure and composition of COS. Ultra high performance liquid chromatography-electrospray ionization/mass spectrometry (UHPLC-ESI-MS) method was used in Stefan Cord-Landwehr's research to quantitatively detect the sequencing of COS (Cordlandwehr et al. 2017). In order to separate the modified chitin oligomer products, Dionex Ultimate 3000RS UHPLC system was used combined with an acquity UPLC BEH Amide column include a precolumn. Chitin oligomers (DP 1–6) were detected by UHPLC system coupled with an Amazon speed ESI-MSn detector. Autoflex Speed MALDI-TOF mass spectrometer equipped with a SmartBeamTM NdYAG-laser (355 nm) was also used to detect the oligomers generated from chitosan (Madhuprakash et al. 2015). MS was also used in our previous work to determine the structure of COS and the results was showed in Fig. 5.7 (Qin et al. 2018a).

5.3.2.2 Infrared Spectrometry

Infrared spectrometer is an instrument to analyze molecular structure and chemical composition by using the absorption characteristics of different wavelength infrared radiations. The main parts of infrared spectrometer are light source, monochromator, detector and computer processing information system. According to the difference of the spectroscopic device, it is divided into dispersive type and interference type. Infrared spectroscopy can be used to study the structure and chemical bond of molecules, such as the determination of force constant and molecular symmetry. By using infrared spectroscopy, the bond length and bond angle of molecules can be measured and the three-dimensional configuration of molecules can

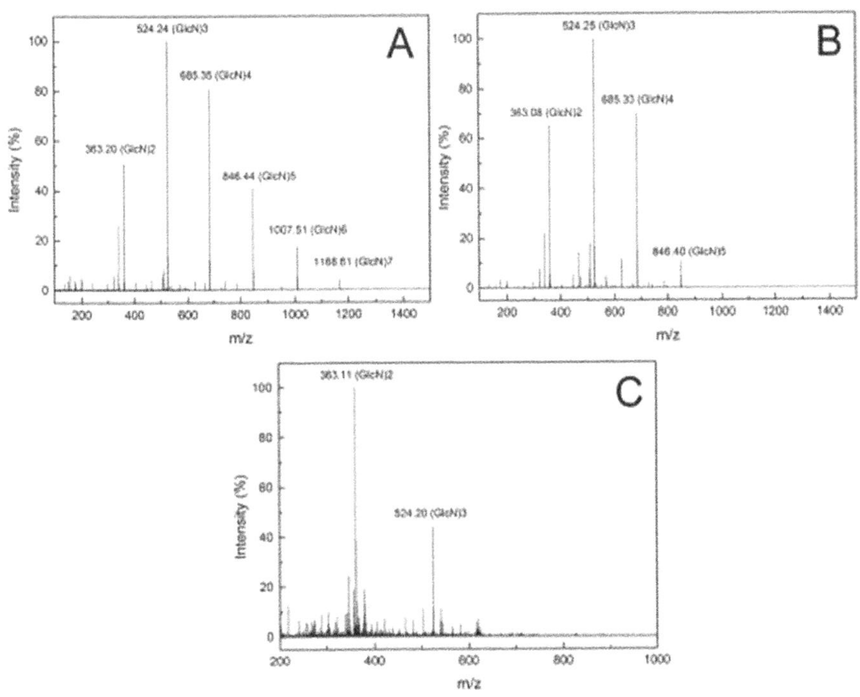

Fig. 5.7 MALDI-TOF MS analysis of different COS mixtures. (**a**) $(GlcN)_2$–$(GlcN)_7$. (**b**) $(GlcN)_2$–$(GlcN)_5$. (**c**) $(GlcN)_2$ and $(GlcN)_3$

be speculated. According to the force constant, the strength of chemical bond can be inferred, and the thermodynamic function is calculated by the simple positive frequency. Certain groups of different compounds or chemical bonds of the band wave number are basically fixed, or change in a small band range only. Therefore, many organic functional groups such as methyl, methylene, carbonyl, cyano, hydroxyl, amino and so on have a characteristic absorption in the infrared spectrum. By infrared spectrometry, people can determine the unknown samples, which exist in the organic functional groups. This laid the foundation for the final determination of the chemical structure of the unknown molecules. Fourier transform infrared spectroscopy (FTIR) was used to determine the degree of deacetylation and characterize of COS. In previous study, FTIR spectrum was used to conform the structures of the synthesized spectrum in compare with the peaks of absorption (Karagozlu et al. 2010).

5.3.2.3 Nuclear Magnetic Resonance

Nuclear magnetic resonance is a physical process in which the spin energy level generates Zeeman splitting under the action of the external magnetic field of the nucleus with non-zero magnetic moment, and the resonance absorbs a certain frequency of radiofrequency radiation. NMR spectroscopy is a branch of spectroscopy that's resonance frequency was within the radio frequency band, and the corresponding transition by the nuclear spin on the nuclear Zeeman energy level.

^{1}H NMR Spectroscopy

^{1}H NMR is an application of NMR spectroscopy to embody the effect of ^{1}H in molecules. It can be used to determine the molecular structure. When a sample contains hydrogen, especially the isotope ^{1}H, the NMR hydrogen spectrum can be used to determine the molecular structure. In previous study, structural analysis of purified chitosan were also analyzed by ^{1}H NMR spectra recorded on a 300 MHz Bruker Advance Spectrometer (Sánchez et al. 2017)

^{13}C NMR Spectroscopy

Most organic molecules are composed of carbon atoms, and ^{13}C NMR is a useful tool for studying organic molecular structures. Since the natural abundance of ^{13}C is only accounted for 1.108%, and the ^{12}C with the greater natural abundance did not have nuclear magnetic resonance signal. ^{13}C NMR signal containing carbon compound was very weak, which led to the extremely limited application of ^{13}C NMR. It was not became a practical test method until the advent of pulse fourier transform spectrometer. In recent years, ^{13}C NMR technology and its application have developed rapidly. It has been widely used in various fields of organic chemistry, research on molecular structure, configuration and conformation, research on synthetic polymer, natural polymer and dynamic process, and has become an indispensable analytical tool in chemistry, biology, medicine and chemical industry. In detecting the structure of COS, ^{13}Carbon (^{13}C) NMR spectra was used to conform that the gallic acid was successfully conjugate onto COS from the aromatic carbon (Sang et al. 2017).

^{15}N NMR Spectroscopy

^{15}N NMR spectrum is a form of NMR spectrum used to detect samples containing ^{15}N. For example, ^{15}N NMR spectroscopy was used to detect the degree of acetylation for chitin/chitosan samples. In previous study, a spectrometer operating at 30–35 MHz equipped with cross-polarization (CP) magic-angle spinning nitrogen NMR (CP-MAS 15N NMR) has been used to record the spectra of chitin/chitosan samples. The CP technique is used to transfer the polarization from the abundant ^{1}H to the rare ^{15}N nuclei thus enhancing the signals (Alvarenga 2011).

5.3.2.4 Thermal Gravimetric Analysis

Thermal gravimetric analysis is a thermal technique that measures the relation between the mass of the sample and the temperature changes under controlled program to study the thermal stability and composition of the material. TGA is a common detection method in both research and development and quality control. Thermo-gravimetric analysis is often used in combination with other analysis methods, in actual material analysis to conduct comprehensive thermal analysis and accurately analyze materials. Thermo-gravimetric analyzer used to examine the thermal properties of COS and chitosan. Analyses are performed using powder sample in a nitrogen gas atmosphere in platinum crucibles from room temperature to 600 °C (Hsiao et al. 2008).

5.4 Summary

It was difficult to separate single DP products from each other, due to the similarity of structure and molecular weight. The functional research of COS was limited by the lack of raw material, especially high DP product. Thus, efficient separation and purification technologies of COS were necessary.

Membrane and chromatographic technology played an important role in separation technology. UF, NF, RO and ED were successfully applied in the separation of COS mixture. Because the separation capacity was limited, membrane technology was used to narrow the distribution of molecular weight, remove the impurities, desalt and concentrate of product, etc. At present, single DP COS could be obtained by chromatographic technology, which obtained size exclusion chromatography, absorption chromatography, affinity chromatography and ion exchange chromatography, etc. While, the purities of COS decreased with the increment of DP. Moreover, the application was limited by the lower yield and productivity. In the future, more in-depth research of separation technology are needed for the production of single DP COS, especially to combine different technologies to design new system for purification of single DP COS. Purification of single DP products will lead to the new innovations toward the applications and functions of COS.

References

Aiba S-i (1994) Preparation of N-acetylchitooligosaccharides by hydrolysis of chitosan with chitinase followed by N-acetylation. Carbohydr Res 265(2):323–328

Aider M, Arul J, Mateescu A-M, Brunet S, Bazinet L (2006a) Electromigration of chitosan d-glucosamine and oligomers in dilute aqueous solutions. J Agric Food Chem 54(17):6352–6357

Aider M, Arul J, Mateescu M-A, Brunet S, Bazinet L (2006b) Electromigration behavior of a mixture of chitosan oligomers at different concentrations. J Agric Food Chem 54(26):10170–10176

Aider M, Brunet S, Bazinet L (2008a) Effect of pH and cell configuration on the selective and specific electrodialytic separation of chitosan oligomers. Sep Purif Technol 63(3):612–619

Aider M, Brunet S, Bazinet L (2008b) Electroseparation of chitosan oligomers by electrodialysis with ultrafiltration membrane (EDUF) and impact on electrodialytic parameters. J Membr Sci 309(1):222–232

Aider M, Brunet S, Bazinet L (2009) Effect of solution flow velocity and electric field strength on chitosan oligomer electromigration kinetics and their separation in an electrodialysis with ultrafiltration membrane (EDUF) system. Sep Purif Technol 69(1):63–70

Alvarenga ESD (2011) Characterization and properties of chitosan. InTech

Amano K-i, Ito E (1978) The action of lysozyme on partially deacetylated chitin. Eur J Biochem 85(1):97–104

Beaudoin ME, Gauthier J, Boucher I, Waldron KC (2005) Capillary electrophoresis separation of a mixture of chitin and chitosan oligosaccharides derivatized using a modified fluorophore conjugation procedure. J Sep Sci 28(12):1390–1398

Blanes L, Saito RM, Genta FA, Donegá J, Terra WR, Ferreira C, do Lago CL (2008) Direct detection of underivatized chitooligosaccharides produced through chitinase action using capillary zone electrophoresis. Anal Biochem 373(1):99–103

Bultel L, Landoni M, Grand E, Couto AS, Kovensky J (2010) UV-MALDI-TOF mass spectrometry analysis of heparin oligosaccharides obtained by nitrous acid controlled degradation and high performance anion exchange chromatography. J Am Soc Mass Spectrom 21(1):178–190

Cao L, Wu J, Li X, Zheng L, Wu M, Liu P, Huang Q (2016) Validated HPAEC-PAD method for the determination of fully deacetylated chitooligosaccharides. Int J Mol Sci 17(10):1699

Chambon R, Despras G, Brossay A, Vauzeilles B, Urban D, Beau JM, Armand S, Cottaz S, Fort S (2015) Efficient chemoenzymatic synthesis of lipo-chitin oligosaccharides as plant growth promoters. Green Chem 17(7):3923–3930

Chen JG (2006) Enzymatic preparation of chitosan oligosaccharides and its antimicrobial activity [D]. Zhejiang University (In Chinese)

Choi WS, Ahn KJ, Lee DW, Byun MW, Park HJ (2002) Preparation of chitosan oligomers by irradiation. Polym Degrad Stab 78(3):533–538

Chuang W-L, McAllister H, Rabenstein DL (2001) Chromatographic methods for product-profile analysis and isolation of oligosaccharides produced by heparinase-catalyzed depolymerization of heparin. J Chromatogr A 932(1):65–74

Cordlandwehr S, Ihmor P, Niehues A, Luftmann H, Moerschbacher BM, Mormann M (2017) Quantitative mass-spectrometric sequencing of chitosan oligomers revealing cleavage sites of chitosan hydrolases. Anal Chem 89(5):2893–2900

Dong H, Wang Y, Zhao L, Zhou J, Xia Q, Jiang L, Fan L (2014) Purification of DP 6 to 8 chitooligosaccharides by nanofiltration from the prepared chitooligosaccharides syrup. Bioresour Bioprocess 1:20

Dong H, Wang Y, Zhao L, Zhou J, Xia Q, Qiu Y (2015) Key technologies of enzymatic preparation for DP 6–8 chitooligosaccharides. J Food Process Eng 38(4):336–344

Einbu A, Grasdalen H, Vårum KM (2007) Kinetics of hydrolysis of chitin/chitosan oligomers in concentrated hydrochloric acid. Carbohydr Res 342(8):1055

El Rassi ZE (1999) Recent developments in capillary electrophoresis and capillary electrochromatography of carbohydrate species. Electrophoresis 20(15–16):3134–3144

Fu Q, Liang T, Zhang X, Du Y, Guo Z, Liang X (2010) Carbohydrate separation by hydrophilic interaction liquid chromatography on a 'click' maltose column. Carbohydr Res 345(18):2690–2697

Gao L, Li C, Wang S (2013) Preparation and analysis of chitooligosaccharide monomers with different degree of polymerization. Chin Mar Med 32(3):21–27 (In Chinese)

Han YP, Lin Q (2012) Purification of oligochitosan by nanofiltration. Membr Sci Technol 32:75–80. (In Chinese)

Hattori T, Anraku N, Kato R (2010) Capillary electrophoresis of chitooligosaccharides in acidic solution: simple determination using a quaternary-ammonium-modified column and indirect photometric detection with crystal violet. J Chromatogr B 878(3):477–480

Hsiao YC, Lin YW, Su CK, Chiang BH (2008) High degree polymerized chitooligosaccharides synthesis by chitosanase in the bulk aqueous system and reversed micellar microreactors. Process Biochem 43(1):76–82

Hu R (2012) Enzymatic hydrolysis of chitosan with specific average molecular weight and narrow distribution [D]. Wuhan University of Technology (In Chinese)

Izume M, Nagae S i, Kawagishi H, Ohtakara A (1992) Preparation of N-acetylchitooligosaccharides from enzymatic hydrolyzates of chitosan. Biosci Biotechnol Biochem 56(8):1327–1328

Jeon Y-J, Kim S-K (2000) Production of chitooligosaccharides using an ultrafiltration membrane reactor and their antibacterial activity. Carbohydr Polym 41(2):133–141

Ji Y, Zhao Y, Zhou X, Weng D, Zhao L (2003) The preparation of chitosan oligomer and the components analysis. Chin J Mod Appl Pharm

Jiang M, Guo Z, Wang C, Yang Y, Liang X, Ding F (2014) Neural activity analysis of pure chito-oligomer components separated from a mixture of chitooligosaccharides. Neurosci Lett 581(1):32–36

Karagozlu MZ, Kim JA, Karadeniz F, Kong CS, Kim SK (2010) Anti-proliferative effect of aminoderivatized chitooligosaccharides on AGS human gastric cancer cells. Process Biochem 45(9):1523–1528

Kuroiwa T, Izuta H, Nabetani H, Nakajima M, Sato S, Mukataka S, Ichikawa S (2009) Selective and stable production of physiologically active chitosan oligosaccharides using an enzymatic membrane bioreactor. Process Biochem 44(3):283–287

Le Dévédec F, Bazinet L, Furtos A, Venne K, Brunet S, Mateescu MA (2008) Separation of chitosan oligomers by immobilized metal affinity chromatography. J Chromatogr A 1194(2):165–171

Li K, Xing R, Liu S, Li R, Qin Y, Meng X, Li P (2012a) Separation of chito-oligomers with several degrees of polymerization and study of their antioxidant activity. Carbohydr Polym 88(3):896–903

Li K, Xing R, Liu S, Qin Y, Li B, Wang X, Li P (2012b) Separation and scavenging superoxide radical activity of chitooligomers with degree of polymerization 6–16. Int J Biol Macromol 51(5):826–830

Li K, Liu S, Xing R, Qin Y, Li P (2013a) Preparation, characterization and antioxidant activity of two partially N-acetylated chitotrioses. Carbohydr Polym 92(2):1730–1736

Li K, Liu S, Xing R, Yu H, Qin Y, Li R, Li P (2013b) High-resolution separation of homogeneous chitooligomers series from 2-mers to 7-mers by ion-exchange chromatography. J Sep Sci 36(7):1275–1282

Li H, Fei Z, Gong J, Yang T, Xu Z, Shi J (2015) Screening and characterization of a highly active chitosanase based on metagenomic technology. J Mol Catal B Enzym 111:29–35

Li K, Xing R, Liu S, Li P (2016) Advances in preparation, analysis and biological activities of single chitooligosaccharides. Carbohydr Polym 139:178–190

Madhuprakash J, El Gueddari NE, Moerschbacher BM, Podile AR (2015) Production of bioactive chitosan oligosaccharides using the hypertransglycosylating chitinase-D from Serratia proteamaculans. Bioresour Technol 198:503–509

Ngo D-N, Lee S-H, Kim M-M, Kim S-K (2009) Production of chitin oligosaccharides with different molecular weights and their antioxidant effect in RAW 264.7 cells. J Funct Foods 1(2):188–198

Ohtakara A, Mitsutomi M (1988) Analysis of chitooligosaccharides and reduced chitooligosaccharides by high-performance liquid chromatography. Methods Enzymol 161:453–457

P S, BT S, A S r, KM V r, SJ H, VG E (2010) Development and application of a model for chitosan hydrolysis by a family 18 chitinase. Biopolymers 77(5):273–285

Pechsrichuang P, Yoohat K, Yamabhai M (2013) Production of recombinant Bacillus subtilis chitosanase, suitable for biosynthesis of chitosan-oligosaccharides. Bioresour Technol 127:407–414

Pechsrichuang P, Lorentzen SB, Aam BB, Tuveng TR, Hamre AG, Eijsink VGH, Yamabhai M (2018) Bioconversion of chitosan into chito-oligosaccharides (CHOS) using family 46 chitosanase from Bacillus subtilis (BsCsn46A). Carbohydr Polym 186:420–428

Purushotham P, Sarma PVSRN, Podile AR (2012) Multiple chitinases of an endophytic Serratia proteamaculans 568 generate chitin oligomers. Bioresour Technol 112:261–269

Qiao X, Li Y, Tang X (2010) Study on the technology of producing chitosan oligosaccharide by enzyme fermentation. J Xiamen Univ (Nat Sci Ed) 49(2):251–255 (In Chinese)

Qin Z, Chen Q, Lin S, Luo S, Qiu Y, Zhao L (2018a) Expression and characterization of a novel cold-adapted chitosanase suitable for chitooligosaccharides controllable preparation. Food Chem 253:139–147

Qin Z, Luo S, Li Y, Chen Q, Qiu Y, Zhao L, Jiang L, Zhou J (2018b) Biochemical properties of a novel chitosanase from Bacillus amyloliquefaciens and its use in membrane reactor. LWT 97:9–16

Sánchez Á, Mengíbar M, Rivera-Rodríguez G, Moerchbacher B, Acosta N, Heras A (2017) The effect of preparation processes on the physicochemical characteristics and antibacterial activity of chitooligosaccharides. Carbohydr Polym 157(6):251–257

Sang VT, Ngo D-H, Bach LG, Ngo D-N, Kim SJ (2017) The free radical scavenging and anti-inflammatory activities of gallate-chitooligosaccharides in human lung epithelial A549 cells. Process Biochem 54:188–194

Semeňuk T, Krist P, Pavlíček J, Bezouška K, Kuzma M, Novák P, Křen V (2001) Synthesis of chitooligomer-based glycoconjugates and their binding to the rat natural killer cell activation receptor NKR-P1. Glycoconj J 18(10):817–826

Sinha S, Chand S, Tripathi P (2014) Production, purification and characterization of a new chitosanase enzyme and improvement of chitosan pentamer and hexamer yield in an enzyme membrane reactor. Biocatal Biotransformation 32(4):208–213

Songsiriritthigul C, Lapboonrueng S, Pechsrichuang P, Pesatcha P, Yamabhai M (2010) Expression and characterization of Bacillus licheniformis chitinase (ChiA), suitable for bioconversion of chitin waste. Bioresour Technol 101(11):4096–4103

Suma K, Podile AR (2013) Chitinase A from Stenotrophomonas maltophilia shows transglycosylation and antifungal activities. Bioresour Technol 133:213–220

Tokuyasu K, Ono H, Hayashi K, Mori Y (1999) Reverse hydrolysis reaction of chitin deacetylase and enzymatic synthesis of β-d-GlcNAc-(1→4)-GlcN from chitobiose. Carbohydr Res 322(1):26–31

Tokuyasu K, Ono H, Mitsutomi M, Hayashi K, Mori Y (2000) Synthesis of a chitosan tetramer derivative, β-d-GlcNAc-(1→4)-β-d-GlcNAc-(1→4)-β-d-GlcNAc-(1→4)-d-GlcN through a partial N-acetylation reaction by chitin deacetylase. Carbohydr Res 325(3):211–215

Wei X, Wang Y, Xiao J, Xia W (2009) Separation of chitooligosaccharides and the potent effects on gene expression of cell surface receptor CR3. Int J Biol Macromol 45(4):432–436

Xiong C, Wu H, Wei P, Pan M, Tuo Y, Kusakabe I, Du Y (2009) Potent angiogenic inhibition effects of deacetylated chitohexaose separated from chitooligosaccharides and its mechanism of action in vitro. Carbohydr Res 344(15):1975–1983

Xu Q, Chen LH, Qin Z, Chen QM, Qiu YJ, Zhao LM (2017) Preparation of chitosan and chitosan monomers and their structure analysis. Food Industry Technol 13:13–18. (In Chinese)

Yang Z, Yi Y, Gao C, Hou D, Hu J, Zhao M (2010) Isolation of inulin-type oligosaccharides from Chinese traditional medicine: Morinda officinalis how and their characterization using ESI-MS/MS. J Sep Sci 33(1):120–125

Yoon JH (2005) Enzymatic synthesis of chitooligosaccharides in organic cosolvents. Enzym Microb Technol 37(6):663–668

Zhai X, Zhao H, Zhang M, Yang X, Sun J, She Y, Dong A, Zhang H, Yao L, Wang J (2018) New stationary phase for hydrophilic interaction chromatography to separate chito-oligosaccharides with degree of polymerization 2-6. J Chromatogr B 1081–1082:33–40

Zhu B-Y, Mant CT, Hodges RS (1991) Hydrophilic-interaction chromatography of peptides on hydrophilic and strong cation-exchange columns. J Chromatogr A 548:13–24

6 Cleaner Production Guide of Chito/Chitin Oligosaccharides and Its Monomer

Xiangzhao Mao, Na Guo, and Jianan Sun

Abstract

Chitin/chito oligosaccharides and its monomer are the depolymerized derivatives of chitin/chitosan with favorable watersolubility and functional biological activities, which are promisingly applied in the pharmaceutical, food, cosmetics and agricultural fields. Traditionally, methods operated for the chitin and its derivatives are chemical processes using corrosive reagents (such as HCl and NaOH), which are acquainted to arouse environmental pollution and additional costs. Nowadays, enterprises are emerging new environment-friendly and clean technologies. Cleaner production is an efficient strategy for resource conservation, corporate conductivity and emission reduction, which manages industrial pollution problems. In this chapter, we summarize the current production situation in chitin industry and introduce methods for cleaner production of chitin, chito/chitin oligosaccharides and its monomer. The concise implementation procedures of cleaner production of chitin and its derivatives have been included. The sustainable development of cleaner production would enable effective, clean, economic, and controllable process for chitin and its derivatives industry.

Keywords

Cleaner production · Sustainable development · Chitin · Oligosaccharide · Monomer

---

X. Mao (✉)
College of Food Science and Engineering, Ocean University of China, Qingdao, China

Laboratory for Marine Drugs and Bioproducts of Qingdao National Laboratory for Marine Science and Technology, Qingdao, China
e-mail: xzhmao@ouc.edu.cn

N. Guo · J. Sun
College of Food Science and Engineering, Ocean University of China, Qingdao, China

L. Zhao (ed.), *Oligosaccharides of Chitin and Chitosan*,
https://doi.org/10.1007/978-981-13-9402-7_6

6.1 Introduction

Hydrolysis of chitin/chitosan for producing chitin/chitosan oligosaccharides can be realized by using physical, chemical and enzymatic methods. Acid hydrolysis was the most widely used traditional method, despite its suboptimal features, including hard controlling, low quality of production, and environmental pollution.

In the past, chitin/chitosan oligosaccharides production with no regard for environmental consequences usually caused serious troubles, such as water and air pollution, soil erosion, global warming, ozone depletion and acid rain. Nowadays, attitudes towards those problems have been switched from environmental control, such as pollution prevention and management, to creation of more sustainable methods of production. Under the circumstances that economic growth and the contradiction between population, resources and environment are increasingly prominent, it is the only way to develop to correctly handle the relationship between man and nature and continue to promote sustainable development. Cleaner production is gradually becoming the theme of healthy development of the chitin/chitosan industry.

Recently, gentle and easily controllable enzymatic hydrolysis of chitin and chitosan has been exploited as an alternative method. Chitosan depolymerized by strong acids (HCl) hydrolysis and commercial enzymatic degradation have been previously compared (Cabrera and Cutsem 2005). The fragments produced by these two methods differed in the degree of polymerization and acetylation, indicating that the enzymatic procedure produced smaller molecules with a higher proportion of fully deacetylated chitooligomers. Here we provide thorough analysis and scheme of emerging cleaner procedures and technologies for the green production in chitin and its derivatives industry.

6.2 Current Situation About Cleaner Production of Chito/Chitin Oligosaccharides and Its Monomer

6.2.1 Concept of Cleaner Production

Cleaner production has been put into effect for several years all over the world, which has been proved to be an efficient tool for resource conservation, corporate conductivity and emission reduction (Kjaerheim 2005). Thereby, it is also a strategy for ensuring sustainability and improving environmental performance, the concept is shown in Fig. 6.1.

With the implementation of Agenda 21, developed by the United Nations Conference on Environment and Development (UNCED) in Rio de Janeiro, Brazil in 1992, a green industrial revolution took place worldwide. Cleaner production is a corresponding to the theme of "Continuous improvement".

As for cleaner production, one of the pioneering European initiatives was given birth by Austria in 1992 by the Ministry for Transports, Innovation and Technology (BMVIT). This generated two initiatives: Prepare and Eco Profit. "PIUS", another

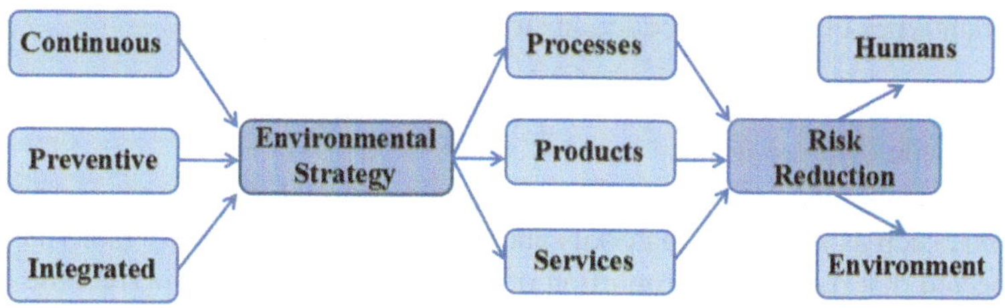

Fig. 6.1 Concept of cleaner production

initiative was set up in Germany in 1999. Since 1994, the United Nations Industrial Development Organization (UNIDO) ran the National Cleaner Production Centre Programme with centres in Central America, South America, Africa, Asia, and Europe (Sakr and Sena 2017). Then, cleaner production as a sustainable strategy has been widely advocated and executed.

The main objectives of cleaner production are to (Schaltegger et al. 2008):

- Minimization of the use, as well as optimization of the reuse and recycling of hazardous and non-hazardous materials.
- More efficient processing for material use, to reduce input amount and non-desired output amount.
- Risk minimization and improvement of human capital through worker hygiene and safety programs.
- Higher monetary returns by lowering energy consumption, decreasing material costs and reducing handling costs. As for this, often capital investment is necessary.

Cleaner production is a new kind of development tactics, which demands varied related theories and technologies, throughout the life cycle of the product in all aspects to take "preventive" measures, the production technology and processes, business management and product and other matters logistics, energy, information and other factors combine, and optimize the operation mode to achieve the minimum environmental impact of energy use the least resources, the best pattern of management and the higher level of economic growth.

Cleaner Production Audit is the process of systematic analysis and evaluation of the running production process by auditors in accordance with certain procedures. Cleaner production Audit is a major technical method and tool for enterprises to implement cleaner production, and is also the basis for implementing cleaner production (Zhou et al. 2010).

Cleaner production audit includes the core part of organizing production or priority sectors, achieving quantitative monitoring of pollution caused by a bid to identify high-consumption, high pollution and further adopt relative strategies to exploit programs to lower and prevent pollutants from the source.

Certain types of cleaner production strategies are shown below:

- To reinforce the production process control as well as management, the general thing is low cost programs.
- To make some changes with respect to raw materials, which meet several requirements, such as non-toxic materials and environmentally-friendly materials, it is expected that reasonable proportion of investment will further perfect the transmission measurement method, and take full advantage of resources and energy, achieving comprehensive utilization or recycling use of raw and auxiliary materials.
- Improved products, to enhance the product yield and quality, to reduce material and energy consumption and to make changes in product design or product packaging, finally to improve product life and alleviate product toxicity as well as environmental hazards.
- Make technological and innovative improvements in order to optimize the process route to better automatic control and updated equipment.
- The recycling use of material and waste.

6.2.2 Current Situation of Production of Chito/Chitin Oligosaccharides and Its Monomer

Chito/chitin oligosaccharides and its monomer as derivatives of chitin, were also called aminosaccharide. Currently, the sorts of chitin/chitosan derivatives products reached approximately 40, including glucosamine, glucosamine sulfate, chitooligosaccharide (COS), N-acetyl-D-glucosamine, etc. Glucosamine is the main aminosaccharide product in commercial, which is produced by chemical treatment traditionally. Statistics showed that the chitin production enterprises in China dropped from hundreds to nearly 100 in the past 20 years due to environmental issues. Nevertheless, the demand and production capacity of chitosan derivatives constantly increased.

The United States, the European Union, Japan, South Korea, Chinese Taipei, other nations and regions, in which the annual demand exceeds 250,000 tons, are main manufacturers and consumers of oligosaccharide products, with the demand for chitosan oligosaccharide of about 60,000 tons (food grade). Chitosan oligosaccharide 20,000 tons, pharmaceutical grade chitosan oligosaccharide 10,000 tons, chitosan oligosaccharide for health food series 15,000 tons, chitosan oligosaccharide for bio-pesticide series 15,000 tons, chitosan oligosaccharide for feed immunization 10,000 tons. In European, American, Korean and Japanese, most manufacturers use oligosaccharides directly in fields of upmarket production, including drugs, health food, pesticides, cosmetics, etc. China is the major exporter of glucosamine products, accounting for 80% of the world production, and plays a pivotal role in the chitin industry. In China, it exported over 6000 tons glucosamine products, consuming 10,000 tons chitin in 2005. Although it has been greatly improved in the production technology, equipment and management level of glucosamine derivatives products in recent years, there is still a gap from the requirements of cleaner production.

The cleaner production of amino saccharide products is through the innovation and optimization of the traditional production process to improve the yield of products. The acidic waste is recycled and utilized to reduce material and energy consumption and minimize the environmental pollutants.

At present, there is few reports related to cleaner production of chito/chitin oligosaccharide and its monomer, the content of this chapter is evaluated and discussed in conjunction with the corresponding policies and regulations and a few examples of enterprises.

6.2.3 Significance of Cleaner Production of Chitin and Its Derivatives

Cleaner production of chitin is the hub of sustainability of chitin industrial development. At present in China, the materials and energy requirement for every ton of chitin production consist of 28–35 t waste from crustaceans, 0.5 t alkali, 8.5 t 30% (wt) HCl, 250–300 t water, 1.5 t coal, etc. Besides, the cost on terminal wastewater treatment amount to 50,000 RMB per ton. The composition of waste water in chitin production is presented in Table 6.1. For producing 1 t COS, it is calculated that 100–130 t water were consumed, resulting in 80–110 t waste water pouring into the environment. The great consumption of materials and energy and the vast cost of environmental treatment induce bottleneck for chitin and its derivatives industry development.

By means of adopting the concept of cleaner production, the cleaner production of chitin and its derivatives renovates conventional process of chitin production and exploits a wide range of processes (including pretreatment of raw materials, acid-base waste closed loop process, boiling alkali recovery, etc.) in a bid to achieve recycling of byproducts (protein, hydrochloric acid and calcium chloride) from raw materials and higher usage rate of energy, reducing waste emission during the process of production and achieving the cleaner production (Ortolano et al. 2012).

More importantly, the environment is usually bound up with economic development. In other word, benign environment will boost economic development, better social activities and support economic activities so that it can further provide the necessary resources and energy, and achieve sustainable development. Therefore, carrying out research on cleaner production of chitin and its derivatives is of great

Table 6.1 Waste water composition in chitin production

Item	Acid waste water	Alkaline waste water	Washing water	Total waste water
COD/(mg/L)	40,000–52,000	70,000–85,000	2500–3500	5000–12,000
SS/(mg/L)	3000–4700	4000–5800	3000–3600	2000
pH	1–2	10–12	5–6	1–2
Chroma	200–400	200–400	100–200	50–100
Waste water/(t/d)	40	10	250	300

significance for realizing the sustainable development of chitin industry, pulling the processing industry and protecting the environment (Wang 1999).

6.3 Cleaner Production of Chitin

Generally, chitin in crustacean byproducts is tightly linked with proteins and minerals. Chemical methods for processing shrimp waste using strong acid and alkali for deproteinization and demineralization have been long used in industries. However, those conventional methods usually lead to hazardous and considerable energy consuming and result in environmental pollution (Gortari and Hours 2013). The previous mode of end-pipe treatment of chitin production wastewater based on physicochemical treatment could not meet the requirements of cleaner production, which inhibits the development of chitin industry.

6.3.1 Technological Keys of Cleaner Production of Chitin

The technological keys of cleaner production needed to be unraveled emergently are as follows.

The first one is the screening and refinement of the cleaner production of chitin. The way to do this is to turn end-of-pipe control treatment into incipient, all-dimensional, whole-process control, extend single physical-chemistry process to comprehensive treatment incorporating biology-chemistry process, physics-chemistry, and recycling of waste water and resources into one process, transform multivariate and intersecting phase into single-partition allocation, which paves the way for the recycling of resources and the following biology-chemistry process of waste water. Ahead of this way, the problem that waste water from chitin production cannot be handled with biology-chemistry approaches because of multiple-dimensional mix, high OD value and high content of chlorine root ions persist all the time.

The second one is the pretreatment process of raw materials. To be specific, It is to pretreat raw materials so as to put an damper on protein's and water's entry into boiling process of alkali.

The third one is to develop reduction or even zero-emission technology for boiling alkali liquor. The main method is the recovery of affluent proteins, degradation products, astaxanthin and grease from boiling alkali liquor, which could fundamentally weaken the emission of organic waste, effectively allow for turning waste into wealth, and eventually decrease the difficulties of terminal waste treatment.

The fourth one is the development of acid-soaked micro-emission technology. In other word, it makes the recycling of acid-soaked waste liquor possible.

The fifth one is comprehensive utilization of waste water. It is required that waste water can reach the standard of discharge and be reused effectively by means of natural settling, physical processing, chemical processing and biological processing. The breakthrough by the cleaning process of chitin production is expected to solve the worldwide problem haunting over it for decades because of waste water processing and at the meantime make the mass recycling of animal protein and other resources possible, which eventually boosting economy and benefiting society evidently.

The last or the most important and prospective one is the synthetic biology technology, which can be manufactured in huge scale by fermentation of microbes with glucose as raw material. This way completely institutes the chemical method and utilizes the glucose instead of shells. The key point of this fermentation way is the construction of the suitable microbe.

It is estimated that, for an enterprise planning to yield 100 t chitin per year, if it puts the cleaner process of chitin production into practice, its economic benefit from the cleaner production will balloon much more than the expense on waste-water processing, the cost of waste-water processing will dropped to 1.0–1.5 RMB from 5 RMB, the mass of recycled protein solid reaches to 150–200 t, hydrochloric acid saved from it totals 100–200 t, the water conserved from it grosses 15,000 t. All those can be explained by reducing waste and energy vastly and further our competiveness of chitin field among nations.

6.3.2 Implementation Process of Cleaner Production Project of Chitin

Through previous research progress and market investigation, this section mainly carries out the following five aspects to achieve cleaner production:

(a) Optimization of production process

The production process of the alkali-cooking after acid leaching in traditional process is adjusted to acid leaching after alkali boiling. It can effectively avoid the problem of multi-cross mixture of protein in chitin of byproducts and its degradation products, astaxanthin and lipids with acid leaching. Besides, it can create favorable conditions for recovery of acid leaching solution and biochemical treatment of subsequent waste water. The bottleneck that originally restricted the treatment of biochemical technology could be solved.

(b) Increasing the pretreatment process for raw materials

Through pretreatment of the raw materials, including washing and pressing dehydration, less protein, water and other substances can be added into the alkali boiling process, which will reduce alkali consumption, energy consumption and washing wastewater in the alkali cooking process.

(c) Recovery of a large amount of protein, astaxanthin and lipids in the alkali boiled waste water

The alkali boiled waste liquid to be discharged is neutralized with the acid liquid, and the protein in the alkali cooking liquid which is combined with astaxanthin is recovered. The product obtained can be used as a high-quality protein feed additive. This not only effectively reduces emissions of organic matter, but also turns waste into treasure. Furthermore, it can bring economic benefits, as well as reduce the difficulty of treatment of integrated wastewater.

(d) Recycling of the acid leaching solution

The previous mode of not reusing the acid leaching solution will result in a large amount of acid solution to be controlled, energy consuming and resources wasting. In the present technology, it can reuse the acid leaching solution by distilling, cooling and crystallizing the acid liquid to recover the $CaCl_2$ product. At the same time, the distillate can be used to neutralize alkali waste liquid, while the remaining liquid of the distillation can be collected for the next round of recovery.

(e) Innovative treatment with integrated wastewater

The integrated wastewater is treated by natural sedimentation, separation and supernatant recovery, and the rest can be treated by anaerobic or aerobic fermentation, and then discharged after flocculation or oxidation.

Besides, membrane technology including alkali-tolerant nanofiltration membrane (NFM) and continuous volumetric diafiltration (CVD) has been applied to separate and concentrate alkali wastewater discharged from chitin-processing selectively (Li et al. 2014; Zhao et al. 2013, 2014) (Figs. 6.2 and 6.3).

6.3.3 Scheme of Cleaner Production

The final cleaner chitin production plan in factories was presented in Table 6.2.

6.3.4 Biotechnologies: Fermentation

Fermentation of crustacean byproducts with protease- or organic acid-producing bacteria can isolate the solid fraction of chitin from liquor containing several kinds of components, such as shrimp proteins, minerals, pigments, and other components (Prameela et al. 2010). In addition, the remained solid fraction after deproteinization is traditionally subjected to organic acid (citric acid) treatment for removing minerals and yielding pure chitin.

With the development of biotechnologies, production lines based on fermentation have been set in few factories in China. The processes have been shown in Figs. 6.4 and 6.5.

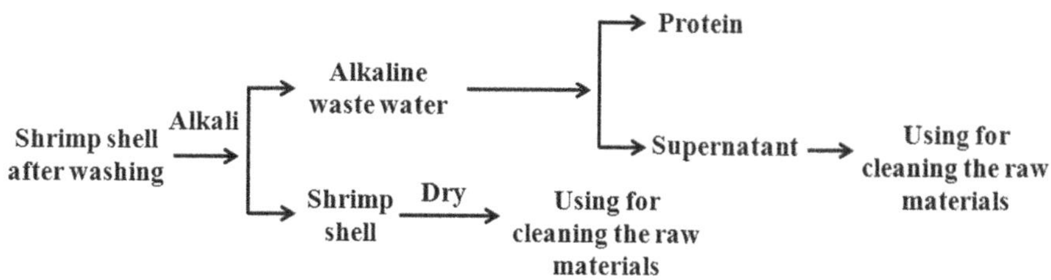

Fig. 6.2 Flow chart of alkali cooking in cleaner production

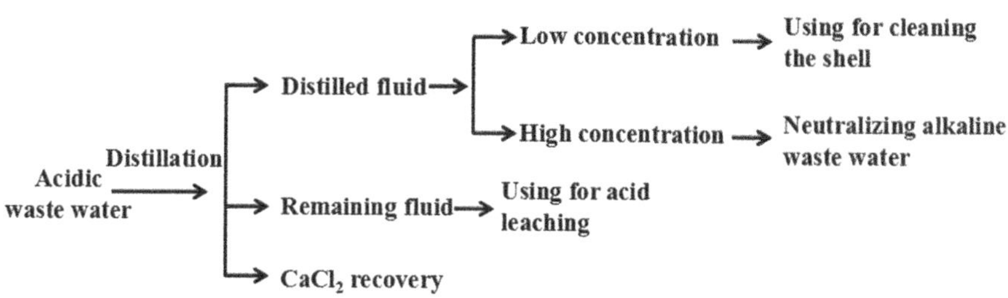

Fig. 6.3 Flow chart of acid leaching in cleaner production

The differences between chemical methods and fermentation technology in chitin production have been generally compared and analyzed, and results were presented in Table 6.3 (Guo et al. 2019). One thousand kilograms chitin were obtained from around 10,000 kg (wet weight) shrimp heads. Traditionally, the quantities of strong acids and alkalis were applied in the removal of protein and minerals for chitin. According to Percot et al., 0.25 M HCl (solid: liquid = 1/40 (w/v)) was applied to achieve complete demineralization within 15 min at room temperature. At 70 °C, deproteinization can be easily achieved in 1 M NaOH within 24 h (Percot et al. 2003). It was calculated that 6700 L concentrated HCl and 1300 kg NaOH were consumed, resulting in over 310 t waste water pouring into the environment. It is certain that this is inappropriate with respect to economic and environmental considerations. While there is 2230 kg organic acid consumption and much less production of harmful effluent wastewater, close to 133 L water was added in fermentation process. Moreover, lower temperature and looser requirement for equipment with the corrosion of strong acid and bases were demanded in fermentation technology, strict requirement for equipment with the corrosion of strong acid and bases.

Comprehensively, it indicated that fermentation, as a novel biotechnology is an economic and cleaner process for chitin extraction from crustacean byproducts.

6.4 Methods for Cleaner Production of Chito/Chitin Oligosaccharide and Its Monomer

Sustainable development challenges the traditional attitude of companies. Either by the profits, services and products they will evaluate or by the impacts they will exert on individuals, collectives or environment. Conjectures usually put forward threats

Table 6.2 Scheme of cleaner chitin production

Description	Scheme	Aim	Expense
Raw materials	Extract with green solvents, such as CO_2 supercritical fluid	Recover astaxanthin and lipids	High
Raw materials	1. Grind and wash material with washing water	1. Precipitation of a portion of protein and lipids, less acid and alkali use	Low
	2. Dry the shell	2. Less the water in shell to be cooked by alkali	
	3. Flocculate the washing water and separate the precipitate	3. Recover part of protein in material, increase the value of products	
Alkali cooking	1. Dry the shell after washing, and then cook with alkali	1. Less the amount of alkali use	Low
	2. Alkali cooking prior to acid leaching(for shrimp head/shell)	2. Separate most protein and lipids, less the amount of acid use in the next procedure	
	3. Acid leaching prior to alkali cooking (for crab head/shell)	3. Separate most mineral and recover $CaCl_2$ products as much as possible, less the amount of alkali in the next procedure	
Waste water after alkali cooking	1. High-concentration waste liquid is used as the recycling alkali	1–2. Reduce the waste water, save water	None
	2. The low-concentration waste liquid is neutralized with the acid leaching waste liquid, and then used to clean the raw materials		
Acid leaching	1. Dry the shell after washing, and then leach with acid	1. Less the amount of acid	Low
	2. The initial volume of the acid is as small as possible	2–3. Reduce loss of acid 4. Save water	
	3. Then supplement low concentration acid or high-concentration acidic waste water	5–6. Recover CO_2	
	4. Use acidic waste water to neutralize the alkaline waste		
	5. Collect the gas production (CO_2) and compress into liquid		
	6. Refine CO_2		
Waste water after acid leaching	Neutralize with the alkaline waste liquid, and then used to clean the raw materials	Reduce the waste water, save water	Low
Integrated wastewater	Biochemical treatment, such as fermentation, flocculation and oxidation	Reuse the wastewater	Medium
Residue	Use as fertilizer	Less waste of resources	Low

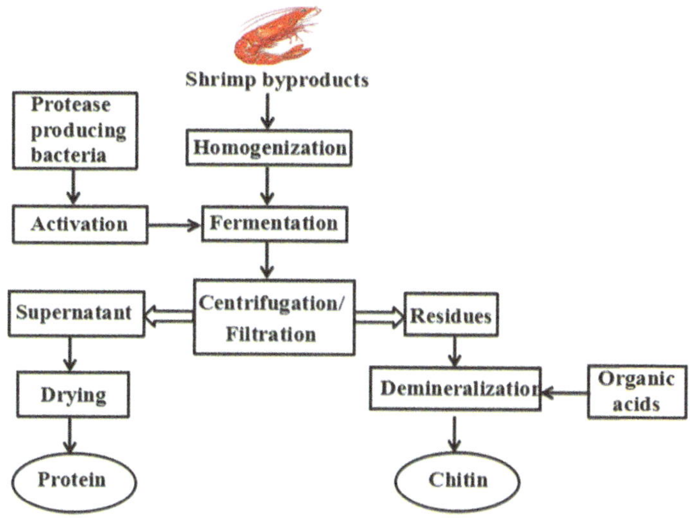

Fig. 6.4 Flow chart of chitin production

Fig. 6.5 Photo of production workshop

and opportunities. Increasing numbers of evidences are proving companies which are acquiring a more sustainable approach are having more positive advantages. There rewards are determined by the areas they are focusing on, the strategies they are carrying out and the phase of evolution. Furthermore, increasing demands are asking companies to take full responsibilities for their social, environmental and economic performances:

Table 6.3 Differences of material parameters between traditional method and fermentation

Index	Traditional method	Fermentation
Chitin production (kg)	1000	1000
NaOH consumption (kg)	1300	0
HCl consumption (L)	6700	0
Organic acids (kg)	0	2230
Water consumption (t)	310	133
Period (h)	12.2	16
The highest temperature in process (°C)	70	50

- Advanced communication technologies allowing for the quick transmission of data and information regarding what are occurring in the world, comprising the deed of different kinds of companies.
- A deeper understanding that companies and social conduct usually give birth to the business threat, especially to brand name and reputation, but also to capital use, functional competence.
- Increasing governmental engagement and activities.
- Reinforcing the authority of international companies – more authority will lead to greater responsibility.
- Growing consciousness that corporate responsibility is closely linked to comportment and action in the supply chain.

6.4.1 Method for Cleaner Production of Glucosamine and Its Derivatives

Glucosamine production refers to process that chitin, after hydrolyzed by hydrochloric acid and other necessary procedures, changes into glucosamine hydrochloride, glucosamine sulfate, N-acetyl glucosamine and other aminosaccharide derivatives (Mao et al. 2017). As a major exporter of aminosaccharide products, which accounts for 80% worldwide yield, China is an indispensable part of the world in chitin field. However, recently China has improved its production technology, facilities and management to the aminosaccharide production, cleaner process still seems to appear a long term.

By the system of cleaner production during the process of production of aminosaccharide renovates conventional process of amino saccharide production and exploits a wide range of processes in order to achieve a variety of aims. Such as higher recycling ration of products, better reuse of wastewater, lower raw material and energy consumption and even the minimum impact to environment. Its technological cruxes needed to be unraveled emergently are as follows.

At first, screen and refine the technologies in aminosaccharide production. In the past, pitfalls exist in aminosaccharide production long include, using ration of hydrochloric acid fails to reach to 1/3 its input, and the hydrochloric and caramel along with acid reaction weigh heavily on products. Besides, some problems are also proved from vacuum evaporation and concentration for mother liquor. It contains increasing energy consumption, considerable operation fees, low recycling ratio and the bad impact on environment because of low-concentration emission of acid waste. Next, the attempt of adopting active carbon to decolorize the products failed due to high cost, energy consumption and bad impact on environment.

Secondly, new technologies should be exploited. Researchers are thinking whether specialized ion-exchange facility is available for the recycling of products and acid waste. Were it to work, we should vastly reduce energy consumption, alleviate labor intensity and promote quality and recycling ratio. In addition, a potential purification technology by nanofiltration (NF) separation is a tool for COS with specific polymerization degree and monosaccharides (Dong et al. 2014; Lei et al. 2018; Zhang et al. 2016).

Thirdly, develop recycling technology by reusing low-concentration acid waste after certain processing.

The final one is to tap crude-product ion-exchange technological potentials. It not only allows for decolorizing product with high efficiency, but makes the recycling of decolorized resin possible.

6.4.2 Process Analysis for Cleaner Production of Glucosamine

The evaluation mainly analyzes the cleaner production process of the project from the aspects of raw and auxiliary materials and energy, techniques and processes, waste treatment measures, environmental management and so on (Fig. 6.6).

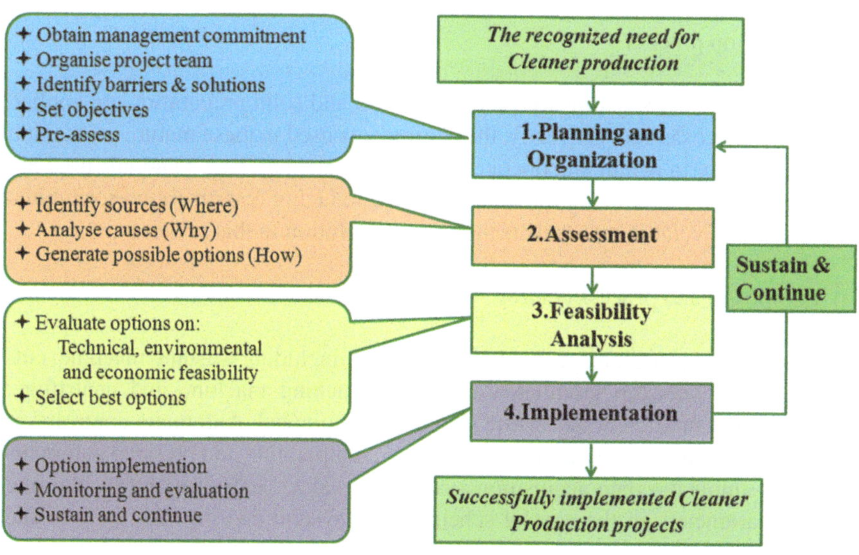

Fig. 6.6 Process analysis of the project

6.4.2.1 Raw Materials and Energy

The raw materials involved in the project are basic chemical materials such as hydrochloric acid and alcohol. Most of these materials are corrosive or inflammable. Therefore, the evaluation of cleaner production level mainly depends on the quality, storage and management of raw and auxiliary materials. Raw and auxiliary materials should be selected with low impurities and high purity of chemical raw materials in order to reduce the amount of pollutants produced in the production process. The storage and transportation equipment raw and auxiliary materials should be selected with good sealing performance of production equipment, to minimize the loss of materials without organization. The raw and auxiliary materials management should be standardizing, and adopting personnel responsibility system. Thereby the raw and auxiliary materials of the project can meet the requirements of cleaner production.

The energy consumption of this project is mainly steam, water and electricity. Therefore, in the design of the project, energy-saving equipment should be selected as much as possible to improve the utilization of water resources.

The project has a specific warehouse for the storage of raw materials. Meanwhile the office, production and raw materials area has been divided into blocks to manage, reducing the incidence of accidents and the storage and making raw and auxiliary materials more in line with the requirements of safe production.

6.4.2.2 Technology Process

Cleaner production is a process of controlling the whole process of production, including saving materials and energy, and reducing their emissions and toxicity as far as possible before discharging the waste produced from production process, and implementing comprehensive utilization of pollutants so as to make waste reuse.

(a) Production process

Hydrochloric acid hydrolysis, plant extraction and solid preparation are applied to produce glucosamine. Currently, the equipments used in these mature production process are easy to install and operate, but the raw and auxiliary materials used are strong corrosive, volatile and flammable. Thus, reducing equipment corrosion and flammable and volatile materials is the primary solution in this process.

(b) Equipment and process control

The production equipment should be advanced, including washing machine, cutting machine, one-step granulator, automatic packing machine and centrifuge, meanwhile the inefficient and waste should be eliminated. Automatic control system in instruments should be adopted to control temperature and pressure automatically, and automatic data acquisition system is used to centralize and display the relevant parameters in the control scheme. This system can give full play to the

Fig. 6.7 Photo of equipments

potential ability of process and equipment, stabilize process operation and reduce man-made errors. It is not only beneficial to strengthen output control, improve product quality and reduce energy consumption, but also reduce the labor intensity of operators. Specifically, the advanced production equipment are selected to improve the conversion rate of raw materials, which is conducive to the production of the post-sequence section, and reduce the amount of pollutants for the sake of environmental protection.

In addition, it is vital that employing the best proportion of raw materials and using temperature and pressure control equipment to stabilize production, improve product quality and avoid serious waste of raw materials. Besides, installment of waste gas absorption devices and waste liquid storage facilities should be executed to avert waste emission (Fig. 6.7).

(c) Pollution control

The waste gas (HCl) produced in the process is suggested to treat by two-stage falling film absorption tower. The alcohol is recovered by alcohol recovery tower. In addition, the recovered or treated products can be sold as byproducts. It can not only create certain additional economic benefits, but also greatly reduce the discharge of air pollutants. Bag collector is used to treat the crush dust and boiler exhaust gas utilizes Venturi water film dust removal technology to improve the removal rate of soot and SO_2 and reduce the discharge of air pollutants.

Boiler waste water can be reused for boiler dust removal to reduce the consumption of fresh water. Furthermore, the produced residual liquor containing alcohol can be mixed into coal for incineration to achieve circular economy, and waste water discharge after treatment by the sewage treatment station to reduce the waste water pollutants.

Table 6.4 List of material balance, energy consumption and pollution emission levels

Production line	Item	Unit	Cleaner production	General level in China
Glucosamine	Hydrochloric acid	Kg/one unit product	1650	1800
	Ethanol	Kg/one unit product	3300	3500
Plant extracts	Ethanol	Kg/one unit product	5000	6000
Energy consumption	Steam	Kg/one unit product	20.8	20.8
	Electricity	Kw·h/one unit product	33	40
	Coal	t/one unit product	2.92	3
	Water	t/one unit product	104	120
Pollutant emission	Waste water	m^3/one unit product	47.2	37
	COD	Kg/one unit product	3.2	3.6

(d) Energy consumption

Since there is no cleaner production norm in China, there is comparison of cleaner production compared with the same industry enterprises in the material consumption, energy consumption and sewage discharge level. The list is shown in Table 6.4.

6.4.2.3 Environmental Management

Developing cleaner production is a new creative concept that continuously applies environmental prevention strategies into the process of production, services and products in order to increase eco-efficiency and reduce man-made risks. From 1980s, global environmental pollution and ecological damage intensified with the increasing development of economic construction. As a result, the economic development has been restricted because of the shortage of resources and energy. Gradually, people have realized the limitation to environment protection by the treatment of controlling pollution. While taking the environmental impact of products and production processes into consideration, it may be more effective to eliminate pollution adopting measures such as improving production processes and strengthening management. Consequently, the practice of cleaner production emerged and popularized around the world.

The principle of cleaner production is prevention, which claims strengthening management and technological progress in the chain of designing products to selection of raw materials, process routes, facilities, waste comprehensive utilization as well as operation management (Klemeš et al. 2012). The idea of prevention reaches the purpose of improving material utilization and reducing or even eliminating the generation of pollutants.

Cleaner production reflects an intensive growth pattern. Enterprises need make such integration:

- Adjust product structure
- Exploit new production technology
- Optimize the condition of production condition
- Perfect technical level of equipment.
- Strengthen management,
- Enhance employee capacity
- Achieve energy conservation and emission reduction
- Reduce pollution
- Increase efficiency
- Allocate resources reasonably

Cleaner production reflects the combination of environmental and economic benefits. Traditional end-of-pipe pollution treatment will cause high investment, high operating cost and difficult governance, while cleaner production can avoid the shortcoming and facilitate the enthusiasm of enterprises to prevent industrial pollution.

6.4.2.4 Services and Management

In practice, all kinds of organizations are promising to make economic profits by adopting cleaner production – the primary and manufacturing sectors, as well as services and knowledge-based organizations. Whilst the emphasis to date has widely been on the manufacturing sector, there are also millions of organizations around the world providing services and management activities, which are also critical in examining moves towards securing sustainable outcomes for businesses and society (Severo et al. 2018).

Enterprises can make adjustments in services and management by introducing cleaner production.

- Develop management regulations and operational guidelines that facilitate cleaner production
- Choose employees with experience and high-level skill, and conduct strict pre-job training for them.
- Strengthen employees' awareness of cleaner production
- Improve a positive attitude towards change and improvement
- Create a "learning organization"
- Enhance relationship between a company and its neighbors as well as between a company and authorities
- Improve occupational health and safety

- Develop incentives and penalties for cleaner production and increase employee engagement in cleaner production
- Promote communication between employees, between companies and their suppliers and between companies and their customers ("Greening the supply chain")

6.4.2.5 Scheme of Cleaner Production

Based on the analysis of production process above, a scheme of cleaner production of glucosamine has been implemented as Table 6.5 shown.

Table 6.5 Scheme of cleaner production

The type of project	Content	Significance and benefits
Raw and auxiliary materials	Use pure materials	Reduce waste emissions
	Use low-sulfur coal	
	Adopt high-grade medicinal materials with less impurities	Less impurity production
	Strengthen the storage management of raw materials	Increase productivity
Technology and equipment	Use energy-efficient equipment (electromechanical equipment, transformers, insulation materials, safety valves, etc.)	Reduce power consumption, save energy and costs
	Use sealed equipment and production technology	Reduce emissions of exhaust gases
	Drain rain and sewage, and recycle wastewater to reduce the amount of fresh water	Reduce pollutant emissions
	Change advanced centrifuge equipment	Improve productivity and product quality
Recycling resources	Recycle HCl by secondary falling film absorption to gain byproducts for profits	Save raw materials and reduce water resources
	Recycle condensate produced in drying steam into boiler for steam production	Reduce wastewater discharge and save water
	Comprehensively utilize of selected medicinal materials waste	Reduce solid waste emissions
	Boiler sewage is reused for dust removal in boiler	Reduce wastewater discharge and save water
Pollution control measures	Adopts "neutralization & A/O process" for wastewater treatment	Reduce the environmental impact of wastewater and ensure the stability of wastewater
	Absorb the exhaust gas from the suction filtration, concentration and secondary filtration section by water circulation	Reduce the environmental impact of waste gas produced by HCl
	Set recovery device for alcohol	Reduce alcohol consumption and save production costs
Equipment	Use automatic control equipment	Save energy, material consumption and save production costs

(continued)

Table 6.5 (continued)

The type of project	Content	Significance and benefits
Production process management	Organize systematic job training for employees	Improve the skills of staff and produce efficiently
	Strengthen the maintenance of equipment	Save costs
	Strengthen the whole process management of raw material transportation, quality inspection, measurement and storage	Reduce material loss
	Add automatic control instruments and strengthen measurement management	Improve the yield of product, save energy and reduce consumption
	Adjust the pipeline layout of water, steam and electricity to strengthen its measurement and testing	Save energy and improve product quality
Continuous cleaner production	Strengthen the cleaner production of the whole site and conduct regular auditing work	Raise the level of cleaner production
	Develop a continuous cleaner production plan so that cleaner production can be carried out in an organized and planned manner in the enterprises	Continuously improve cleaner production level

6.5 Future Challenges and Trends in Cleaner Production

Methods applied for the industrial chitin/chitosan oligosaccharide are chemical procedures using corrosive reagents, which are inevitable to cause environmental pollution and resource wastage and increase the additional costs. However, novel environment-friendly and cleaner technologies in laboratory-scale remain to be further emerged in future.

The biotechnologies referred in previous discussions are potential to efficient production with lower cost, reduced labor, and greater safety than traditional methods (acid hydrolysis). However, to date, there are limitations and outstanding bottlenecks in developing biotechnologies into industrial scale. The technical challenges of biotransformation, enzymatic synthesis and fermentation methods include limited reuse and stability of enzymes, difficulty in sustaining continuous reactions, and small total sample capacities, and high efficiency "intelligent" microorganisms construction. Therefore, in-depth study on "intelligent" microorganisms construction, enzyme immobilization, rational enzyme design, solid-state fermentation, and continuous mixing bioreactor systems are required to achieve biocatalyst reactions with high efficiency, stability, specificity, and total yield.

6.6 Conclusion

The trend toward sustainable development of environment and economy has resulted in large-scale discussions on the cleaner production of chitin/chitosan oligosaccharide. Combination of environmental and economic benefits, process adjustment, byproducts recycling and integrated control based on current technology should be conducted to achieve cleaner production.

In particular, in recent years, biotechnologies have become the mainstream of future research and manufacture instead of hazardous chemical routes. A simple, fast, effective, clean, economic, and controllable bioprocess for the chitin/chitosan oligosaccharide production needs to be further investigated.

Moreover, adapting laboratory-scale biotechnological methods for ecologically safe industrial-scale bio-manufacture of oligosaccharides from chitin/chitosan is a major challenge of future research.

References

Cabrera JC, Cutsem PV (2005) Preparation of chitooligosaccharides with degree of polymerization higher than 6 by acid or enzymatic degradation of chitosan. Biochem Eng J 25(2):165–172

Dong H, Wang Y, Zhao L, Zhou J, Xia Q, Jiang L, Fan L (2014) Purification of DP 6 to 8 chitooligosaccharides by nanofiltration from the prepared chitooligosaccharides syrup. Bioresour Bioprocess 1(1):20

Gortari MC, Hours RA (2013) Biotechnological processes for chitin recovery out of crustacean waste: a mini-review. Electron J Biotechnol 16(3):1–18

Guo N, Sun J, Zhang Z, Mao X (2019) Recovery of chitin and protein from shrimp head waste by endogenous enzyme autolysis and fermentation. J Ocean Univ China 18(3):719–726

Kjaerheim G (2005) Cleaner production and sustainability. J Clean Prod 13(4):329–339

Klemeš JJ, Varbanov PS, Huisingh D (2012) Recent cleaner production advances in process monitoring and optimisation. J Clean Prod 34:1–8

Lei Y, Zhen Q, Chen Q, Zhao M, Zhao H, Ahmad W, Fan L, Zhao L (2018) Insights into the nanofiltration separation mechanism of monosaccharides by molecular dynamics simulation. Sep Purif Technol 205:48–57

Li J, Zhao L, Wang Y, Chen C, Zhou J, Qiu Y, Du H (2014) Analysis of membrane fouling by proteins during nanofiltration of chitin alkali wastewater. J Water Reuse Desalination 4(4):253–262

Mao X, Guo N, Sun J, Xue C (2017) Comprehensive utilization of shrimp waste based on biotechnological methods: a review. J Clean Prod 143:814–823

Ortolano L, Cushing KK, Warren KA (2012) Cleaner production in China. Ecol Econ 19(5):431–436

Percot A, Viton C, Domard A (2003) Optimization of chitin extraction from shrimp shells. Biomacromolecules 4(1):12–18

Prameela K, Mohan CM, Hemalatha KPJ (2010) Extraction of pharmaceutically important chitin and carotenoids from shrimp biowaste by microbial fermentation method. J Pharm Res 3(10):2393–2395

Sakr D, Sena AA (2017) Cleaner production status in the Middle East and North Africa region with special focus on Egypt. J Clean Prod 141:1074–1086

Schaltegger S, Bennett M, Burritt RL, Jasch C (2008) Environmental management accounting (EMA) as a support for cleaner production. In: Eco-efficiency in industry and science, vol 24. Springer, pp 3–26

Severo EA, de Guimarães JCF, Dorion ECH (2018) Cleaner production, social responsibility and eco-innovation: generations' perception for a sustainable future. J Clean Prod 186:91–103

Wang J (1999) China's national cleaner production strategy. Environ Impact Assess Rev 19(5–6):437–456

Zhang S, Zhou J, Fan L, Qiu Y, Jiang L, Zhao L (2016) Investigating the mechanism of nanofiltration separation of glucosamine hydrochloride and N-acetyl glucosamine. Bioresour Bioprocess 3(1):34

Zhao L, Zhao H, Nguyen P, Aoran LI, Jiang L, Xia Q, Rong YI, Qiu Y, Zhou J (2013) Separation performance of multi-components solution by membrane technology in continual diafiltration mode. Desalination 322(8):113–120

Zhao L, Li J, Wang Y, Fan J, Li A (2014) Mechanism underlying the nanofiltration of protein hydrolysates in chitin alkali wastewater. Desalination Water Treat 52(37–39):6936–6946

Zhou J, Gou X, Shi Z (2010) Overview of cleaner production audit. In: Key engineering materials, vol. 439–440. Trans Tech Publications, Switzerland, pp 1533–1536

Modification of Chitosan/Chitin and Its Oligosaccharides

7

Xiaoli Liu and Qixing Jiang

Abstract

In view of rapidly growing interest in the amino polysaccharide chitosan/chitin and its oligosaccharides as functional biopolymer, a recent progress of basic and application studies in chitosan/chitin oligosaccharides chemistry is reviewed as well as some basic aspects of this specialty biomass resource. A special emphasis is placed on the controlled modification reactions to prepare chitin derivatives with well-defined structures and thereby to construct sophisticated molecular architecture having various advanced functions. The reactions discussed here include acylation, quaternary ammonium salt, carboxyalkylation, graft copolymerization, quaternary salt formation, Schiff base formation, reductive alkylation, microwave modification. For conducting modification reactions in a facile and controlled manner, some soluble chitosan/chitin oligosaccharides derivatives are convenient. To fully explore the high potential of these specialty biopolymers, basic and application researches are being made extensively.

Keywords

Polymer modification · Chitosan/chitin oligosaccharides · Biopolymers

7.1 Introduction

Chitosan/chitin and its oligosaccharides, which are considered the most promising, flexible source for the preparation of advanced antibacterial biopolymers, are being investigated by several groups (Konwar et al. 2016). These materials are

X. Liu (✉) · Q. Jiang (✉)
State Key Laboratory of Food Science and Technology, School of Food Science and Technology, Collaborative Innovation Center of Food Safety and Quality Control in Jiangsu Province, Jiangnan University, Wuxi, China

L. Zhao (ed.), *Oligosaccharides of Chitin and Chitosan*,
https://doi.org/10.1007/978-981-13-9402-7_7

biodegradable, have low toxicity and are compatible with mammals' tissues. One characteristic of the chitosan/chitin and its oligosaccharide is that they possess cationic sites (positively charged amino groups) that have affinity towards negative sites of the cell walls.

In spite of their advantageous properties, their relatively low biological activity and solubility have limited their applications. Three reactive groups are found in chitosan/chitin and its oligosaccharide (amino group at C_2, secondary OH group at C_3, and primary OH group at C_6), which make this molecule behave in a unique way. They can be conveniently derivatized and can be readily derivative to improve the biological activity and solubility (Liu et al. 2014). Several synthetic strategies have been introduced where the C_2 amino group, C_6 position and C_3 position OH groups of polymer are the strongest activity common target for modification (Geisberger et al. 2013; Huang et al. 2013b) Thus, different COS derivatives were synthesized by acylation, alkylation, Schiff base formation (Badawy and Rabea 2014), reductive alkylation, carboxy alkylation, heterocyclic aldehydes (Sahariah et al. 2014), quaternary reaction, and the carboxymethyl reaction (Bodnar and Perreault 2015; He et al. 2015). These derivatives showed then improved antibacterial properties. In is necessary to have effective preparative procedures for synthesizing well-defined structures in order to create novel types of polysaccharide-bases new materials with specific functions. So, the efforts have been focused on the development of well-controlled, high-yield derivation reactions under mild conditions. Some important reactions are discussed in the next sections.

7.2 Acylation

The acylation of chitosan/chitin oligosaccharide derivative is the earliest reaction in the chemical modification of them. Chito/chitin oligosaccharide can be introduced aliphatic or aromatic acyl groups of different molecular weight on macromolecular chain through the reaction of acyl chloride or anhydride. There are both hydroxyl and amino groups in the molecular chain of chitosan. Therefore, the acylation reaction can not only produce ester on hydroxyl group (O-acylation) but also on amino group (N-acylation) to form amide. The introduction of acyl group can destroy the hydrogen bond between chitosan/chitin oligosaccharide and its macromolecules, change the crystal structure of chitosan/chitin oligosaccharide, and enhance the solubility in common organic solvents.

7.2.1 Chitin Acylated Derivative

The intra- and inter-molecular hydrogen bonds of chitin are hard to break by molecular solvents, and there is a strong micelle structure. Hence, the acylation reaction is very difficult to react. Generally, use anhydride as acylation reagent, the corresponding strong acid as a solvent or catalyst. There are more than ten kinds of reaction methods reported. The main difference of these methods is the use of different solvents and catalytic conditions. Chitin and chitosan can introduce aliphatic or aromatic acyl groups of different molecular weight on macromolecular chains by reacting with acyl chloride and anhydride. Acylation reaction can be carried out on hydroxyl group (O-acylation) or amino group (N-acylation). The presence of acyl groups can destroy inter- and intramolecular hydrogen bonds of chitin and its derivatives, change its crystalline structure, and make the resulting products soluble in organic solvents.

Chitin is not soluble in organic solvents normally used for acetylation, and the reaction is very slow under heterogeneous conditions (Singh et al. 2012). Extensive acetylation may be realized only under strong conditions; for example, acetylation of chitin was achieved with acetic anhydride and hydrogen chloride (Pillai et al. 2009).

Acetic anhydride achieves complete acetylation of chitin in the solvent methanesulphonic acid (Kurita 2006). The reaction mixture was heterogeneous initially, but the acetylated derivative went into solution as the degree of acetylation increased.

Another solvent system (trichloroacetic acid +1,2-dichloroethane) dissolves chitin and acetylation is done in solution. All these strongly acidic solvents cause partial degradation of the polymer backbone. Also, trichloroacetic acid chlorinates chitin (though to a low extent) (Huang et al. 2013a).

β-chitin swells in many solvents like methanol, although it is insoluble in most of them. Thus, the amino groups in β-chitin (Scheme 1) react with acetic anhydride in methanol giving poly (N-acetyl D-glucosamine), with no free amino groups. Under these conditions α-chitin cannot be acetylated. Also, β-chitin could be readily and totally acetylated in pyridine with acetic anhydride and the catalyst 4-dimethylaminopyridine (Fig. 7.1) (Kurita 2001). These results indicate that β-chitin may be used to conquer the problems caused by the intractability of α-chitin. Thus, it is a more versatile starting material than the more common α-chitin (Kurita et al. 1994).

Chitin and chitosan may be derivatized with a variety of acylation reactions. Products soluble in chloroform may be synthesized with chlorides of long chain aliphatic carboxylic acids like hexanoyl, dodecanoyl, and tetradecanoyl chlorides.

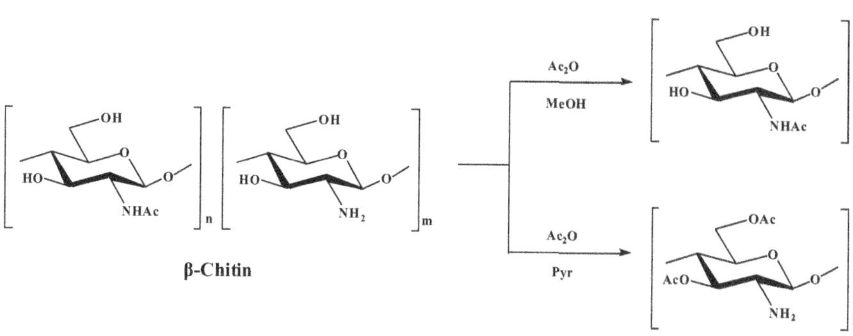

Fig. 7.1 β-chitin treated with acetic anhydride in pyridine in the presence of 4-dimethylaminopyridine as the catalyst. (Modified from Kurita 2001)

Such products show a degree of acylation of 3 (Fujii 1980). Fourier Transform Infrared Spectroscopy (FTIR) and thermal analyses data have revealed that chitosan with a higher degree of deacetylation was more susceptible for acylation owing to a decrease in hydrogen bonding. Derivatives of chitosan prepared with lactones (cyclic esters) like β-propiolactone or γ-butirolactone produce derivatives with N-hydroxyalkanoyl groups (Loubaki et al. 1989).

In the acetylation reaction, early studies used dry HCl and saturated acetic anhydride to acetylate chitin, but the reaction time was long and the degradation of polymer was a serious problem. It was found that strong acid solvents such as HCl, $HClO_4$, methane sulfonic acid and mixed solvents such as dichloroethane trioxy acetate, N, N-dimethyl acetamide, anhydrous LiCl were suitable for the reaction of chitin. In recent years, it has been found that meso sulfonic acid can be acylated instead of HCI. For the acylation reaction of chitin, masonic acid is both a solvent and a catalyst. The reaction can be carried out in homogeneous phase, and the acylation degree of the product is high. Mixed solvents such as dichloroethane trioxy acetate and lithium chloride N. N-dimethylacetamide can dissolve chitin directly and make the reaction in a homogeneous phase thus the derivatives with high degree of substitution and uniform distribution can be prepared. Hirano et al. (Hirano and Zhang 2000) reported that the filament surface and inside chitosan fibers were N-acylated by reaction with different carboxylic acid anhydrides at room temperature in methanol. It was demonstrated that the mechanical properties of chitin filaments such as tenacity and elongation was little affected by N-acylation. Treatment of chitin fiber and chitin–cellulose mixed fiber with 40% NaOH at 95–100 °C for 4 h in suspension afforded a chitosan fiber and a novel cellulose–chitosan mixed fiber, respectively. Novel N-acetyl, N-propionyl, N-butyryl, N-hexanoyl, and N-octanoyl chitosan fibers were obtained. The fibers were not soluble in water, acidic, or basic solutions.

The characteristic absorption band of amide I band at 1660 cm^{-1} and 1628 cm^{-1}. The amide II band, amide III, and CH_3-CO vibrational absorption band were also found in 1557 cm^{-1}, 1313 cm^{-1} and 1379 cm^{-1}, respectively, in Chitin. Characteristic absorption bands are also found at 1158 cm^{-1} and 895 cm^{-1}. Comparing the IR spectra of chitin and acetylated chitin, it was found that the characteristic absorption band of C=O stretching vibration of ester bond appeared in the infrared spectra of acetylated chitin at 1743 cm^{-1}, 1230 cm^{-1} and 1380 cm^{-1}, and with the increase of the amount of acetic anhydride. Its strength is gradually increasing. The disappearance of the characteristic absorption band at 895 cm^{-1} of the carbohydrate unit linked to the β-(1,4) glycoside bond of chitin further indicates that the acetylation reaction destroys the ordered structure of the chitin chain. This can be further confirmed by the fact that acetylated chitin dissolves in 88% formic acid but chitin does not dissolve in formic acid.

Benzylated, hex acylated, decyl, dodecanoylated and a series of p-substituted benzylated chitin were also successfully prepared by using the corresponding acyl chloride and chitin system. Acylation degree depends on the amount of acyl chloride, usually to obtain high degree of substitution products, need to add excessive reactants. In general, 1 mol acetylglucosamine was required with 4–6 mol (acyl

Table 7.1 The relationship between reaction conditions and acylation degree

Sample	The amount of dodecyl chloride[a]	Reaction condition	Degree of dodecyl chloride
1	0.6	0 °C, 2 h+ −20 °C, 12 h	0.15
2	1	0 °C, 2 h+ −20 °C, 12 h	0.8
3	5	0 °C, 2 h+ −20 °C, 12 h	1.5
4	5	10 °C, 2 h+ −20 °C, 12 h	1.1
5	3	0 °C, 4 h+ −20 °C, 12 h	1.5
6	2	0 °C, 2 h+ −20 °C, 12 h	1.7
7	4	0 °C, 3 h+ −20 °C, 12 h	1.7
3'	/	Shunt precipitation[b]	1.9
6'	/	Shunt precipitation[b]	1.9

[a]The amount of dodecanoyl chloride is equal to the amount of N-acetylglucosamine units
[b]3′ and 6′ is obtained by freezing the fractionation product of hot methanol

chloride or anhydride), and the reaction conditions were still stirred at 0 °C for 2 h and overnight at −20 °C. The influence of reaction conditions on the reaction is listed in the following table (Table 7.1) with the example of dodecanoylation. Because of the steric resistance effect, it is difficult to obtain the high substitution product with the increase of the chain length of the substituent. The complete acylation product can be obtained by boiling reaction of excessive hexyl chloride, decyl chloride or dodecyl chloride in anhydrous pyridine chloroform.

7.2.2 Chitosan Acylated Derivative

Chitosan acylation reaction is easier than chitin, generally no catalyst, the reaction medium commonly used methanol or ethanol. The early acylation of chitosan was carried out in anhydride or acyl chloride. The reaction conditions were mild, but the reaction time was long, and the degradation of polymer was serious. Like the acylation of chitin, mesylate is also a good solvent for the acylation of chitosan, and has a higher acylation degree of the product. Chitosan can be dissolved in acetic acid solution, adding the same amount of methanol does not precipitate. Therefore, acylated chitosan can be prepared by acetic acid-methanol solvent. Acylated chitosan reaction usually takes place on the amino to obtain the N-acylated products, but the reaction cannot be completely selectively on the amino group, and the o-acylated chitosan will also take place, but the formation of O-acylated chitosan is more difficult. In general, the straight chain aliphatic derivatives (formyl, acetyl, hexanoyl, dodecyl, tetradecyl, etc.) can be prepared in methanol or pyridine/chloroform solvents. The branched aliphatic acyl derivatives (N-isobutyl, N-trimethyl acetyl, N-isoamyl) can react in formamide solution. Aromatic acyl rowers (benzoyl, benzosulfonyl, etc.) are often prepared in methylene sulfonic acid solvents. The acylation degree of chitosan obtained is more than 1.8, and most of them are dissolved in a

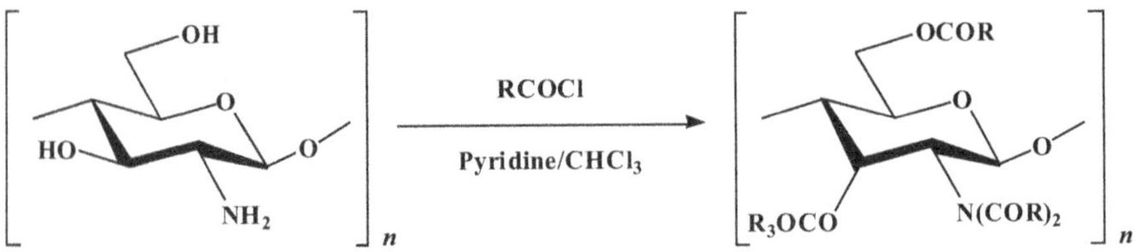

Fig. 7.2 Structural formulas for complete acylated chitosan derivatives

variety of organic solvents (Jenkins and Hudson 2001). The structural formula of fully acylated chitosan derivatives is shown in Fig. 7.2.

7.2.2.1 N-acylated Chitosan Derivative

The structure and degree of substitution of N-acyl chitosan have important effects on its properties. For N-acyl chitosan containing saturated fat chains, the length and degree of substitution of N-acyl fat chains are important factors determining its biological function, biodegradability and selective aggregation of certain cancer cells. Therefore, it is of great significance to study the preparation and control of the extent of substitution in biomedical materials.

Concerning the preparation of derivatives with definite structure and functional materials with better properties by chitosan, it is very important to find a method to control the reaction easily. In recent years, the selective reaction of N-phthalic chitosan has attracted much attention.

In 1930, Karrer first prepared N-acyl chitosan by reaction of chitosan and anhydride at high temperature (Karrer and White 1930). Hirano improved the method by adding formamide, methanol, tetrahydrofuran and other solvents into the system to improve the yield and degree of substitution of N-acyl chitosan (Hirano et al. 1976). Kurita took acetic acid and acetone as the media and added maleic anhydride to prepare N-maleylated chitosan, which is a kind of excellent water-soluble N-acylated chitosan (Kurita et al. 1982).

One of the major functions of N-acylation modification of chitosan is to prepare chitosan acetylated by 50%. When the extent of acetylation or of deacetylation of chitosan is 50%, the amino and acetylamino groups in the sugar chain are in a completely irregular arrangement, and chitosan is completely water-soluble. The completely water-soluble chitosan by adding pyridine to acetic acid-ethanol system and controlling the ratio of chitosan to acetic anhydride 1:2.

The N-acylation modification of chitosan not only improves its solubility in water, but also improves its forming and processing properties, and expands the application range of chitosan. Kaifu K prepared chitin with butylated chitin, and found that the compound had good anticoagulant activity and could be used in medical materials (Kaifu et al. 1981). N- formylated chitosan, N- acetylated chitosan and chitin mixed into fiber, can be used for surgical sutures, sutures of this kind of tensile strength and retention rate are better than the absorption of polyethyl lactide suture Dexon. Chitosan N-maleated at various extents had higher disinfectant power and moisturizing performance than hyaluronic acid. Meanwhile, the price of such compounds was much lower than that of very expensive hyaluronic acid products at present.

In acetic acid solution or in high swelling pyridine gel, the reverse point of N acetylation is easy to occur in chitosan, and 50% N-acetylated chitosan may be obtained adjusting the process conditions. Because it can form gel in organic solvent, it has better reactivity activity. Therefore, it can also be used as raw material for secondary modification. If the soluble chitin aqueous solution is mixed with pyridine, Dimethyl Formamide (DMF) or like solvents, a highly swelled gel can be obtained. N-acylation reaction can occur with amino groups using phthalic anhydride and tetracarboxylic anhydride (Kurita et al. 1982). Acylation reaction in acetic acid/methanol medium, the concentration of acetic acid has a significant effect on the degree of substitution of N-acyl group of chitosan. Chitosan has a higher degree of N-acetyl substitution at low acetic acid concentration. With the increase of acetic acid concentration, the degree of N-acetyl substitution decreases gradually. When the concentration of acetic acid increases to 10%, the degree of substitution is almost no longer affected by the change of acetic acid concentration. The acetic acid concentration effect on the degree of substitution of N-acyl groups showed a similar regularity in the N-propionyl, N-butyryl and N-hexyl reactions of chitosan. A nucleophilic addition-elimination mechanism occurs in the N-acylation of chitosan with anhydride. The degree of substitution of N-acyl group is determined by the nucleophilic ability of amino group and the space shielding of amino group in chitosan polycationic polymer chain. When the concentration of acetic acid is low, the degree of amination is lower, and its nucleophilic property is relatively strong. The repulsive effect of protonated amino groups at low concentration makes the polycationic polymer chains more extensive, resulting in less special shielding of amino groups. Therefore, the N-acyl reaction is easy to occur and has a high degree of substitution of N-acyl group. With the increase of the concentration of acetic acid, the degree of protonation of amino groups deepens, the concentration of acetate ion increases accordingly and the opposite electric field is formed around the chain of chitosan polycationic polymer, which weakens the repulsive action between protonated amino groups in the chain. The chain expansion is weakened, the polymer chain becomes more curled up and has a greater shielding effect on the amino group, so the N-acyl group reaction is relatively difficult to occur, and the degree of substitution of N-acyl group is relatively low.

Generally, the methods of preparing acetylated chitosan (total acetylated chitin) are as follows: A dispersion of fully dried 1 g chitosan powder in 150 mL methanol was prepared, and excessive acetic anhydride was added. The mixture was stirred for 16 h at room temperature, filtered, washed with methanol twice, soaked in 50 mL 0.5 mol/L ethanol and KOH solution for 16 h, then filtered, washed fully with methanol, dehydrated with ether and dried in air, the yield was quantitative. This method of preparation is important because it is difficult to prepare fully acetylated chitin from chitin, and chitin is decomposed, which is much easier to prepare from chitosan and the product has better properties and lower cost.

Chitosan was suspended in DMF, heated to 120–130 °C, and reacted with excessive phthalic anhydride, the phthalates formylation product can be dissolved in

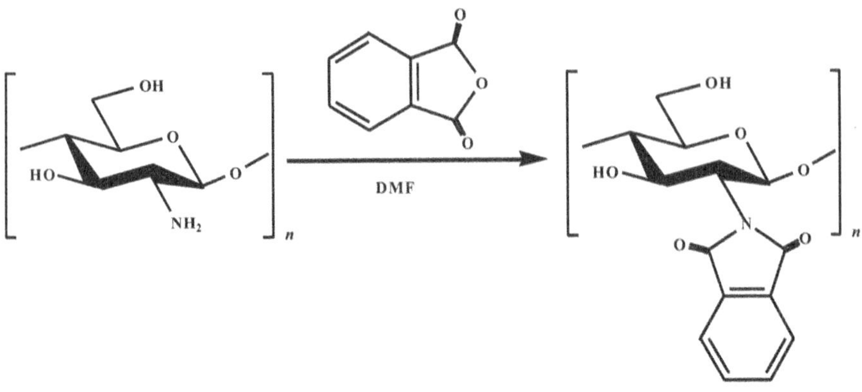

Fig. 7.3 Preparation of N- phthalic chitosan. (Modified from Li et al. 2004)

dimethylsulfoxide (DMSO) (Fig. 7.3) (Liu et al. 2004). Some O-phthalic acylation also takes place in this reaction, but phthalic amide is sensitive to alkaloids. In the presence of methanol and sodium, transesterification occurs, and the departure of O-acyl groups only produces N-phthalic.

The degree of substitution of acylated chitosan derivatives depends directly of the ratio of the added substances. When the other reaction conditions remain unchanged, the sample whose degree of substitution is more than 0.38 can be dissolved in water by changing the mass ratio of the material. Because the hydrogen bond between chitosan molecules makes it insoluble in water, the 2-carboxylbenzoyl group leads to the destruction of the dense grade structure of chitosan molecules, which greatly reduces the crystallinity and almost becomes amorphous.

The higher the degree of substitution, the more amorphous phase, and the better the solubility. The experimental results show that the higher the mass ratio, the higher the degree of substitution, and the higher the intrinsic viscosity and relative molecular weight. The reason is that the acylation reaction can be carried out in the solution system with high viscosity due to the presence of pyridine. However, the intrinsic viscosity and the relative molecular weight of the product decreased obviously when the ratio of the added material was 2, and the color of the product deepened. That's because the chain of chitosan molecules breaks as the reaction goes on. The more anhydride, the more acidic the reaction medium, the greater the degree of sucrose chain breakage. According to Mark Houwink equation, the relative molecular weight of the product is proportional to the intrinsic viscosity of the product. Therefore, when the ratio of the added material is 1.25, the high viscosity of phthalic chitosan is obtained, and the relative molecular weight of the sample is the largest.

7.2.2.2 O-Acylated Chitosan Derivative

In the study of acylation of chitosan, the chemical structure of acyl group has great influence on the O-acylation reaction of acylated organisms. The degree of substitution of large volume acyl group is generally less than 0.5, otherwise it takes a long time to obtain a higher degree of substitution. It is difficult to obtain O-acylated chitosan because the amino group is more reactive than the hydroxyl group, and the acylation reaction first occurs on amino group. Therefore, the amino groups of chitosan are usually protected by benzaldehyde, then acylated under mild conditions, and then the protective groups are removed after the reaction (Aytemir and Ozcelik 2011).

Acylated chitosan reactions usually occur on amino groups. However, the reaction cannot be completely selective on the amino group, there will also be an O-acylation reaction (Rúnarsson et al. 2010). N-acetylation of chitosan occurs easily in highly swelling pyridine gels or in water/acetic acid solutions. With a careful control of the reaction conditions, 50% N- acetylated chitosan can be obtained. Because it can form gel in organic solvent and has good reactivity, it can also be used as the raw material for secondary modification. The addition of an aqueous solution of chitin (soluble in water) is added to solvents like pyridine and dimethylformamide, a highly swelling gel can be obtained, and N-acylation reaction can occur with amino groups using phthalic anhydride and tetracarboxylic anhydride.

N-phthalic chitosan can undergo many selective modification reactions under homogeneous conditions (Fig. 7.4). For example, in pyridine, the C6 hydroxyl group first performs triphenyl methylation reaction, and after the reaction is complete, C_3 performs acetylation reaction, and finally C_6 removes the triphenylmethyl group to obtain the free hydroxyl group (Ning Ma 2004). These reactions can be carried out smoothly and quantitatively in the solvent (Li et al. 2011). Thus, N-phthalic acid group plays a protective role in the selective substitution reaction of amino group.

Triphenylmethyl chitosan can form sulfonate, and C_2 and C_3 sulfonated chitosan derivatives can be obtained after triphenylmethyl is removed. When triphenylmethyl chitin derivative was sulfated, the derivative of O-sulfonic acid could only be obtained at C_3 position. The derivative showed strong antiviral activity and had a good inhibitory effect on AIDS virus. The O- sulfonate chitin at C_6 has

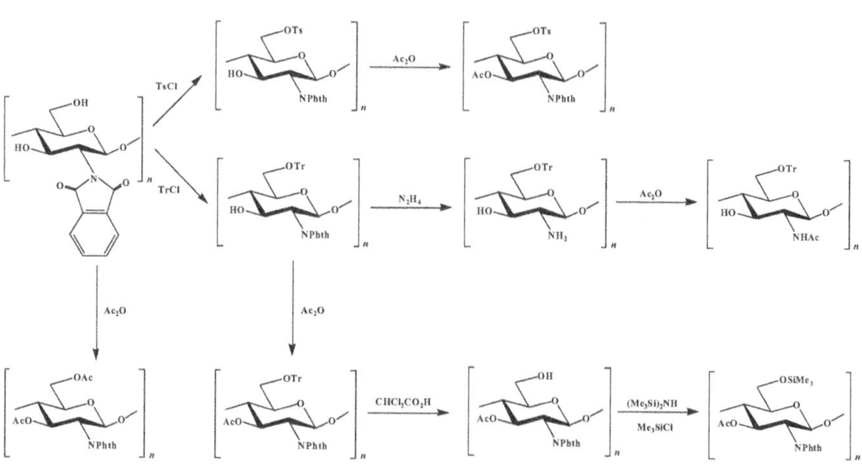

Fig. 7.4 The selective reaction of phthaloyl chitosan

anticoagulant function. It can be seen that the effect of sulfonate derivatives on AIDS virus is related to the sulfonate groups at specific sites, rather than the degree of substitution (Nishimura et al. 1998).

In the acylation reaction of chitosan, the adsorption of metal ions is not the higher the degree of substitution, the better the performance. The lower the extent of substitution of acetylated chitosan, the higher the adsorption amount of Cu (II). This is because the presence of a small amount of acyl group, on the one hand, will destroy the crystal structure of chitosan, and on the other hand, occupy less position of functional group amino group, so the adsorption of metal increases. The effect of nonyl groups is more pronounced than that of acetyl groups, because nonyl groups are larger in size and more hydrophobic (Sahariah et al. 2015).

Octanal, benzoyl and lauroyl chitosan derivatives were prepared from acetic acid aqueous solution and methanol mixed solvent with corresponding acyl chloride. The gel is a good adsorbent of amino acids and has a higher adsorption capacity for L-type amino acids than D-type amino acids. This gel can be used as the stationary phase of liquid chromatography to effectively split the optical isomers of amino acids, and the lower the degree of substitution, the better the separation effect. At a lower temperature, chitosan reacts with benzoyl chloride to obtain benzylated chitosan (Uragami et al. 2016). It is made into a thin film which can be used to separate the mixture of phenyl cyclohexylamine.

7.2.2.3 N, O-Acylated Chitosan Derivative

The swelling completely deacetylated chitosan was added to the pyridine solution of phthalic anhydride, the N-, and O-phthalic acid chitosan was obtained with a total degree of substitution between 0.25 and 1.81. The chitosan was dissolved in dimethyl sulfoxide, dichloroacetic acid and formic acid to form a solubilized liquid crystal, and its critical concentration was basically unaffected by the change of the degree of substitution.

6-triphenylmethyl chitosan is soluble in organic solvents and can be obtained by removing phthalic acid group with hydrazine, so it is an important raw material for reaction. If the reaction conditions are controlled, hexadecyl chitosan derivatives with double substitution and trisubstitution can be prepared. The product can be further sulfonated, which is a kind of amphiphilic molecule that can form Langmuir layer (Shin-Ichiro et al. 1993).

Ten millimoles Chitosan was dissolved in 20 mL methane sulfonic acid at room temperature, and then acyl chloride was added for 1 h. After stirring for 5 h at room temperature, the reaction was terminated by adding 30 g of crushed ice. The excess acid was removed by dialysis (1 day), and $NaHCO_3$ was added to neutral pH. Finally, the mixture was additionally dialyzed for at least 3 days and freeze-dried to obtain N, O-acyl chitosan (Badawy et al. 2004; Ning Ma 2004). The synthesis route is shown in Fig. 7.5.

Acylated chitosan with long fat chain was prepared by the reaction of dodecyl chloride, tetradecyl chloride and hexadecyl chloride with chitosan using chloroform and pyridine as the reaction medium. The results showed that under the condition of high pyridine/chloroform volume ratio, the average acylation degree of the initial

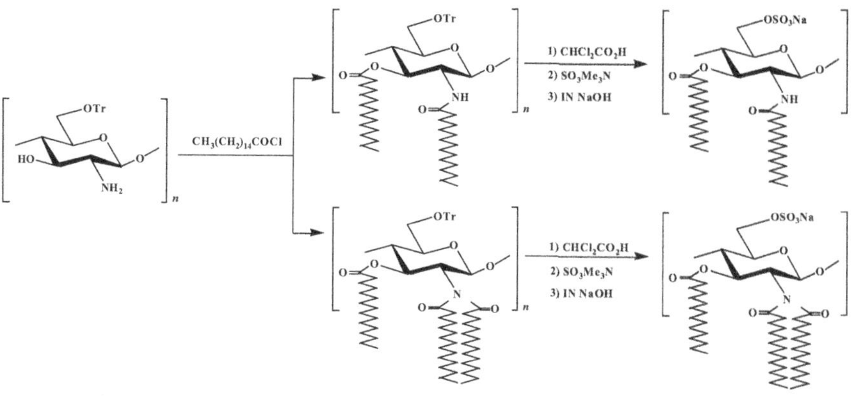

Fig. 7.5 The palmitoylated derivative of chitosan

products was high, and the acylation degree decreased with the decrease of the volume ratio. This is because the boiling point of different solvent ratio is different, and the reflux temperature is high under the condition of high pyridine/chloroform volume ratio, which is conducive to the acylation reaction.

7.2.3 Solubility of Acylated Products

The acyl groups in acylated chitin and its derivatives destroys the oxygen bond between the macromolecules of chitin and its derivatives, change their crystal structure and improve the solubility of chitin. By acylation, the solubility of acylated products in organic solvents was improved, especially in carboxylic acids like methanoic acid and dichloroacetic acid. Some products with high acylation degree were also soluble in common organic solvents, such as high substituted chitin benzylated in phenylmethanol/dimethyl sulfoxide; Highly substituted chitin can be dissolved in benzene, phenol, tetrahydrofuran and dichloromethane. In addition, the processability of acylated chitin and its derivatives has been greatly improved.

Various acyl chitin/chitosan derivatives are widely used in textile, pharmaceutical, cosmetics, environmental protection and materials due to their good water-solubility and other properties. However, the acylated chitosan was prepared by heterogeneous or homogeneous reactions. In heterogeneous reactions, solvents such as acetone, dimethylformamide and dimethyl sulfoxide have the disadvantage that many preparation processes require inorganic acid or alkali-breaking catalysts. Catalyst cannot be reused. The salt produced is difficult to separate from the product, and the equipment is corroded and polluted seriously. The disposal capacity of the three wastes is large: it is difficult to realize the control of the target product and the product is uneven. It has brought many difficulties to industrial production. Homogeneous reaction. The main use of acetic acid as solvent, some also added formamide/methanol, formamide/ethanol, methanol, ethanol, pyridine and other organic solvents. The disadvantages are that the reaction time is long, the reaction process is easy to form a gel, and the post-treatment is complex, which leads to the breakdown of the chitosan skeleton and reduction of the molecular weight. Therefore, it has become an urgent problem in the further processing of chitosan to find a green solvent that can effectively dissolve chitosan and make the preparation of its derivatives under the condition of homogeneous reaction.

7.3 Alkylation

Under different reaction conditions, chemical modification of chitin/chitosan can produce N-, O- and N, O-substitutions products. Chitosan can be modified by halogenated alkanes with different carbon chain lengths to prepare ethyl chitosan, butyl chitosan, octyl chitosan and hexadecyl chitosan. The alkylation reaction of chitosan occurs mainly in $-NH_2$, substitutions can also occur at C_3 and C_6 OH. In the O- site alkylation reaction, the reaction conditions are harsh because chitin has strong intermolecular forces. Therefore, most of the alkylation reactions were reported by chitosan. In the alkylation of chitosan, the reaction time, reaction temperature, reaction medium, the amount of alkali and amount of modifier directly affect the physical and chemical properties of the modified products. The hydrogen bond between molecules was significantly weakened when chitosan was introduced into alkyl groups. Therefore, the biological properties of burnt chitosan could be dissolved in water.

However, if the alkyl chain was too long, its derivatives were not completely soluble in water or even in acidic aqueous solution.

7.3.1 Chitin Alkylation Derivative

The alkylation of chitin takes place on hydroxyl groups. Chitin alkali is the intermediate of the alkylation reaction of chitin, which is usually made by the reaction of chitin and concentrated NaOH solution at low temperature, and then the alkylation product is generated by the reaction of chitin alkali with halogenated hydrocarbon or sulfuric acid ester (Kurita et al. 2002). Reaction between chitin base and chloroethane in autoclave, ethyl chitin with a substitution extent higher than 1 and soluble in a variety of organic solvents can be prepared, and the product can be used to control the release of sustained-release capsules.

Chitin remains stable in nature because it is soluble only fluorinated solvents, N, N-dimethylacetamide/LiCl, and methanol/$CaCl_2$. Meanwhile, randomly 50% deacetylated chitin and chitin derivatives with tosyl, iodol, trimethylsilyl, and glycosyl groups show solubility in water and organic solvents (Kurita et al. 2002). The process of N – alkylation has potential for preparation of simple chitin derivatives with higher solubility and lower crystallinity. The structure bonded to C_2 is similar to that of N, N-dimethylacetamide (DMAC), which is an excellent solvent for many substances. Many experiments were done to make polymers by ring-opening polymerization of 2-oxazolines, which have DMAC units in their backbone.

At low temperature, alkali chitin was prepared by treating chitin with NaOH/sodium dodecyl sulfate, and then reacted with haloalkanes of different lengths to obtain alkylated chitin products. The degree of dissolution and expansion of the product in water depends on the length and volume of the alkyl carbon chain. Partial destruction of chitin molecular crystal structure is the main reason for the increase of hydrophilicity. Alkyl chitin fiber was prepared by injecting alkylated chitin into the mixed solution of formic acid/dioxo acetic acid and then injecting human ethyl acetate into the nozzle. The hygroscopicity of the fiber is much higher than that of chitin or N-acetyl chitosan fiber. The results show that the hydroxyl group at the C_6 position is more prone to alkylation than the hydroxyl group at the C3 position. The chitin and benzyl chloride in 0–5 °C for 1 h reaction, then the reaction at room temperature 20 h benzyl chitin can be obtained from the door (Ravi Kumar 2000).

The process of reaction can be accelerated by microwave treatment. Chitin is treated with hexadecyltrimethylammonium bromide and NaOH to prepare alkalified chitin, and then added bromo-n-butane into the alkylated chitin, which is well mixed and placed in the center of microwave oven. After radiation at a certain power for a certain time, the reaction mixture was cooled and neutralized with dilute acetic acid until the pH was neuter. After washing with distilled water for many times until the detection of dehydration of acetone and drying to constant weight. Octyl chitin, decyl chitin, dodecyl chitin and hexadecylformin can be prepared by the same method (Dong 2001). The preparation process is shown in Fig. 7.6.

Derivatives of chitin may be prepared by treatment of alkali chitin with alkene oxides. For the preparation of hydroxyethyl chitin (also known as glycol chitin), ethylene oxide is the reagent. Side reactions occur, like the N-deacetylation of chitin and the graft copolymerization of ethylene oxide under the alkaline conditions. Hydroxyethyl chitin is soluble in water and is used for the assay of chitinolytic enzymes. The reaction with propylene oxide gives hydroxypropyl chitin and its

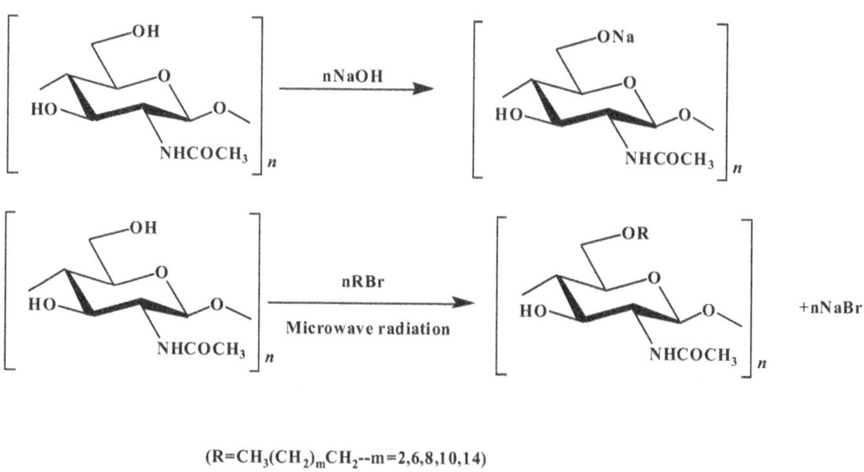

Fig. 7.6 Preparation of chitin by alkylation

course depends upon the pH. The substitution occurs preferentially at the hydroxyl groups in alkaline medium and at the amino groups in neutral medium. Hydroxy alkylations with 2-chloroethanol (giving the same product as ethylene oxide) and 3-chloropropane-1,2-diol were also reported. With N, N-diethylaminomethyl chloride, diethylaminomethyl chitin was produced.

7.3.2 Chitosan Alkylation Derivative

Alkylation can be carried out on the hydroxyl group of chitosan (O-alkylation) or the amino group of chitosan (N-alkylation). N-alkylation is more likely to occur. Chitin base usually reacts with halogenated hydrocarbon or sulfuric acid ester to form alkylation product. Ethyl chitosan, Butyl chitosan, Octyl chitosan and hexadecyl chitosan can be prepared by modifying chitosan with halogenated alkanes of different chain lengths (Zargar et al. 2015). Alkylation of chitosan mainly occurs on -NH_2 of C_2, and substitution reactions can also occur on -OH of C_3 and C_6.

In the alkylation of chitosan, the reaction time, reaction temperature, reaction medium, amount of alkali and amount of modifier directly affect the physical and chemical properties of the modified products. In general, in order to produce high degree of substitution and high viscosity of derivatives, the reaction time is 2–4 h, the reaction temperature with degree of 40–60 °C advisable (Popescu et al. 2016).

7.3.2.1 O-Alkylated Chitosan Derivative

O-alkylation generally has three methods: Schiff base method, N-phthalic acid method, metal template synthesis method. These three methods are the first to protect the amino group, in O-alkylation reaction, and then remove the protective group, has been alkylation products. Alkylated chitosan has the characteristics of liquid crystal, good film-forming property, good adsorption and chelating ability for metal ions, and has been widely used in medicine, environmental protection, textile and other fields. Chitin or chitosan reacts with monochloroacetic acid under alkaline conditions to form O-carboxymethylation. The reaction is greatly affected by temperature and alkali concentration. Some researchers used chitosan as raw material to prepare carboxymethyl chitosan with different degree of substitution. The results showed that the best preparation conditions were: reaction time 4 h, mass ratio of chloroacetic acid-sodium hydroxide 9:10, and the addition of 2% sodium dodecyl sulfate and a small amount of sodium borohydride. Schiff base amino protection method is that chitosan is reacted with aldehydes to form Schiff base, then alkylated by haloalkane, and the protective group is removed in alkyd solution, which is only substituted at O position. Schiff bases derived from chitosan may be prepared by reaction of the available amino groups of chitosan with aldehydes or ketones. This reaction occurs readily in a composed of water, methanol, and acetic acid. Initially the mixture is homogeneous, but a gel is often formed due to the low solubility of these derivatives (Ravi Kumar 2000). A glass transition was observed with the product from decanal due to the long side chain (Kurita et al. 2005). In neutral or basic solutions, the imine linkage C=N is to a degree stable. In acid medium, however, it is readily hydrolyzed. Thus, when it is desired to chemically modify the free hydroxy groups, the reaction with aldehydes or ketones is useful for the protection de amino groups. Glutaraldehyde (1,5-Pentanedial) is often used as a cross-linking agent for chitosan. Regioselective N-substitution of these imine linkages may be accomplished with the reagent sodium borohydride or sodium cyanoborohydride. Introduction of sugar groups was realized by reductive alkylation by sugars like glucose, galactose, N-acetylglucosamine, lactose and cellobiose (Matzuoka et al. 2000). These branched chitosan derivatives were soluble in water and/or dilute acids, and this solution had interesting properties.

7.3.2.2 N-Alkylated Chitosan Derivative

N-alkylated chitosan is usually synthesized by reaction of aldehydes with $-NH_2$ in chitosan molecules to form Schiff base and then reduced by $NaBH_3CN$ or $NaBH_4$ to obtain target derivatives. Ethanylated chitosan can be obtained by reaction of acetaldehyde with chitosan after reduction (Kurita 2001) (Fig. 7.7). The derivatives of methyl, ethyl, propyl and aromatic compounds were induced by this method, and have good adsorption or chelating ability to various metal ions (Avadi et al. 2005; Bratskaya et al. 2012; Sangeetha et al. 2015).

Because of their amphiphilic properties, long chain N-alkylated chitosan derivatives can be prepared by self-assembled pharmaceutical microcapsules, but modified by Schiff base reaction with higher fatty acid aldehyde, because of the two-phase reaction, the degree of substitution is low. The substitution degree of n-alkylated chitosan can be improved by adding sodium decalkyl sulfonate. But there are also problems of long reaction time and low efficiency. The method of microwave irradiation can make the reaction faster and more efficient. After stirring and dissolving chitosan at 40 °C, the pH is adjusted to about 7 by NaOH solution, and then the 30 min alkalization is continuously stirred, and the acetic acid solution with 5% volume fraction is added. After stirring at 40 °C for 1 h, lauric aldehyde and catalyst sodium decyl sulfonate were added in turn, and microwave reaction was carried out for a certain time after full stirring and dissolving. pH was modified with NaOH solution, then reduced with 10% NaBH solution, and then stirred for 2–3 h, acetone precipitated, and oil was overheated. N-dodecylated chitosan was obtained by washing with water and acetone.

7.3.2.3 N, O-Alkylated Chitosan Derivative

Under alkaline conditions, chitosan can react directly with halogenated alkyl to prepare derivatives that can be substituted at N, O-position at the same time. The solubility of the product is different with different reaction conditions. The reaction

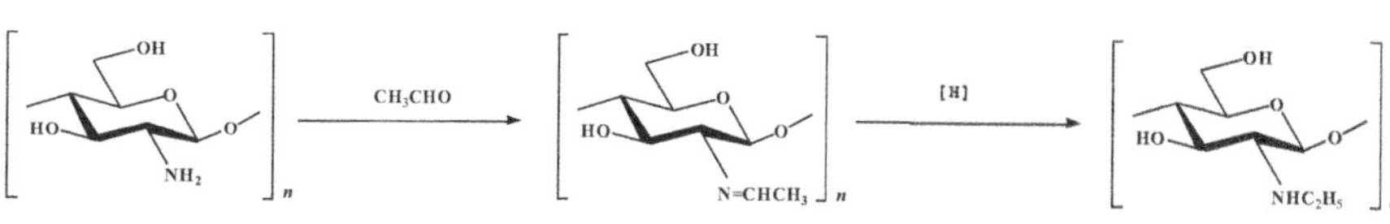

Fig. 7.7 Synthesis of N-ethyl chitosan derivatives. (Modified from Kurita 2001)

process involves adding chitosan to isopropanol containing NaOH, stirring for 30 min, adding halogenated alkanes, adjusting pH to neutral for 4 h, precipitation, filtration, washing and drying.

The deacetylated chitosan, benzaldehyde and anhydrous methanol were added to the three bottles, under magnetic stirring at room temperature reaction after 20 h suction filter flocculent white solid, placed in a Soxhlet extractor using anhydrous methanol extraction 4 h, with steaming water washing, solids with $NaBH_4$ reduction, after filtering, vacuum drying to light yellow solid N,O- benzyl chitosan (Ravi Kumar 2000).

7.4 Quaternary Ammonium Salt

The study of quaternary ammonium salt of chitin/chitosan is an important direction of chemical modification of chitin/chitosan. The quaternary ammonium salt reaction can be carried out on the amino and hydroxyl groups of chitin/chitosan. Due to the high steric hindrance and strong hydration ability of the quaternary ammonium salt group introduced, the hydrogen bond between chitin/chitosan molecules can be weakened to a large extent, thus increasing the solubility in water. In addition, the properties of chitin/chitosan derivatives can be improved by quaternary ammonium salting, making the chitin/chitosan derivatives an ideal advanced material.

7.4.1 Chitin Quaternary Ammonium Salt Derivative

Chitin reacted with water-soluble glycidyl trimethylammonium chloride to produce quaternary ammonium salt to treat staple fiber monofilament. After 1 month's use of the fabric, the weight loss rate of *Staphylococcus epidermidis* was 9.8–100% (Bhatnagar and Sillanpää 2009). The quaternary ammonium salt chitin which is substituted on the C_6-site of chitin can maintain the original ion characteristic of chitin. In water system, hydroxypropyl trimethyl chitin derivatives can be prepared by continuous constant temperature stirring reaction at 80 °C for 24 h. By comparing the reaction activity of α-chitin with β-chitin, it is found that the substitution degree of quaternary ammonium salt derivatives obtained from different crystalline chitin under the same reaction conditions is quite different. The product substitution degree in solid phase and liquid phase of β-chitin hydroxypropyl trimethyl quaternary ammonium salt is higher than that in solid phase and liquid phase of α-chitin hydroxypropyl trimethyl quaternary ammonium salt, and the product substitution degree in biological liquid phase of the same crystal chitin quaternary ammonium salt is higher than that in solid phase. Therefore, the reaction activity of β -chitin is higher than that of a-chitin.

7.4.2 Chitosan Quaternary Ammonium Salt Derivative

The study of quaternary ammonium salt of chitosan is an important direction of the molecular functional alterations of chitosan. Due to a better water solubility than chitosan and its unique properties such as surface activity, it has been paid more and more attention recently. The quaternization of chitosan can be divided into the following three methods: (1) according to the properties of chitosan soluble in various dilute organic acids, chitosan was dissolved in acetic acid, which was formed by salt chain quaternary ammonium salt. (2) chitosan reacts with alkyl halides such as methyl iodide to form N, N-trimethyl ammonium salts of chitosan; (3) chitosan reacts with quaternary ammonium compounds. At present, the latter two methods are more widely used in the quaternary ammonium process of chitosan.

The quaternary ammonium salt of chitosan is a kind of amphoteric polymer. In general, chitosan with more than 25% substitution can be soluble in water. Chitosan quaternary ammonium salt can be obtained by the reaction of excess haloalkane with chitosan. Because of the high activity of iodoalkane, it is a common halogenation reagent. Chitosan quaternary ammonium salts can also be synthesized by reaction of chitosan with aldehyde, then reduced by $NaBH_3CN$ or $NaBH_4$, and then synthesized by iodomethane reaction. The typical synthesis route is shown in Fig. 7.8.

Chitosan reacts with aldehyde to get Schiff base, which is reduced by $NaBH_4$, and then reacts with excess iodine methane to get chitosan quaternary ammonium salt, and studies its antibacterial activity. The bacteriostatic experiments of *Escherichia coli* were carried out with 0.25% and 0.50% quaternary ammonium chitosan, respectively. It was found that the bacteriostatic ability of *Escherichia coli* increased with the increase of the concentration of quaternary ammonium salt, and its acid solution was superior to the aqueous solution (Jia et al. 2001). In this case, the quaternary ammonium salt containing hydroxyl group was obtained using quaternary ammonium salt containing epoxy alkanes as the reactant with chitosan. Hydroxypropyl trimethylammonium chloride chitosan was synthesized with glycidyl trimethylammonium chloride and chitosan, and its water solubility increased with the increase of degree of substitution. Quaternary ammonium chitosan derivatives with different carbon chain lengths can also be prepared by using ethyl sulfate. The quaternary ammonium degree of N- methyl quaternary ammonium salt can reach 35%, and the physical and chemical properties of chitosan quaternary ammonium salt can be improved obviously. The quaternary ammonium salt obtained by the reaction of glycidyl trimethylamine halide with chitosan can be used in shampoo.

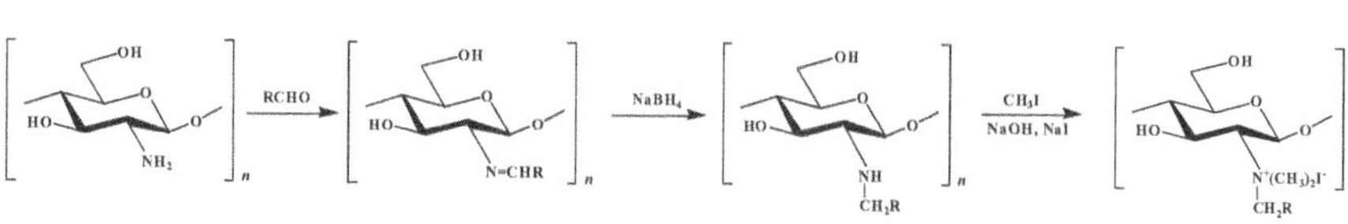

Fig. 7.8 Synthesis route of quaternary ammonium chitosan by halogenation

7.5 Carboxyalkylation

So far, among the reported derivatives of chitin/chitosan, carboxylated chitin/chitosan is the earliest and the most studied, because the introduction of carboxyl group can obtain fully water-soluble polymers on the one hand, and more importantly, amphoteric chitosan derivatives containing anions on the other hand. The reaction of chitin/chitosan with chloroalkanic acid or glyoxylic acid can introduce carboxyalkyl groups on the hydroxyl or amino groups of chitin/chitosan. The reductive alkylation of chitosan produces N-carboxymethyl derivatives that have a well-defined structure, because this reaction is regioselective at the amino group. This is in contrast with the above-mentioned conventional preparations like those with chloroacetic acid that give not-well-defined structures. Thus, the reductive alkylation is necessary for the inclusion of carboxymethyl groups in the chitosan structure.

7.5.1 Chitin Carboxymethyl Derivative

Chitin was first frozen in NaOH solution overnight, and then reacted with chloroacetic acid to prepare carboxymethyl chitin. The reaction process is shown in Fig. 7.9. After fully swelling chitin in dimethyl sulfoxide, carboxymethylation of chitin was carried out under the alkaline strip, carboxymethylated chitin sodium salt was obtained. If the product is further stripped of the acyl group, carboxymethylated chitosan is obtained. Factors such as reaction temperature, concentration of alkali and reaction time have great influence on the physicochemical properties of carboxymethyl chitin products (Sini et al. 2005).

At room temperature, carboxymethyl chitin can be prepared by the reaction of monochloroacetic acid with chitin for 72 h. There are many hydroxyl groups, acetyl amino groups and small amounts of amino groups in the repeated monolayers of chitin molecules, which can form intramolecular and inter-molecular hydrogen bonds. As a result, the aggregation state structure is close and the crystallinity is higher. In the process of chitin carboxymethylation, chitin was dissolved by concentrated alkali. Because of the action of NaOH and water, the oxygen bond between molecules was destroyed, and the crystal plane distance was increased, which made the carboxymethylation substitution reaction easy. After the reaction,

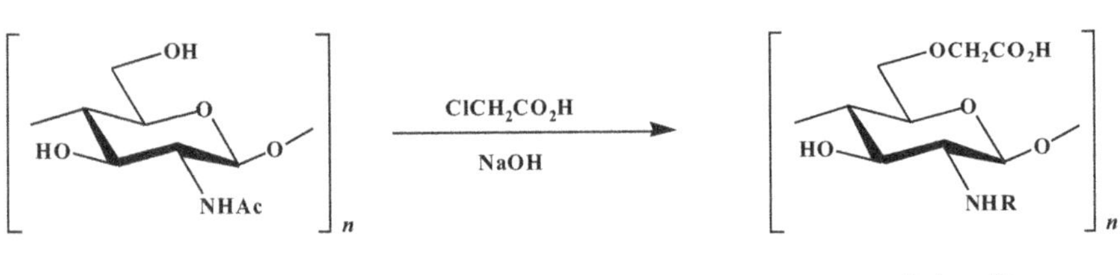

Fig. 7.9 Reaction of carboxymethyl chitin

the -CH_2COONa group was induced to the molecular repeating unit, which reduced the crystallinity of the whole molecule. The higher the degree of substitution, the more -CH_2COONa, the lower crystallinity of carboxymethyl chitin has better moisture retention, surface absorption and certain bacteriostatic effect.

The reaction of chitin with chloroacetic acid and a strong alkali like sodium hydroxide occurs similarly to the carboxymethylation of cellulose. The reaction is favored by using isopropanol as a cosolvent. The preferred position for substitution is the C_6 hydroxyl group of chitin, as was confirmed by hydrolysis. N-deacetylation occurs at the same time, so the polymer product contains free amino and carboxyl groups. Aqueous alkali dissolves this polymer, carboxymethylated chitin.

7.5.2 Chitosan Carboxymethyl Derivative

Carboxymethyl substitution can take place on the hydroxyl groups of C_3, C_6, and amino groups of C_2 positions of chitosan. The substitution product on the hydroxyl group is called O-carboxymethyl chitosan, the substitution product on the amino group is called N-carboxymethyl chitosan, and the substitution product on the amino and hydroxyl groups is called N, O-carboxymethyl chitosan. Because of steric hindrance effect of C_3 site and intermolecular hydrogen bond between C_3 and C_2, it is difficult for C_3 site to have substitution reaction, while C_6 site O carboxymethyl chitosan is the main one. For C_6-OH and C_2-NH_2, the substitution activity of carboxymethyl on the hydroxyl group is higher than that of the amino group under alkaline conditions. Only when the degree of substitution is close to 1 and higher than 1, substitution will occur on the amino group at the same time to form N, O-carboxymethyl chitosan. The degree of substitution of carboxymethyl was determined by potentiometric titration. The preferred order of substitution was $C_6-OH > C_3-OH > C_2-NH_2$, as shown by the results. Therefore, different carboxymethylated chitosan derivatives can be preferentially synthesized changing the conditions (Wu 2017) (Fig. 7.10).

7.5.2.1 O-Carboxymethyl Chitosan

O-carboxymethyl chitosan preparation is usually based on chitin as raw material, the first preparation of O-carboxymethyl chitin, and then the deacetylation reaction, the final generation of O-carboxymethyl chitosan. The specific preparation method is that chitin is immersed in 40–60% NaOH solution, soaked at a certain temperature for a few hours, then slowly added chloroacetic acid in the stirring process. The reaction was carried out at 70 °C for 0.5–5 h, the mass ratio of acid and base was controlled at 1.2–1.61:1, and the reaction mixture was kept at 0–80 °C for 36 h, then neutralized with HCI or acetic acid. The separated product was washed with 75% ethanol solution and dried at 60 °C. The preparation process of O-carboxymethyl chitosan is shown in Fig. 7.11.

This preparation method has less pollution, but the reaction medium needs to be changed twice in the preparation process, and the properties of carboxymethyl chitosan products are not uniform. O-carboxymethyl chitosan can also be prepared by

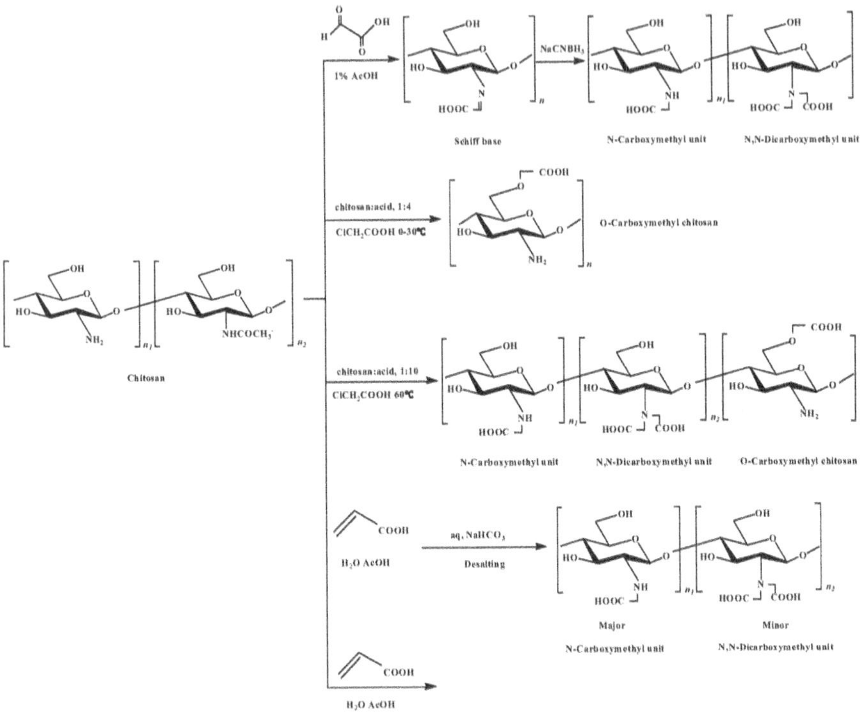

Fig. 7.10 Synthesis of carboxymethylation and carboxyethylation of chitosan

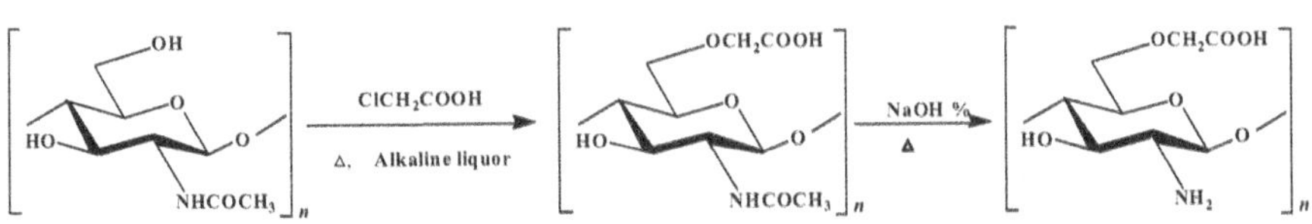

Fig. 7.11 Synthesis of carboxymethylation and carboxyethylation of chitosan

direct reaction of chitosan and chloroacetic acid. Carboxymethyl chitosan with uniform properties, high molecular weight and high viscosity can be prepared by batch reaction in a medium.

In the KOH isopropanol system, the chitosan derivative was prepared by the reaction of chitosan and chloroacetic acid. The effect of temperature on the structure of the product was related to the mass ratio of chitosan to chloroacetic acid. When the mass ratio of chitosan to oxyacetic acid and KOH to chloroacetic acid was 2:1 and 2.3:1, respectively, the O-carboxymethyl chitosan with high degree of substitution could be prepared by reaction at room temperature for 5 h. The preparation method is simple, rapid, low reagent dosage, low product cost, suitable for a certain scale of production (LogithKumar et al. 2016).

The promoting effect of ultrasonic wave on heterogeneous reaction is related to the "cavity effect" produced by ultrasonic wave, which forms a high energy center in the reaction system that can initiate or accelerate the reaction. And the secondary effects of ultrasonic such as mechanical oscillation, diffusion, emulsification, crushing are also conducive to the full mixing of reactants, greatly promoting the reaction (Wang et al. 2018). This method is of great significance for large-scale production of water-soluble chitosan derivatives. Water soluble O-carboxymethyl chitosan can also be prepared by ultrasonic radiation. When other reaction conditions are the same, the degree of substitution and reaction time of carboxymethyl can be significantly increased by ultrasonic radiation than by mechanical stirring.

The carboxymethyl groups included as substituents in the chitosan chain will be negative charged in basic media. Similarly to the synthesis of carboxymethyl cellulose, the derivatization of chitosan is done with alkalis chitosan and chloroacetic acid (Li et al. 2011). The derivatization is favored with the addition of a suitable organic solvent. When the derivatized polymer chain is hydrolyzed, the study of the structure of the monomers obtained indicates that carboxymethylation proceeds preferentially at C_6. The strongly alkaline conditions, however, cause an important degree of N-deacetylation (that may reach a value of 0.5), resulting in the formation of polymers with amino and carboxyl groups (Kurita 2001) (Fig. 7.12). In addition, the free amino groups formed may suffer themselves carboxymethylation. Given the uncertainties in the structure of the polymer formed, it is difficult to discuss the resulting structure property relationships. However, carboxymethyl chitin is still useful as an easy to prepare derivative. Its utilization is possible in many areas, and applications as a biomaterial, for example as a drug carrier, have been proposed (Zargar et al. 2015)

7.5.2.2 N-Carboxymethyl Chitosan

Schiff base was obtained by reaction of glyoxylic acid with the C_2 amino group of chitosan in acidic medium. Then N-carboxymethyl chitosan was reduced by $NaBH_4$ to obtain N-carboxymethyl chitosan. The specific method is that chitosan was dissolved in glyoxylic acid aqueous solution, Schiff base is formed, pH was adjusted to 6 with NaOH, then stirred with sodium borate at room temperature for several

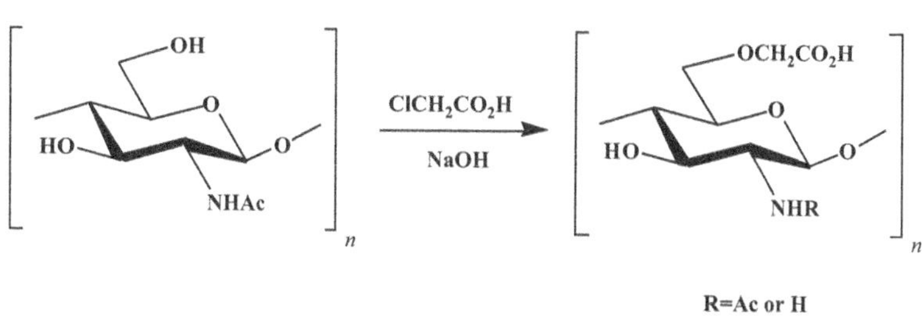

Fig. 7.12 Synthesis of amphoteric chitosan polymers. (Modified from Kurita 2001)

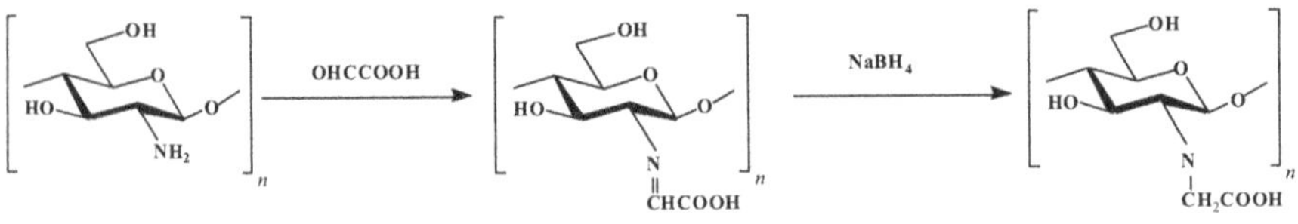

Fig. 7.13 Preparation of N- carboxymethyl chitosan

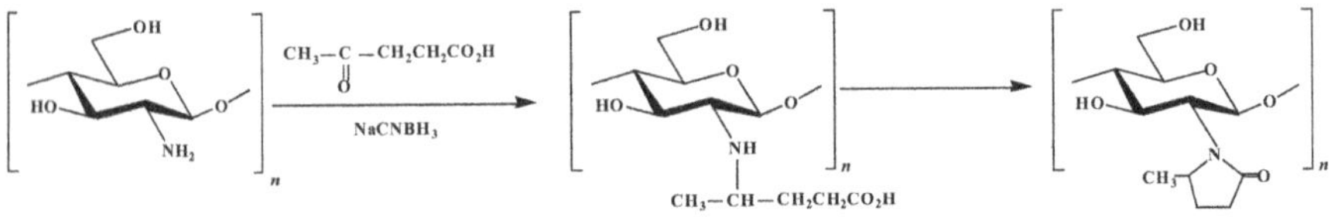

Fig. 7.14 The synthesis of N-carboxybutyl chitosan

hours, Schiff base is reduced and the reaction solution is poured into ethanol. White precipitates will be precipitated, filtered, washed with anhydrous ethanol and acetone, then dried in vacuum to produce N-carboxymethylated chitosan from white powder (Muzzarelli et al. 1982). The reaction process is shown in Fig. 7.13. The characteristic of the reaction is that it does not need to be heated and the reaction is easy to carry out, but the post-treatment is cumbersome. Therefore, on the basis of the above method, N carboxymethyl chitosan can also be prepared by adding glacial acetic acid to dissolve the raw material and then adding glyoxylic acid to the reaction (Kurita 2006).

The synthesis of N-carboxyalkyl derivatives was done by the reaction of chitosan with pyruvate, β-hydroxypyruvate, α-ketoglutaric acid and 2-carbonylpropionic acid under similar conditions. N-carboxybutyl chitosan can be prepared from levulinic acid (Singh and Ray 2000) (Fig. 7.14). Under certain conditions, 5-methylpyrrolidone chitosan can also be obtained. N-carboxybutyl chitosan and 5-methylpyrrolidone chitosan have good bioactivity and compatibility (Wang et al. 2018). Therefore, as a new biological material, it has been used as a coating agent for wound healing and a growth-promoting agent for tissue repair (Kurita 2006).

It is generally prepared by aldehyde reaction. Glyoxylic acid as carboxymethylation reagent to synthesize N-carboxymethyl chitosan with high and low viscosity from chitosan with different viscosity. The results showed that n-carboxymethyl chitosan with high and low viscosity could be used as intestinal absorption promoter for anionic macromolecules (Thanou et al. 2001).

N-carboxymethyl chitosan can prevent tissue adhesion, gel or membrane used in surgical wounds, can prevent blood from clotting on the tissue surface, thus preventing adhesion. N-carboxymethyl chitosan curing enzyme may be used in the treatment of leukemia. N-carboxymethyl chitosan can inhibit spore's germination and fungal spores' formation of toxic fungi in plants and thus reduce the formation of aflatoxin which is harmful to human body. These properties are not found in N, O-carboxymethyl chitosan and O-carboxymethyl chitosan. The carboxyl and amino groups in N-carboxymethyl chitosan can also coordinate with metal ions to form

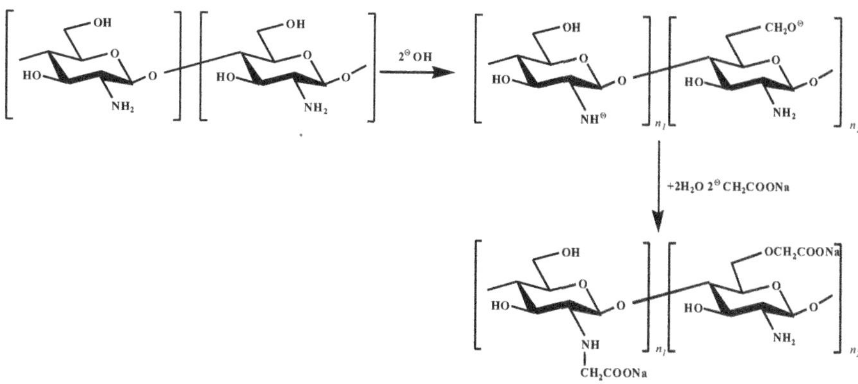

Fig. 7.15 Reaction mechanism of N, O-carboxymethyl chitosan

stable complexes, which have been applied in environmental protection and chemical industry.

7.5.2.3 N, O-Carboxymethyl Chitosan

N, O-carboxymethyl chitosan can be prepared by direct reaction of chitosan and chloroacetic acid (Lin et al. 2005) (Fig.7.15).

According to the structure analysis of monomer obtained from hydrolysis, carboxymethylation mainly occurs on C_6. But because the reaction is in the strong alkali, both deacetylation side reaction and N-carboxymethylation reaction occur. Especially when products with higher degree of substitution are obtained, or when the reaction time is longer, the appearance of N-site products will be more obvious, and the reaction process will be easily affected by temperature fluctuations. Although the biological structure obtained by this method is not very clear, it is still one of the most widely used chitosan derivatives. Chloroacetic acid can be substituted with C_6-site or C_2-position amino group of chitosan in alkaline condition, so the final product is a mixture of methyl chitosan and N, O-carboxymethyl chitosan.

At present, the research of water-soluble N, O-carboxymethyl chitosan is generally in the presence of alkali, using chloroacetic acid to modify chitosan carboxymethylation, but because of the poor solubility of chitosan, it has a direct impact on the reaction results. Therefore, N, O-carboxymethyl chitosan can be prepared by phase transfer catalyst (Rahmani et al. 2016). On the basis of investigating the effects of catalyst type, reaction temperature, reaction time and ratio of water to alcohol on the substitution and solubility of carboxymethyl chitosan, it was found that the reaction temperature was 55 °C with triethyl benzyl ammonium chloride as catalyst. When the reaction time was 4 h and the ratio of water to alcohol was 1:4 (volume ratio), the degree of substitution of carboxymethyl chitosan could reach 1. 12 (Acet et al. 2018; Rahmani et al. 2016).

7.6 Graft Copolymerization

Graft copolymerization is not a theme that has been largely researched with chitin and chitosan. The field, however, is rapidly expanding, and there is a strong increase in the activity in the number of publications and papers during the last decade. Recently, reviews have been published. Methods for polymer synthesis and analysis are presently being developed, specifically methods for the efficient modification of chitin. With this kind of reaction, specially designed polymers can be created, tailored for different purposes. The properties of graft copolymers can be controlled by the number and properties (structure and length) of the grafted chains. Possible applications are separations (analytical and preparative), water purification, cation adsorption, agriculture, and many others. Graft copolymerization is one of the most promising methods for chitin/chitosan modification. Under certain conditions, C_6 primary hydroxyl group, C_3 secondary hydroxyl group and C_2 amino group of chitin/chitosan can be grafted into the side chain of polymer, thus the application range of chitin/chitosan can be widened. By changing the structure, chain length and the number of branched chains, the properties of chitin/chitosan could be maintained and the solubility, temperature resistance and brittleness of chitin/chitosan could be improved. At present, chitin/chitosan graft copolymers have been applied in many fields such as tissue engineering materials, medical materials, adsorbents, flocculants, ion exchange resins and biodegradable plastics. There are two main ways to graft copolymerization of chitin/chitosan:

(a) The production of free radicals on the polymer skeleton of chitin/chitosan initiated another monomer polymerization. For example, the graft copolymerization of acrylic acid and methacrylic acid onto chitosan was initiated by cerium ammonium nitrate.
(b) By coupling the reactive functional groups on the chitin/chitosan polymer chain with other polymer chains, such as the graft copolymer of chitosan and polyethylene glycol. The types of graft copolymerization of chitin/chitosan include free radical graft copolymerization, coupling graft copolymerization, ion grafting copolymerization and orientation graft copolymerization.

7.6.1 Grafting onto Chitin

Cerium (IV) a reagent useful for grafting onto cellulose, may also be used for doing the same with chitin. Chitin in aqueous suspension was grafted with acrylic acid and acrylamide (vinyl monomers). The resulting polymers showed increased solubilities and swelling. Those grafted with polyacrylate were soluble in dichloroacetic acid, and those grafted with polyacrylamide swelled in this solvent. Both these polymers were more hygroscopic than chitin (Aly and El-Bisi 2018).

Two similar monomers, methyl methacrylate and methyl acrylate, were also grafted into chitin. The swelling behavior of the methyl methacrylate derivative depended upon the percentage of grafting. Above 400%, the swelling became pronounced to give transparent gels from which films could be cast (Kurita 2001). The differential scanning calorimetry method was used to find the glass transition temperature (130 °C) of the grafted derivative. Methyl methacrylate was grafted with tributylborane in water with low efficiency. Another grafting method is with redox initiators (Thanou et al. 2001).

The grafting of styrene on chitin is initiated byγ-Ray irradiation in the presence of water (as is done with cellulose) with low yields. Another method is photo-induced initiation, which was used with methyl methacrylate and chitin or oxychitin (oxidized chitin). DMF could be used to shorten the induction period and increase the percent grafting. Graft copolymerization induced by light in the absence of catalysts was smooth compared with that in the presence of 2,2-azobisisobutyronitrile or hydrogen peroxide (Jenkins and Hudson 2002).

7.6.2 Grafting onto Chitosan

Graft copolymerization with chitosan is generally done with the reagents 2,2′-azobisisobutyronitrile, Ce(IV), or a radical-generating redox system.

At present, the graft copolymerization of chitosan and vinyl monomer initiated by Ce(IV) salt has been studied. The graft polymerization of chitosan and alkene monomers was studied by adding ammonium cerium nitrate or ammonium cerium sulfate in the presence of water in heterogeneous conditions (Setia 2018). Studies have shown that Ce(IV) is a highly effective initiator with mild conditions, and the main influencing factors are reaction time, reaction temperature, monomer ratio and initiator concentration, etc. If the initiator is below a certain content, the reaction can hardly be carried out, because chitosan is a reducing polysaccharide, and its reductive terminal group consumes a quantitative initiator. Only above this amount can the graft copolymerization be initiated. The initiation mechanism of graft copolymerization of chitosan and ethylene-monomer initiated by Ce(IV) can be established by kinetic study and polymer chain structure analysis (Wei et al. 1993). The process is shown in Fig. 7.16.

Persulfate such as potassium persulfate and ammonium persulfate are common initiators and are also important free radical initiators for grafting vinyl monomer onto natural polymers. With potassium persulfate and ammonium persulfate as initiators, the reaction conditions were mild, the operation process was simple, the reagent was cheap, and there was no residue in the graft copolymer. However, there are still many controversies about the initiation mechanism of the mechanical initiators in the reaction process. When the graft copolymerization of methyl methacrylate and chitosan is initiated by potassium persulfate, the viscosity of the solution decreases significantly, which may be because the graft copolymerization follows the following initiation mechanism: first, $S_2O_8^{2-}$ decomposed into SO_4^- under heating. Because the free radical is close to the C4 of the pyran ring, the hydrogen atom which stores the C_4 position transfers the free radical to C_4, which leads to the breaking of the C-O-C bond in the main chain of chitosan. Chitosan is decomposed into two parts: some containing carbonyl terminal and the other containing free

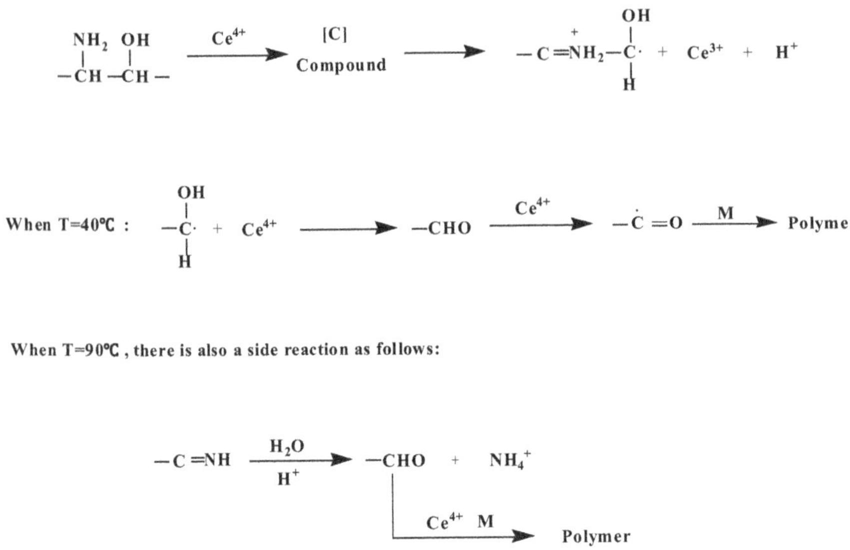

Fig. 7.16 Ce^{4+} -initiated graft copolymerization of chitosan and vinyl monomer

radicals in the disconnected position. The formation of macromolecular free radicals leads to monomer polymerization. It can be seen that potassium persulfate is not only the initiator but also the degradation agent of chitosan molecular chain in the reaction. During the reaction, chitosan is not only the reactant, but also can be used as surfactant to accelerate the reaction. However, the end carbonyl group formed by the disconnection of the main chain can also inhibit the reaction by terminating the free radical. At the same time, if the degradation degree of chitosan is high, the negatively charged $S_2O_8^{2-}$ and SO_4^-. It is easy to be surrounded by positively charged chitosan segments, resulting in a "cage effect", which reduces the reaction rate (Shoda et al. 2016).

Styrene can also be grafted onto chitosan powder or film by γ-Ray irradiation. The adsorption of chitosan-g-polystyrene copolymer on bromine is better than chitosan itself. Compared with chitosan film, the copolymer film has less swelling and better ductility in water. When the graft copolymerization of hydroxyethyl methyl acrylate and chitosan film was initiated by ^{60}Co γ-ray irradiation, it was found that the tensile strength of the copolymer film decreased with the increase of graft ratio, but the thermal stability of the film was improved (Huang et al. 2013c). The copolymer has good blood compatibility. After the graft copolymerization of chitosan with N-dimethylaminoethyl methacrylate, the tensile strength, crystallinity and expansion ratio of the product decreased with the increase of grafting rate, but the thermal stability was improved after the graft copolymerization of chitosan and N-dimethylaminoethyl methacrylate initiated by ^{60}Co γ-ray irradiation (Bardajee et al. 2012).

7.7 Microwave Modification

Microwave is an electromagnetic wave with a frequency of about 300–300 GHz and a wavelength of 1 mm–1 m. It lies between the infrared radiation (spectrum) of the electromagnetic spectrum and the radio waves. The civil frequency is between 2450 MHz and 915 MHz. The radar has a band between 1 and 25 cm and the rest is used for telecommunication transmission. In the late 1980s, microwave radiation was widely developed as a new organic synthesis technology with the function of promoting organic reaction. Gedye and Griguere (Gedye et al. 1986) compared with conventional conditions in the microwave oven, vinegar, hydrolysis, oxidation and the result of the substitution reaction, found that permanganate potassium oxidation of toluene in the microwave oven for the reaction of benzoic acid is five times as fast as conventional reflux, and 4-cyano phenolate and the reaction of benzyl chloride is 240 times faster, the discovery is the sign of microwave synthesis of organic chemistry began. From 1986 to the following decades, the research on Microwave Induced Organic Reaction has developed into a new and attractive field-MORE chemistry (Microwave Induced Organic Reaction Enhancement Chemistry). Because microwave radiation heating technology has the advantages of fast heating speed, high efficiency, strong permeability and even heating, the grafting reaction induced by microwave radiation has become the latest technology for the synthesis of grafted copolymers. In addition, in the grafting reaction initiated by microwave irradiation, the microwave irradiation can only stimulate the polar bonds of organic compounds, causing the polar bonds to break and form free radicals. And the C-C bond, which is the backbone of the organic chain, is a non-polar bond and will not be excited by microwaves. Microwave irradiation modification method has the advantages of short reaction time, simple operation and high grafting rate.

Chitin/chitosan can be used as a carrier for chromatography or enzyme fixation. Vanillin grafted chitosan was synthesized by microwave irradiation and the products were reduced to study the adsorption of Ni^{2+}, Cr^{6+} and Mn^{2+}. The adsorption capacity of modified Chitin/Chitosan on Ni^{2+}, Mn^{2+} and Cr^{6+} was significantly improved compared with that of chitosan, and the adsorption capacity of the modified chitosan on the three metal ions could reach 116.2, 63.83 and 47.27 mg. g^{-1}, respectively. Another example is Pd(II) catalyst supported by chitosan Schiff base applied to the microwave assisted Suzuki coupling reactions of aryl halides with phenyl boronic acids, in the absence of solvents, for 4 min at 50 °C. The catalyst loading was low (0.02 mol%) and can be used again, without loss of activity, up to seven times. Also, the values of turnover number and turnover frequency were greater than in previous similar studies (Baran et al. 2016).

7.8 Modification of Chitosan Oligosaccharide

Functional groups related to chemical properties on the molecular chain of (COS) include C_6 primary hydroxyl group on the glucosamine unit, C_3 secondary hydroxyl group, C_2 amino or acetylamino group, and glycoanhydride bond. Among them, C_3

secondary hydroxyl group, C_6 primary hydroxyl group and C_2 amino group are the ideal sites for chemical modification of COS. COS is easily soluble in water and organic solvents, and is prone to various chemical reactions. At present, there are few reports about modification of COS, but most of them focus on the improvement of related activities. Therefore, new active groups can be introduced through molecular design to form various functional derivatives, so as to broaden the application range of COS and improve their biological properties. At present, studies on chemical modification of COS mainly include the following reaction types, such as carboxymethylation, alkylation, quaternization, etherification, Schiff base, esterification, hydrolysis and graft copolymerization. COS was modified with kojic acid (Liu et al. 2014, 2018a, b), have shown no toxicity and can be used as a food preservative, and the antibacterial activity of these COS derivatives has been effectively improved. Because of its unusual properties, COS shows great potential to be a biopolymer to improve the performance of polyurethanes. Base on the good water solubility of COS, Jia groups prepared novel hemocompatible waterborne polyurethanes using COS as an extender via the self-emulsion polymerization method (Ravi Kumar 2000). The COS-based waterborne polyurethanes emulsion showed satisfactory freeze/thaw stability, and the films cast from the emulsions exhibited excellent mechanical properties and good anticoagulating character. The active amino and hydroxyl groups contained in the structure of COS/polyurethanes could carry out further modification, which made it an excellent candidate for wide application in biomedical field. Stearic acid (SA) grafted chitosan oligosaccharide (COS–SA), which was synthesized by an 1-ethyl-3-(3-dimethylaminopropyl) carbodiimide (EDC)-mediated coupling reaction, was demonstrated to form micelle like structure by self-aggregation in aqueous solution (Hu et al. 2006). The critical micelle concentration (CMC) of CSO–SA with 15.4% amino substituted degree of COS was about 0.035 mg/mL. The low cytotoxicity of the present COS–SA micellar vector, having an unhampered superior DNA condensation capacity, may be beneficial for non-virus gene transfer carrier.

7.9 Conclusions

It is now demonstrated that chitin/chitosan are functional biopolymers with potential for many important applications. Also, there is a huge annual generation. However, their utilization has not been extensively developed. This can be explained by the intractable nature of these materials. Considering their molecular structures with amino and hydroxyl groups, controlled reactions could be used to produce advanced designs with specified properties. Extensive studies have been done with this purpose. It must be pointed out that the understanding of the chemistry of these polysaccharides at a fundamental level is having a rapid progress. Thus, many applications have been suggested, like in toiletry, water absorption and moisture retention, in medicines, in viricidal, bactericidal and fungicidal applications, biomedical materials, supports for enzyme reagents, cation adsorbents, and in agriculture.

References

Acet Ö, Baran T, Erdönmez D, Aksoy NH, Alacabey İ, Menteş A, Odabaşi M (2018) O-carboxymethyl chitosan Schiff base complexes as affinity ligands for immobilized metal-ion affinity chromatography of lysozyme. J Chromatogr A 1550:21–27

Aly AA, El-Bisi MK (2018) Chapter 11 – grafting of polysaccharides: recent advances. In: Thakur VK (ed) Biopolymer grafting. Elsevier, Cambridge, MA, pp 469–519

Avadi MR, Jalali A, Sadeghi AMM, Shamimi K, Bayati KH, Nahid E et al (2005) Diethyl methyl chitosan as an intestinal paracellular enhancer: ex vivo and in vivo studies. Int J Pharm 293(1):83–89

Aytemir MD, Ozcelik B (2011) Synthesis and biological activities of new Mannich bases of chlorokojic acid derivatives. Med Chem Res 20:443

Badawy MEI, Rabea EI (2014) Synthesis and antifungal property of N-(aryl) and quaternary N-(aryl) chitosan derivatives against Botrytis cinerea. Cellulose 21(4):3121–3137

Badawy MEI, Rabea EI, Rogge TM, Stevens CV, Smagghe G, Steurbaut W, Höfte M (2004) Synthesis and fungicidal activity of new N,O-acyl chitosan derivatives. Biomacromolecules 5(2):589–595

Baran T, Açıksöz E, Menteş A (2016) Highly efficient, quick and green synthesis of biarlys with chitosan supported catalyst using microwave irradiation in the absence of solvent. Carbohydr Polym 142:189–198

Bardajee GR, Hooshyar Z, Zehtabi F, Pourjavadi A (2012) A superabsorbent hydrogel network based on poly((2-dimethylaminoethyl) methacrylate) and sodium alginate obtained by γ-radiation: synthesis and characterization. Iran Polym J 21(12):829–836

Bhatnagar A, Sillanpää M (2009) Applications of chitin- and chitosan-derivatives for the detoxification of water and wastewater – a short review. Adv Colloid Interf Sci 152(1):26–38

Bodnar ED, Perreault H (2015) Synthesis and evaluation of carboxymethyl chitosan for glycopeptide enrichment. Anal Chim Acta 891:179–189

Bratskaya SY, Azarova YA, Matochkina EG, Kodess MI, Yatluk YG, Pestov AV (2012) N-(2-(2-pyridyl)ethyl)chitosan: synthesis, characterization and sorption properties. Carbohydr Polym 87(1):869–875

Fujii S (1980) Preparation of poly (acyl) chitosans. Carbohydr Res 83:389–393

Gedye R, Smith F, Westaway K, Ali H, Baldisera L, Laberge L, Rousell J (1986) The use of microwave ovens for rapid organic synthesis. Tetrahedron Lett 27(3):279–282

Geisberger G, Gyenge EB, Hinger D, Käch A, Maake C, Patzke GR (2013) Chitosan-thioglycolic acid as a versatile antimicrobial agent. Biomacromolecules 14(4):1010–1017

He J, Wang J, Zhong H, Ding J, Zhang L (2015) Cyanoethylated carboxymethyl chitosan as water soluble binder with enhanced adhesion capability and electrochemical performances for LiFePO4 cathode. Electrochim Acta 182:900–907

Hirano S, Zhang M (2000) Cellulose–acidic glycosaminoglycan blend fibers releasing a portion of the glycosaminoglycans in water. Carbohydr Polym 43(3):281–284

Hirano S, Ohe Y, Ono H (1976) Selective N-acylation of chitosan. Carbohydr Res 47(2):315–320

Hu F-Q, Zhao M-D, Yuan H, You J, Du Y-Z, Zeng S (2006) A novel chitosan oligosaccharide–stearic acid micelles for gene delivery: properties and in vitro transfection studies. Int J Pharm 315(1–2):158–166

Huang J, Chen W-w, Hu S, Gong J-Y, Lai H-W, Liu P et al (2013a) Biochemical activities of 6-carboxy β-chitin derived from squid pens. Carbohydr Polym 91(1):191–197

Huang J, Jiang H, Qiu M, Geng X, Yang R, Li J, Zhang C (2013b) Antibacterial activity evaluation of quaternary chitin against Escherichia coli and *Staphylococcus aureus*. Int J Biol Macromol 52:85–91

Huang W, Wang Y, Zhang S, Huang L, Hua D, Zhu X (2013c) A facile approach for controlled modification of chitosan under γ-ray irradiation for drug delivery. Macromolecules 46(3):814–818

Jenkins DW, Hudson SM (2001) Heterogeneous chloroacetylation of chitosan powder in the presence of sodium bicarbonate. J Polym Sci A Polym Chem 39(23):4174–4181

Jenkins DW, Hudson SM (2002) Heterogeneous graft copolymerization of chitosan powder with methyl acrylate using trichloroacetyl–manganese carbonyl co-initiation. Macromolecules 35(9):3413–3419

Jia Z, shen D, Xu W (2001) Synthesis and antibacterial activities of quaternary ammonium salt of chitosan. Carbohydr Res 333(1):1–6

Kaifu K, Nishi N, Komai T, Tokura S, Somorin O (1981) Studies on chitin. V. formylation, propionylation, and butyrylation of chitin. Polym J 13:241

Karrer P, White SM (1930) Polysaccharide XLIV. Weitere Beiträge zur Kenntnis des Chitins. Helv Chim Acta 13(5):1105–1113

Konwar A, Kalita S, Kotoky J, Chowdhury D (2016) Chitosan–iron oxide coated graphene oxide nanocomposite hydrogel: a robust and soft antimicrobial biofilm. ACS Appl Mater Interfaces 8(32):20625–20634

Kurita K (2001) Controlled functionalization of the polysaccharide chitin. Prog Polym Sci 26(9):1921–1971

Kurita K (2006) Chitin and chitosan: functional biopolymers from marine crustaceans. Mar Biotechnol 8(3):203

Kurita K, Ichikawa H, Ishizeki S, Fujisaki H, Iwakura Y (1982) Studies on chitin, 8. Modification reaction of chitin in highly swollen state with aromatic cyclic carboxylic acid anhydrides. Die Makromol Chem 183(5):1161–1169

Kurita K, Ishii S, Tomita K, Nishimura S-I, Shimoda K (1994) Reactivity characteristics of squid β-chitin as compared with those of shrimp chitin: high potentials of squid chitin as a starting material for facile chemical modifications. J Polym Sci A Polym Chem 32(6):1027–1032

Kurita K, Mori S, Nishiyama Y, Harata M (2002) N-alkylation of chitin and some characteristics of the novel derivatives. Polym Bull 48(2):159–166

Li C, Tian H, Rong N, Liu K, Liu F, Zhu Y, Jiang Y (2011) Chitosan grafted with macrocyclic polyamines on C-2 and C-6 positions as nonviral gene vectors: preparation, characterization, and in vitro transfection studies. Biomacromolecules 12(2):298–305

Lin Y-H, Liang H-F, Chung C-K, Chen M-C, Sung H-W (2005) Physically crosslinked alginate/N,O-carboxymethyl chitosan hydrogels with calcium for oral delivery of protein drugs. Biomaterials 26(14):2105–2113

Liu L, Li Y, Li Y, Fang Y-E (2004) Rapid N-phthaloylation of chitosan by microwave irradiation. Carbohydr Polym 57(1):97–100

Liu X, Xia W, Jiang Q, Xu Y, Yu P (2014) Synthesis, characterization, and antimicrobial activity of Kojic acid grafted chitosan oligosaccharide. J Agric Food Chem 62(1):297–303

Liu X, Jiang Q, Xia W (2018a) One-step procedure for enhancing the antibacterial and antioxidant properties of a polysaccharide polymer: Kojic acid grafted onto chitosan. Int J Biol Macromol 113:1125–1133

Liu X, Xia W, Jiang Q, Yu P, Yue L (2018b) Chitosan oligosaccharide-N-chlorokojic acid mannich base polymer as a potential antibacterial material. Carbohydr Polym 182:225–234

LogithKumar R, KeshavNarayan A, Dhivya S, Chawla A, Saravanan S, Selvamurugan N (2016) A review of chitosan and its derivatives in bone tissue engineering. Carbohydr Polym 151:172–188

Loubaki E, Sicsic S, Le Goffic F (1989) Modification chimique du chitosane avec la δ-gluconolactone, la β-propiolactone et le glycidol. Eur Polym J 25(4):379–384

Muzzarelli RAA, Tanfani F, Emanuelli M, Mariotti S (1982) N-(carboxymethylidene)chitosans and N-(carboxymethyl)chitosans: novel chelating polyampholytes obtained from chitosan glyoxylate. Carbohydr Res 107(2):199–214

Ning Ma AW (2004) Progress in chemical modification of chitin and chitosan. Prog Chem 16(4):643–653. In Chinese

Nishimura S-I, Kai H, Shinada K, Yoshida T, Tokura S, Kurita K et al (1998) Regioselective syntheses of sulfated polysaccharides: specific anti-HIV-1 activity of novel chitin sulfates. Carbohydr Res 306(3):427–433

Pillai CKS, Paul W, Sharma CP (2009) Chitin and chitosan polymers: chemistry, solubility and fiber formation. Prog Polym Sci 34(7):641–678

Popescu V, Muresan A, Popescu G, Balan M, Dobromir M (2016) Ethyl chitosan synthesis and quantification of the effects acquired after grafting it on a cotton fabric, using ANOVA statistical analysis. Carbohydr Polym 138:94–105

Qizhi Dong ZG (2001) O-alkylation of chitosan under microwave irradiation. Nat Sci J Xiangtan Univ 23(2):57–59. in Chinese

Rahmani S, Mohammadi Z, Amini M, Isaei E, Taheritarigh S, Rafiee Tehrani N, Rafiee Tehrani M (2016) Methylated 4-N,N dimethyl aminobenzyl N,O carboxymethyl chitosan as a new chitosan derivative: synthesis, characterization, cytotoxicity and antibacterial activity. Carbohydr Polym 149:131–139

Ravi Kumar MNV (2000) A review of chitin and chitosan applications. React Funct Polym 46(1):1–27

Rúnarsson ÖV, Holappa J, Malainer C, Steinsson H, Hjálmarsdóttir M, Nevalainen T, Másson M (2010) Antibacterial activity of N-quaternary chitosan derivatives: synthesis, characterization and structure activity relationship (SAR) investigations. Eur Polym J 46(6):1251–1267

Sahariah P, Gaware V, Lieder R, Jónsdóttir S, Hjálmarsdóttir M, Sigurjonsson O, Másson M (2014) The effect of substituent, degree of acetylation and positioning of the cationic charge on the antibacterial activity of quaternary chitosan derivatives. Mar Drugs 12(8):4635

Sahariah P, Benediktssdóttir BE, Hjálmarsdóttir MÁ, Sigurjonsson OE, Sørensen KK, Thygesen MB, Másson M (2015) Impact of chain length on antibacterial activity and hemocompatibility of quaternary N-alkyl and N,N-dialkyl chitosan derivatives. Biomacromolecules 16(5):1449–1460

Sangeetha Y, Meenakshi S, SairamSundaram C (2015) Corrosion mitigation of N-(2-hydroxy-3-trimethyl ammonium)propyl chitosan chloride as inhibitor on mild steel. Int J Biol Macromol 72:1244–1249

Setia A (2018) Chapter 1 – Applications of graft copolymerization: a revolutionary approach. In: Thakur VK (ed) Biopolymer grafting. Elsevier, Cambridge, MA, pp 1–44

Shin-Ichiro N, Yoshiaki M, Longdi R, Manabu S, Akihiko Y, Norio N et al (1993) An efficient method for the syntheses of novel amphiphilic polysaccharides by regio- and thermoselective modifications of chitosan. Chem Lett 22(9):1623–1626

Shoda S-i, Uyama H, Kadokawa J-i, Kimura S, Kobayashi S (2016) Enzymes as green catalysts for precision macromolecular synthesis. Chem Rev 116(4):2307–2413

Singh DK, Ray AR (2000) Biomedical applications of chitin, chitosan, and their derivatives. J Macromol Sci C 40(1):69–83

Singh V, Kumar P, Sanghi R (2012) Use of microwave irradiation in the grafting modification of the polysaccharides – a review. Prog Polym Sci 37(2):340–364

Sini TK, Santhosh S, Mathew PT (2005) Study of the influence of processing parameters on the production of carboxymethylchitin. Polymer 46(9):3128–3131

Thanou M, Nihot MT, Jansen M, Verhoef JC, Junginger HE (2001) Mono-N-carboxymethyl chitosan (MCC), a polyampholytic chitosan derivative, enhances the intestinal absorption of low molecular weight heparin across intestinal epithelia in vitro and in vivo. J Pharm Sci 90(1):38–46

Uragami T, Matsuoka Y, Miyata T (2016) Removal of dilute benzene in water through ionic liquid/poly(vinyl chloride) membranes by pervaporation. J Membr Sci Res 2(1):20–25

Wang Y, Zhang X, Qiu D, Li Y, Yao L, Duan J (2018) Ultrasonic assisted microwave synthesis of poly (chitosan-co-gelatin)/polyvinyl pyrrolidone IPN hydrogel. Ultrason Sonochem 40:714–719

Wei L, Zhaoyang L, Wenshen L, Feng X-D (1993) Chemical modification of biopolymers-mechanism of model graft copolymerization of chitosan. J Biomater Sci Polym Ed 4(5):557–566

Wu D (2017) Synthesis and characterization of O-carboxymethyl chitosan grafted colorants and supported catalysts. PhD thesis. Tianjin University

Zargar V, Asghari M, Dashti A (2015) A review on chitin and chitosan polymers: structure, chemistry, solubility, derivatives, and applications. Chem Bio Eng Rev 2(3):204–226

Part III

The Application of Chito/Chitin Oligosaccharides and Its Monomer

Biological Activities and Potential Application in Food Industry

8

Mengyao Zhao and Liming Zhao

Abstract

Currently, the number of natural products associated with possible health benefits with regard their functional effects are increasing rapidly. Previous studies have indicated that chitooligosaccharides (COS) possess numerous beneficial properties such as antibacterial, anti-diabetic, immunity-enhancing, anti-obesity, and anti-tumor advantages. In this chapter, the protective effects of COS were presented in various models, and their critical underlying molecular mechanisms, as well as their prevention and intervention effects, were summarized. Moreover, COS display significant diversity regarding its utilization in the food industry, and their extraordinary benefits exceed that of chitosan. The advantages presented by COS not only relate to extending the shelf-life of food and retaining its quality but also possess nutraceutical properties that are beneficial in functional daily supplements and food for special medical purposes (FSMP). Based on the current studies, the various and potential application of COS in the food industry was also summarized in this chapter. Furthermore, the limitations derived from the existing research was illustrated, and additional possibilities were considered for the active intervention and development of a new application direction.

Keywords

Chito/chitin oligosaccharides · Bioactivity · Food application · Daily supplements · Food for special medical purposes

M. Zhao · L. Zhao (✉)
State Key Laboratory of Bioreactor Engineering, School of Biotechnology, East China University of Science and Technology, Shanghai, China
e-mail: zhaoliming@ecust.edu.cn

L. Zhao (ed.), *Oligosaccharides of Chitin and Chitosan*,
https://doi.org/10.1007/978-981-13-9402-7_8

8.1 Introduction

Chronic diseases such as diabetes mellitus, cardiovascular disease, some types of cancer, and more showed high morbidity and mortality in worldwide (World Health Organization 2014). The molecular mechanism of these diseases represents multi-faceted problems. The initiation of these diseases can be associated with unhealthy habits, lacking of exercise, family history and mutations in regulatory gene pathways, and more (Franz 2016; Sechler et al. 2013; Spaans and Goss 2014). Traditionally, therapy for chronic diseases is necessary for medication to be taken for extended periods of time, potentially leading to dependency and various other side effects, and inevitably reducing in life quality of patients (Adefegha 2018; Costa et al. 2017a; Gangemi et al. 2016). Therefore, the prevention of chronic disease largely depends on the successful formulation of safe, efficient and high-quality treatment under this condition.

Recently, phytochemical combined therapy methods have received a great deal of attention. Chitooligosaccharides (COS) and other derivatives from chitin and chitosan generally exhibit characteristics of low molecular weight, high water solubility, high bioavailability, and various biological activities (Muanprasat and Chatsudthipong 2017; Zou et al. 2016a), all of which have attracted the attention of consumers, industries and academic researchers. In the first section of this chapter, we summarized COS possess numerous beneficial properties such as anti-bacterial, anti-diabetic, immunity-enhancing, anti-obesity, and anti-tumor advantages, etc. as well as their critical underlying molecular mechanisms.

The reported scientific evidence could form a link between phytochemical and human health status. Food products like functional foods, food supplements, nutraceuticals and FSMP which contained COS and other derivatives from chitin and chitosan are designed to satisfice the new therapeutic alternatives for the prevention of chronic diseases in the developed societies. Additionally, COS and other derivatives from chitin and chitosan also could be used as antistaling agent due to their excellent anti-bacterial properties (Liaqat and Eltem 2018). In the second section of this chapter, we concluded the application of COS and other derivatives from chitin and chitosan in food industry such as preservation, dietary supplements and FSMP. The scope of this chapter is aimed to clarifying the framework of their biological activities and potential application in food industry. Especially, it could give a better understanding for their biological activities and potential applications in chronic diseases intervention and management.

8.2 Biological Activities of COS

8.2.1 The Anti-oxidative Activity of COS

Numerous pathological and physiological processes require the effective operation of reactive oxygen species (ROS) and free radicals (Di Meo et al. 2016). Therefore, it was evident that COS relied on a concentration-dependent technique to remove various unwanted impurities including hydrogen peroxide, alkyl radicals, superoxide radicals, carbon-centered radicals, hydroxyl radicals, and 2,2-diphenyl-1-picrylhydrazyl (DPPH) free radicals (Je et al. 2004a, b; Li et al. 2012a, b; Park et al. 2003). Each report presented different conclusions on which degree of polymerization (DP) or molecular weight (MW) of COS possessed the optimal antioxidant capacity (Table 8.1). Chen et al. found that both chitobiose and chitotriose could scavenge hydroxyl radicals, but only chitobiose is able to scavenge superoxide radicals (Chen et al. 2003). Park et al. claimed that COS possess a strong ability to scavenge hydroxyl and superoxide radicals, while it is less able to scavenge alkyl and DPPH radicals. Furthermore, COS with an average MW of 1–3 kDa displays the highest capacity for scavenging radicals among Mw of below 1, 3–5, and 5–10 kDa (Park et al. 2003). Je et al. claimed that COS (Mw 1–5 kDa) exhibits the highest scavenging activity on DPPH, hydroxyl, superoxide, and carbon-centered radicals among COS with Mw below 1 and 5–10 kDa (Je et al. 2004a, b). Furthermore, Chang et al. obtained six different degraded chitosans (Mw 156.0, 72.1, 29.2, 7.1, 3.3, and 2.2 kDa), and found that the scavenging activity on hydrogen peroxide and DPPH radicals appeared significantly higher in conjunction with a decline in MW (Chang et al. 2018). Li et al. found that COS with low DP had displayed a higher scavenging capacity of hydroxyl radicals and in reducing power when compared to high DP levels. In contrast, the superoxide radical scavenging activity of COS exhibited high levels when the DP increased (Li et al. 2012b). The reason for the diverse effects of various antioxidants might be related to the resource variety of the raw materials.

More intensive research relating to cell-level experiments proved that COS possesses antioxidant abilities. In B16F1 murine melanoma cells, COS not only provides protection against the oxidative damage of purified genomic DNA but also inhibits nitric oxide (NO) induced by lipopolysaccharide (LPS). Additionally, COS down-regulates the nuclear factor kappa-B (NF-κB) gene expression caused by H_2O_2, but had no significant effect on the glutathione (GSH) level (Mendis et al. 2007). In the ECV304 human umbilical vein endothelial cells, COS restores the H_2O_2 induced damage, reducing ROS and the production of lipid peroxidation. Furthermore, the activities of endogenous antioxidants were restored, including those of superoxide dismutase (SOD) and glutathione peroxidase (GSH-Px). Additionally, the levels of NO and nitric oxide synthase (NOS) increased as well (Liu et al. 2009). Similarly, COS protected the P22 bacteriophage against deactivation and the erythrocytes against hemolysis by scavenging $H_2O_2^-$ and $AAPH^-$ radicals (Fernandes et al. 2010a). In the pancreatic islet cells in rats with diabetes caused by streptozotocin, COS improved the activity of SOD and decreased the malondialdehyde (MDA) content (Yuan et al. 2009). The nuclear factor erythroid-2-related factor-2 (Nrf2) signaling is an essential intracellular technique to defend against oxidative stress. Meanwhile, the transcriptional activation of a Nrf2 route was utilized for the induction of several antioxidant genes including SOD, hemeoxygenase1 (HO-1), and NAD(P)H dehydrogenase [quinone] 1 into ethanol-stimulated L02 cells. This process allowed COS to impede the manifestation of lipid peroxidation and ROS, as well as the depletion of GSH (Luo et al. 2014a). Similarly, the

Table 8.1 A summary of anti-oxidative activities of COS, ChOS and its derivatives

No.	Average molecular weight	Degree of deacetylation	Dose	Model	Ending point	Proposed mechanism	References
1	/	/	100 mM	Hydroxyl radicals assay	Superoxide radicals↓	Oxidation-reduction	Chen et al. (2003)
				Superoxide radicals assay	Hydrogen radicals↓		
				Hydrogen peroxide assay			
2	5–10 kDa	93%	/	Hydroxyl radical assay	(Dependent on concentration and molecular weight)	Oxidation-reduction	Park et al. (2003)
	3–5 kDa			Superoxide radical assay	Hydrogen peroxide↓		
	1–3 kDa			Alkyl radical assay	Superoxide radicals↓		
	<1 kDa			DPPH radical assay	DPPH radical↓		
					Alkyl radical ↓		
3	5–10 kDa	90%	/	Scavenging effect on DPPH radical	90-MMWCOS, the highest scavenging activity	Oxidation-reduction	Je et al. (2004a)
	5–1 kDa	75%		Hydroxyl radical scavenging activity	DPPH radical↓		
	<1 kDa	50%		Superoxide anion radical scavenging activity	Hydroxyl radical ↓		
				Assay for carbon-centered radical	Superoxide anion radical ↓		
					Carbon-centered radical↓		

4	5–10 kDa	90%	/	Reducing power	The 90-MMWCOS, 90% deacetylated the highest	Oxidation-reduction	Je et al. (2004b)
	5–1 kDa	75%		DPPH scavenging activity	DPPH radical↓		
	<1 kDa	50%					
5	COS DP 2–3	/	300 mg/kg	Male Wistar strain rats (Carbon tetrachloride-induced)	Malondialdehyde and 4-hydroxy-2-alkenals↓	Chitobiose and chitotriose have the ability to exert a protective action against CCl_4-induced acute hepatoxicity, probably by their antioxidant activity	Chen et al. (2005)
					Plasma aspartate transaminase↓		
					Alanine transaminase↓		
6	1–3 kDa	90%	0–500 μg/mL	B16F1 cells	Low molecular weight is important for activities	COS was observed to be successively participated in suppression of NF-κB gene promoter activity suggesting its capability to prevent oxidative stress related disease complications	Mendis et al. (2007)
	<1 kDa				Hydroxyl radical↓Superoxide radical↓		
					Carbon-centered radicals↓		
					DNA damage↓NO↓NF-κB↓		
7	229.21–593.12 Da	<10%	1000 μg/mL	HL-60 cells	Myeloperoxidase↓	NA-COS act as a potent antioxidant in cells	Ngo et al. (2008)
				Raw 264.7 cells	Membrane protein oxidation↓		
					DNA protection activity		
					Cellular radical↓		

(continued)

Table 8.1 (continued)

No.	Average molecular weight	Degree of deacetylation	Dose	Model	Ending point	Proposed mechanism	References
8	COS DP 2–6	>95%	25–200 μg/mL	HUVECs (ECV304 cells) (H_2O_2-induced)	ROS↓	COS can effectively protect HUVECs against oxidative stress by H_2O_2	Liu et al. (2009)
					Decrease in SOD and GSH-Px activities was restored		
					NO↑NOS activity↑		
					Apoptosis↓		
					Lipid peroxidation↓		
9	NA-COS	<10%	10–1000 μg/mL	RAW 264.7 cells	(Dose-dependent manner)	NA-COS act as a potential scavenger against oxidative stress in cells	Ngo et al. (2009)
	1–3 kDa				Membrane protein oxidation↓		
	<1 kDa				DNA protective activity		
					Cellular direct radical scavenging effect		
					Intracellular glutathione↑		
10	/	90%	500 mg/L	Pancreatic islet cells in diabetic rats (streptozotocin-induced)	Apoptosis↓ Ntioxidant capacity ↑	The possible mechanism of COS' protective function in the pancreas is that they have strong in vivo antioxidative properties	Yuan et al. (2009)
					Superoxide dismutase↑ malondialdehyde↓		
11	COS DP 2–8	90%	0.1–0.4 mg/mL	L02 cells	Radical ↓	The underlining mechanism of the protective effects of COS was partly attributed to its antioxidant activity and upregulation of anti-apoptotic protein Bcl-XL	Xu et al. (2010)
					Apoptosis↓(PARP cleavage ↓ Bcl-XL expression↑)		
					Cellular uptake (plasma membrane and nucleolus)		

12	5–10 kDa	/	500 μg/mL	Pancreatic β-cells	Radical↓apoptotic↓	Chitooligosaccharides protect β-cells from oxidative-stress-induced cellular damage, presumably due to radical scavenging activity	Karadeniz et al. (2010)
	3–5 kDa						
	1–3 kDa						
	<1 kDa						
13	<3 kDa	80–85%	0.05, 0.01, 0.005 mg/mL	Erythrocytes and bacteriophages	Hemolytic↓DNA damage↓	COS can be used as antioxidants in biological systems	Fernandes et al. (2010a)
	<5 kDa				H_2O_2-radicals↓AAPH-radicals↓		
14	COS DP 6–16	Fully deacetylated	/	Superoxide radical scavenging assay	DP 10–12 exhibited higher scavenging activity	Oxidation-reduction	Li et al. (2012b)
					Superoxide radical↓		
15	COS DP 2–12	82%	/	Hydroxyl radical scavenging assay	Low DP: better hydroxyl radical scavenging activity and reducing power	Oxidation-reduction	Li et al. (2012a)
				Superoxide radical scavenging assay	High DP: better superoxide radical scavenging activity		
				Measurement of reducing power			
16	<1 kDa	≥95%	0.25, 0.5, 1.0 mg/mL	L02 cells (ethanol-induced)	Dose-dependent protect cells	COS-mediated activation of Nrf2 and reduction of MAPK phosphorylation may be important for its hepatoprotective action	Luo et al. (2014a)
					ROS↓		
					Lipid peroxidation and GSH depletion↓		
					Antioxidant gene (HO-1, NQO-1, SOD)↑		
					The nuclear translocation of Nrf-2		
					MAPK phosphorylation↓		

(continued)

Table 8.1 (continued)

No.	Average molecular weight	Degree of deacetylation	Dose	Model	Ending point	Proposed mechanism	References
17	/	/	100, 200, 400, 800 mg/L	SH-SY5Y cells (Cu^{2+}-induced)	Caspase-3↓	Nrf2 activation might be involved in the protection of COS against Cu^{2+}-induced cellular oxidative damage	Huang et al. (2015a)
					pSer40-Nrf2 protein↑		
					HO-1↑		
18	COS DP 2 3 5 6	N-acetylated	/	Superoxide radical scavenging assay	Superoxide radical↓	Oxidation-reduction	Salgaonkar et al. (2015)
				Hydroxyl radical scavenging assay	Hydroxyl radical↓		
				H_2O_2 scavenging assay	H_2O_2 ↓		
				DPPH radical scavenging assay	DPPH radical↓		
				Assessment of protection of DNA damage	DNA damage↓		
19	/	/	0.00–0.01%	DPPH and hydroxyl radicals assay	Radical scavenging activity↑	The formation of 5-hydroxymethylfurfural, trans-2-nonenal and phenylacetaldehyde were decreased	Yang et al. (2017)
20	1.5 kDa	/	0.50%	Female ICR mices	Superoxide dismutase↑	COS has potent antioxidant activity that can protect mice from oxidative stress	Qu and Han (2016)
	COS DP 2–6				Catalase↑		
					Glutathione peroxidase↑		
21	/	/	200, 350, 500 mg/kg	Broiler	Glutathione peroxidase↑	Dietary chitosan oligosaccharides supplement could improve the antioxidant activity	Li et al. (2017b)
				Ileum mucosa of broiler	Superoxide dismutase↑		
					Hydroxy radical↓		

22	<1 kDa	>95%	100 mg/kg	Weaned pigs	Superoxide dismutase↑	COS supplementation can accelerate weaned pig growth through enhancing antioxidant	Wan et al. (2017)
	COS DP 2–8				Catalase↑		
					Total antioxidant capacity↑		
23	156.0, 72.1, 29.2, 7.1, 3.3, 2.2 kDa	95%	/	Hydrogen peroxide scavenging assay	Molecular weights dependent manner	Oxidation-reduction	Chang et al. (2018)
				DPPH radical-scavenging assay	Hydrogen peroxide↓		
				Ferrous ion chelating assay	2, 2-diphenyl-1-picryl hydrazyl radical↓		
					Chelating ferrous ion activity		

Nrf2 related pathway alteration was further verified in Cu^{2+}-induced SH-SY5Y human neuroblastoma cells (Huang et al. 2015a).

COS are hydrolyzed products of chitosan which is a deacetylated form of chitin, meaning some types of COS are partly deacetylated (ChOS). ChOS displayed the ability to scavenge hydroxyl, superoxide, and DPPH radicals (Salgaonkar et al. 2015). In HL-60 human myeloid cells and Raw 264.7 mouse macrophages, NA-COS was able to inhibit Myeloperoxidase (MPO) activity, protein oxidation and DNA oxidation, and increase the GSH level (Ngo et al. 2008, 2009). However, the antioxidant capacity of COS is higher than that of NA-COS (Chen et al. 2003). Hydroxyl and amine groups within the chitosan molecule were responsible for the radical scavenging activities. Therefore, higher deacetylated chitosan led to greater antioxidant capacity.

Animal experiments were conducted based on in vitro studies. COS protected rats and mice from CCl_4-induced lipid peroxidation and 2,3,7,8-tetrachlorodibenzo-p-dioxin-induced lipid peroxidation, and was related to its antioxidant capacity (Chen et al. 2005). In a high-fat diet (HFD)-mouse model, COS not only increased the levels of SOD, catalase (CAT), and GSH-Px in the stomach, liver, and serum of mice but also improved the intestinal integrity (Qu and Han 2016). Furthermore, in the broiler and weaned pigs model, a similar conclusion was obtained in Li et al. (2017b) and Wan et al. (2017).

In a word, COS was able to alter the ROS metabolism in mitochondria, and the potential mechanism is summarized in Fig. 8.1. COS could quench the additional ROS formation and promote simultaneous oxygen eliminating enzyme activity such as SOD, CAT and GSH-Px.

8.2.2 The Anti-inflammatory Activity of COS

COS is regarded as a potential immunostimulatory compound for use in anti-inflammatory activated macrophages, microglia cells in vitro and modulated multiple signaling pathways in animal inflammation models (Table 8.2).

Macrophages played a vital role in the immune system. They could phagocytose and kill intracellular parasites, bacteria, tumor cells, and dead cells. Moreover, macrophages could exert influence on the body's immune defense, homeostasis, and surveillance (Franken et al. 2016; Schwartz and Hirschberg 1995). COS could decrease tumor necrosis factor-α (TNF-α), interleukin-6 (IL-6), NO, inducible nitric oxide synthase (iNOS), cyclooxygenase 2 (COX-2), and prostaglandin (PGE2) in LPS-induced RAW 264.7 macrophage cells (Lee et al. 2009c; Yoon et al. 2007). Furthermore, Ma et al. indicated that COS decreased IL-6 and TNF-α in LPS-induced macrophages by repressing not only the phosphorylation of p38, extracellular signal regulated kinase 1/2 (ERK1/2), c-Jun N-terminal kinase (JNK), phosphoinositide 3-kinase (PI3K), and protein kinase B (Akt) but also the activation of NF-κB and AP-1 (Ma et al. 2011). According to Hyung, J.H. et al., COS was responsible for the expression of HO-1, as well as the reduction of COX-2 and iNOS in RAW264.7 cells activated by LPS. Furthermore, the activation of the

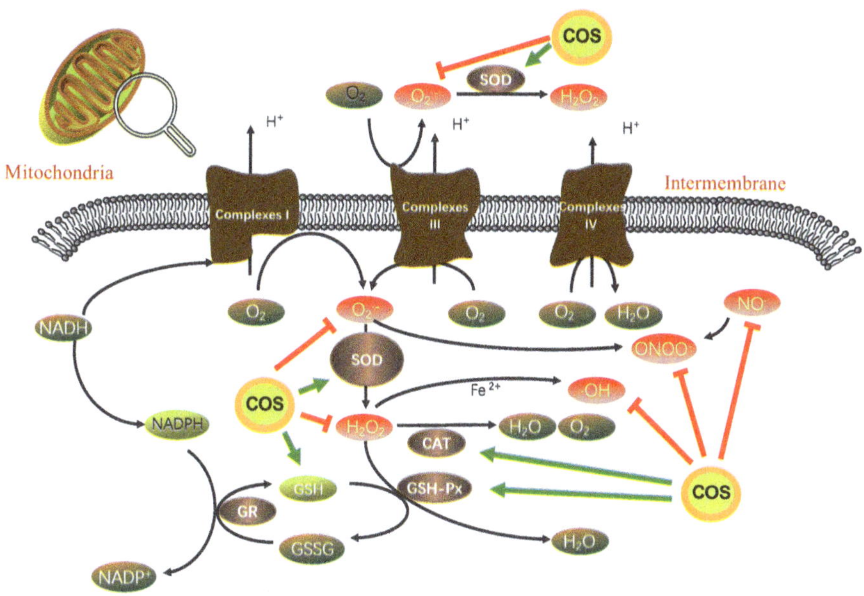

Fig. 8.1 The mechanism leading to ROS metabolism alteration via COS. The mono-electronic reduction of O_2 is responsible for the formation of superoxide radical anions (O_2), a process that primarily occurs at Complexes I and III of the respiratory chain. Dismutation via Cu, Zn-SOD and Mn-SOD is responsible for converting O_2 into H_2O_2 in the intermembrane space and in the matrix, respectively. GSH is employed to enable enzymes such as mitochondrial CAT or thiol peroxidases like GSH-Px to remove H_2O_2. (The figure was modified from Kowaltowski et al. 2009)

nuclear translocation of Nrf2, exerted an arbitrary effect on this process. It is reasonable to assume that the Nrf2 activation could be ascribed to the COS incited phosphorylation of ERK1/2, JNK, and p38 MAPK (Hyung et al. 2016). Notably, both sulfated COS (SCOS) and highly n-acetylated COS exhibited anti-inflammatory properties. SCOS inhibited IL-6/TNF-α in LPS-induced RAW264.7 macrophage cells via mitogen-activated protein kinase (MAPKs) pathways, which were dependent on NF-κB activation (Kim et al. 2014c). Highly N-acetylated COS not only decreased IL-6, TNF-α, ROS, and NO, but also inhibited the phosphorylation of PI3K and Akt in LPS-induced RAW 264.7 macrophage cells (Xu et al. 2017a).

Microglial cells are members of the mononuclear phagocytic cell family and are widely recognized as the primary immune effector in the central nervous system. Microglia and its mediated neuro-inflammation are essential to the treatment of central nervous system injury and disease (Rothhammer et al. 2018; Streit et al. 1998). In LPS-induced BV2 microglia cells, COS attenuated the productions of NO and PGE2 by inhibiting iNOS and COX-2 expressions, while TNF-α, IL-6 and IL-1β were decreased by suppressing the phosphorylation of JNK and p38 MAPK (Qiao et al. 2011). Furthermore, in LPS-induced N9 microglial cells, COS decreased NO by suppressing iNOS expression and inhibited not only phosphorylation of p38 MAPK and ERK1/2, but also the activation of nuclear NF-κB and AP-1 (Wei et al. 2012).

Table 8.2 A summary of anti-inflammatory activities of COS, ChOS and its derivatives

No.	Average molecular weight	Degree of deacetylation	Dose	Model	Ending point	Proposed mechanism	References
1	3.5 kDa	/	5.1 g/day	Clinical trial	IL-1β↓ TNF-α↓IL-12↑ IFN-γ↑	The oral intake of COS may have beneficial effects on specific cell-mediated immunity	Kim et al. (2006)
2	<10 kDa	90–95%	0.05%, 0.1%, 0.5%	RAW 264.7 cells (LPS-induced)	NO↓IL-6 ↓ TNF-α↓	Via the stimulus of TNF-α	Yoon et al. (2007)
3	COS DP 3–9	>95%	50–150 μg/mL	Rabbit neutrophils (PMA-induced)	The cell viability ↑	COS was demonstrated that it was a positive regulator of resting neutrophils	Dou et al. (2007)
					Production of NO and O_2^- of resting neutrophils↑	COS was a negative regulator of PMA-activated neutrophils	
					Production of NO and O_2^- of PMA-activated neutrophils↓		
					The degranulation of PMA-activated neutrophils↓		
					The adhesion to fibronectin of PMA-activated neutrophils↓		
4	5–10 kDa	90%	1, 2.5, 5 mg/mL	RAW264.7 cells (LPS-induced)	90-HMWCOS: TNF-α↓ IL-6 ↓iNOS ↓ COX-2↓PGE2↓	90-HMWCOS may have anti-inflammatory effect via down-regulation of transcriptional and translational expression levels of TNF-α, IL-6 and iNOS and COX-2	Lee et al. (2009c)
	1–5 kDa	50%					
	<1 kDa						

5	1.2 kDa (COS3)	80–85%	50–1000 mg/kg	Balb/c mice (Carrageenan-induced)	Dose-dependent and molecular weight dependent	Cyclooxygenase inhibition and reduction of prostaglandins	Fernandes et al. (2010b)
	5.3 kDa (COS5)				Edema ↓		
6	COS DP 2–6	>95%	50–200 μg/mL	Human umbilical vein endothelial cells (LPS-induced)	IL-6↓	This can be regulated by at least two parallel signalling pathways: one via p38 MAPK pathway independent of NF-κB activation and one via ERK1/2 pathway dependent on NF-κB activation	Liu et al. (2010)
7	COS DP 2–6	95%	50–200 μg/mL	RAW264.7 cells (LPS-induced)	p38↓ERK1/2↓JNK↓PI3K↓Akt↓NF-κB↓AP-1↓	Chitosan oligomers inhibited the elevated expression of IL-6/TNF-α in LPS-induced macrophages, regulated by MAPKs and PI3K/Akt pathways dependent on NF-κB/AP-1 activation	Ma et al. (2011)
					IL-6 ↓ TNF-α↓		
8	COS DP 2–6	95%	100 mg/kg	BALB/c mice (LPS-induced)	Rate of septic↑	COS can protect mice from the LPS damage by anti-inflammatory effects	Qiao et al. (2011)
					Organ damage and dysfunction↓		
					Cytokine levels(neutrophil infiltration, TNF-α, IL-1β)↓		
					Oxidative tissue damage↓		
					JNK ↓ p38↓		

(continued)

Table 8.2 (continued)

No.	Average molecular weight	Degree of deacetylation	Dose	Model	Ending point	Proposed mechanism	References
9	<1 kDa 1–3 kDa 3–5 kDa 5–10 kDa	/	10–500 μg/mL	BV2 microglia (LPS-induced)	NO↓PGE2↓IL-6↓TNF-α↓ IL-1β↓	(NO) (PGE2): inhibiting inducible NO synthase (iNOS) and cyclooxygenase-2 (COX-2) expressions (TNF-α) (IL-6) (IL-1β): the suppressive effects on the phosphorylation of JNK and p38 mitogen-activated protein kinase (MAPK) by COS	Pangestuti et al. (2011)
10	1–3 kDa 3–5 kDa 5–10 kDa	/	1000 μg/mL	RBL-2H3 cells	TNF-α↓, IL-1β↓, IL-4↓, IL-6↓	The suppressive effects on the phosphorylation of extracellular signal-regulated kinases (MEK/ERK) and p38 kinase	Vo et al. (2011)
11	<1 kDa	100%	500, 1000, 1500 μg/mL	RBL-2H3 cells	Antigen-stimulated degranulation, cytokine generation↓	COS can abrogate the symptoms of asthma via anti-inflammatory activities by down regulating Th2 cytokines and proinflammatory cytokine in vitro and in vivo	Chung et al. (2012)
			16.5 mg/kg	Allergic asthma model mice	IL-4↓IL-5↓IL-13↓TNF-α ↓		

13	<1 kDa	≥95%	50–200 μg/mL	N9 murine microglial cells (LPS-induced)	NO↓	COS could suppress the production of NO in LPS-induced N9 microglial cells, mediated by p38 MAPK and ERK1/2 pathways	Wei et al. (2012)
14	5–10 kDa	>90%	10–20 mg/kg/day	Mice (DSS-induced)	NF-κB↓Apoptosis↓TNF-α↓IL-6↓	COS may be effective in the treatment of inflammatory bowel disease through inhibition of NF-κB signaling and apoptosis of intestinal epithelial cells	Yousef et al. (2012)
			20, 100, 500 μg/mL	Human colonic epithelial cells (T84 cells)			
15	/	/	5, 10 mg/kg	Lewis rats	TNF-α↓iNOS↓MCP-1↓RANTES↓Fractalkine↓ICAM-1↓	COS treated EAAU by inhibiting the activation of NF-κB and reducing the expression of inflammatory mediators	Fang et al. (2014)
					IκB↑p65↓		
16	<1 kDa	>95%	100 μg/mL	Endothelial cells	IL-1β↓MCP-1↓E-selectin↓	COS decreased OGT-dependent O-GlcNAcylation of NF-κB and thereby attenuated LPS-induced vascular endothelial inflammatory response	Li et al. (2014b)
			1 mg/mL in normal drinking water	C57BL/6 J mice (LPS-induced)			
17	COS DP 3–7	>95%	50–200 mg/mL	Porcine iliac artery endothelial cells (PIECs) (LPS-induced)	E-selectin↓ ICAM-1↓	COS downregulate the expression of E-selectin and ICAM-1 by inhibiting the phosphorylation of MAPKs and the activation of NF-κB in LPS-treated PIECs	Li et al. (2014c)

(continued)

Table 8.2 (continued)

No.	Average molecular weight	Degree of deacetylation	Dose	Model	Ending point	Proposed mechanism	References
18	Sulfated-COS	90%	10, 50, 100 μg/mL	RAW264.7 cells	NO↓IL-6↓TNF-α↓	The sulfated-COS suppressed inducible nitric oxide synthase (iNOS), phosphorylation of JNK and translocation of p65, a subunit of NF-κB, into the nucleus by inhibiting degradation of IκB-α	Kim et al. (2014c)
						Sulfated-COS inhibit IL-6/TNF-α in LPS-induced macrophages, regulated by mitogen-activated protein kinases (MAPKs) pathways dependent on NF-κB activation	
19	COS DP 2–8	/	2%	C57BL/6 mice (DSS-induced)	NF-κB↓COX2↓iNOS↓TNF-α↓IL-6↓	The anti-inflammatory action of COS occurs via down-regulation of transcriptional and translational expression levels of TNF-α, IL-6, iNOS and COX-2	Azuma et al. (2015)
20	5–14 kDa	>90%	100 μg/mL	Intestinal epithelial cells (IEC)	Activation of AMPK	Activating AMPK via CaSR-PLC-IP3 receptor channel-mediated calcium release from endoplasmic reticulum	Muanprasat et al. (2015)

21	/	/	5, 10 mg/kg	Mice (retinal I/R injury)	NF-κB↓ JNK↓ ERK↓ p38 ↑	COS prevented retinal I/R injury by blocking the activation of NF-κB, JNK, and ERK but promoting the activation of p38 activation	Fang et al. (2015)
22	/	/	300 μg/kg	Piglets (LPS-induced)	TNF-α↓IL-6↓IL-8↓	COS has the potential to reduce the intestinal inflammatory response, which is concomitant with the activation of CaSR and the inhibition of NF-κB signaling pathways under an inflammatory stimulus	Bo et al. (2016)
					Pro-inflammatory cytokine mRNA↓		
					Anti-inflammatory cytokine mRNA ↑		
					CaSR↑PLCβ2↑		
					p-NF-κBp65↓IKKα/β↓IκB↓		
23	5 kDa	90%	100 μg/mL	Rabbit and human synoviocytes	Activation of AMPK	The mechanism of COS-induced AMPK activation involves an increase in the ADP/ATP ratio but not calcium/calmodulin-dependent protein kinase kinase beta (CaMKKβ). Activation involves an increase in the ADP/ATP ratio but not calcium/calmodulin-dependent protein kinase kinase beta (CaMKKβ)	Kunanusornchai et al. (2016)
			5 mg/kg (L-COS) or 10 mg/kg (H-COS)	Rabbit osteoarthritis model	iNOS↓COX-2↓	COS suppressed the TNFα-induced iNOS and COX-2 expression via an AMPK-dependent mechanism	

(continued)

Table 8.2 (continued)

No.	Average molecular weight	Degree of deacetylation	Dose	Model	Ending point	Proposed mechanism	References
24	3.5 kDa	90%	4 mg/mL	RAW264.7 cells	iNOS↓COX-2↓Nrf2↑ERK1/2↑JNK↑p38 MAPK↑	COS exert an anti-inflammatory effect through Nrf2/MAPK-mediated HO-1 induction	Hyung et al. (2016)
25	1.5 kDa	90%	0.25, 0.5,1.0 mg/mL	Caco-2 transformed human intestinal cells (LPS-induced)	TNF-α↓ PGE2↓COX-2↓TLR4↓NF-κB↓ Apoptosis↓ caspase-3↑ bcl-2↑	COS in the control of inflammatory bowel diseases, which may be due to the downregulation of TLR4/NF-κB pathway	Yang et al. (2016)
26	1.5 kDa	90%	Low (125 mg/kg) Middle (250 mg/kg) High (500 mg/kg)	Female C57BL/6J mice (DSS-induced)	TLR4/NF-B-k/COX-2↓	The COS administration prevented the DSS-induced activation of TLR4/NF-K/COX-2 signaling cascade in mouse colon	Tong et al. (2016)
27	<1 kDa (DP 2–7)	>90%	50, 75, 100, 150, 200 ng/mL	IPEC-J2 (TNF-α-induced)	IL-8↓MCP-1↓Claudin-1 ↓	COS had anti-inflammatory activity	Yang et al. (2018a)
28	ChOS DP 2–6	Highly N-acetylated	50–200 μg/mL	RAW 264.7 cells (LPS-induced)	IL-6 ↓ TNF-α↓ROS↓ NO↓ PI3K↓Akt↓	ChOS inhibit LPS-induced inflammatory response in macrophages	Xu et al. (2017a)
29	<1 kDa	≥93%	50 uL (1 mg/mL)	Rat osteoarthritis model	Prevent and treat osteoarthritis OPG ↑ RANKL↓	COS treatment upregulated the ratio of OPG/RANKL and downregulated the RANKL/RANK	Zhang et al. (2017)

Moreover, various additional immune cells displayed inflammatory properties, which were successfully impeded by COS. Considering basophilic leukemia RBL-2H3 cells in rats, both the cytokine production and antigen-induced degranulation were suppressed by COS (Chung et al. 2012). In calcium ionophore A23187 plus phorbol 12-myristate 13-acetate (PMA)-induced RBL-2H3 rat basophilic leukemia cells, COS decreased various types of cytokines including TNF-α, IL-1β, IL-4 and IL-6 by suppressing the MEK/ERK and p38 kinase (Vo et al. 2011).

Besides the immune-related cell models described above, both the endothelial cell model and the intestinal epithelial model are commonly used in models related to in vitro inflammatory cells. Liu et al. found that COS decreased IL-6 in LPS-induced HUVECs via the p38 MAPK pathway independent from NF-κB activation, as well as via the ERK1/2 pathway which relied on the activation of NF-κB (Liu et al. 2010). Moreover, COS attenuates LPS-induced vascular endothelial inflammation through the obstruction of MAPK phosphorylation and successfully in reducing ICAM-1 and E-Selectin. Furthermore, NF-κB activation in LPS-stimulated endothelial cells located in the porcine iliac artery exhibited a significant decrease (Li et al. 2014b). Yousef et al., confirmed that COS decreased NF-κB activation, TNF-α, IL-6, and oxidative stress in LPS-induced T84 human colonic epithelial cells (Yousef et al. 2012). Research indicates that COS displays a capacity for stimulating AMPK via a calcium-sensing receptor-phospholipase C-IP_3 (CaSR-PLC-IP_3) receptor channel-mediated calcium release from endoplasmic reticulum (Muanprasat et al. 2015). Furthermore, COS suppressed the inflammatory response and apoptosis in LPS-induced Caco-2 cells by decreasing TNF-α, PGE2, COX-2, toll-like receptor 4 (TLR4), and NF-κB (Yang et al. 2016). Additionally, COS attenuated the mRNA expression of IL-8 and MCP-1 induced by TNF-α in the IPEC-J2 cells (Yang et al. 2018a).

The anti-inflammatory effect of COS was further confirmed in various animal inflammation models. In a mice model of acute colitis induced by dextran sulfate sodium (DSS), COS suppressed the NF-κB activation, as well as the TNF-α and IL-6 levels (Yousef et al. 2012). COS inhibited the activation of the TLR4/NF-κB/COX-2 signaling cascade and could contribute in determining the related NF-κB flow in mice models (Tong et al. 2016). Observation of the model of inflammatory bowel disease in mice suggested that COS reduced the shortening of colon and tissue injury induced by DSS. COS not only suppressed the activation of MPO, NF-κB, COX-2, and iNOS but also decreased TNF-α and IL-6 levels (Azuma et al. 2015). When considering the model dealing with sepsis in mice, COS attenuated LPS-induced organ damage and dysfunction by reducing the levels of TNF-α and IL-1β (Qiao et al. 2011). When observing allergic and asthmatic mice, COS improved Ovalbumin-induced lung inflammation by decreasing IL-4, IL-5, IL-13, and TNF-α (Chung et al. 2012).

Similarly, in a specific rat model, COS protected rats against experimental autoimmune anterior uveitis (EAAU) by reducing TNF-α, iNOS, MCP-1, RANTES, fractalkine, and ICAM-1. Moreover, COS decreased the IκB degradation and p65 presence in the ciliary body by inhibiting NF-κB/DNA binding activity (Fang et al. 2014). Both reperfusion injury, and retinal ischemia were prevented due to the ability of COS to reduce the manifestation of inflammatory mediators. During this process, the activation of ERK, NF-κB, and JNK were restricted, while the production of p38 was stimulated (Fang et al. 2015). In a rat osteoarthritis model, COS improved the cartilage damage by up-regulating the ratio of OPG (osteoprotegerin)/RANKL (receptor activator of NF-κB ligand) and down-regulating the RANKL/RANK ratio (Zhang et al. 2017).

In other animal models, COS induced AMPK activation by increasing the ADP/ATP ratio but not calcium/calmodulin-dependent protein kinase kinase beta (CaMKKβ). Therefore, iNOS and COX-2 prompted by TNF-α were suppressed in the rabbit model of anterior cruciate ligament transection-induced osteoarthritis (Kunanusornchai et al. 2016). In piglets with induced LPS, COS alleviated the intestinal damage and decreased TNF-α, IL-6, and IL-8 levels. Additionally, the intestinal abundance of pro-inflammatory cytokine mRNA was reduced, while the mRNA expression of anti-inflammatory cytokine increased. Considering the underlay mechanism, COS increased the expression of both CaSR and PLCβ2 and reduced the expression of p-NF-κB p65, and IKKα/β in the intestines (Huang et al. 2016).

This section examines the underlying mechanism relating to the anti-inflammatory activity of COS. COS may be beneficial in the prevention of inflammation via multiple signal suppression, especially in NF-κB pathway, JNK/ERK/p38 signals, PI3K/AKT signals, and NOS formation pathways (Fig. 8.2).

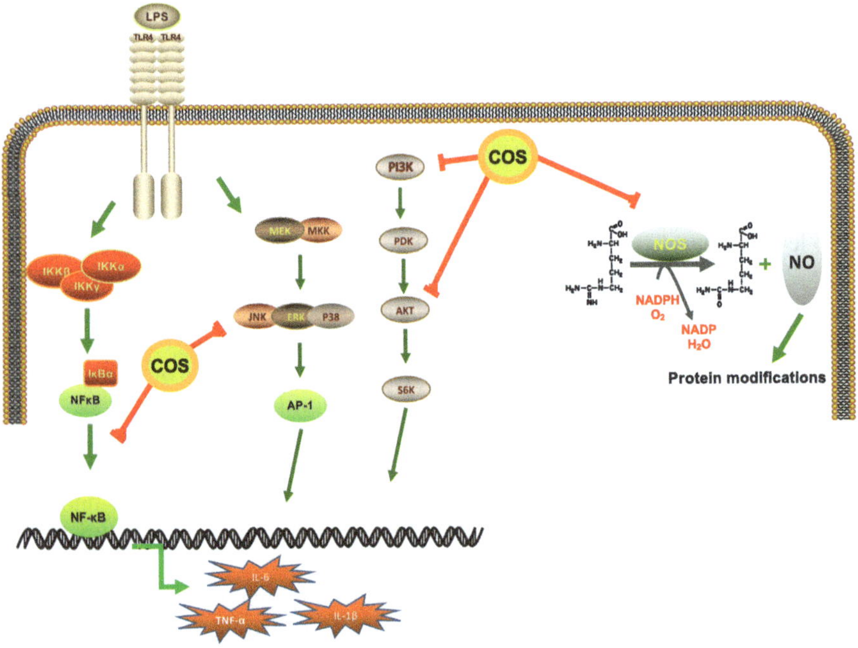

Fig. 8.2 The multiple signaling pathways contribute to the anti-inflammatory effects of COS. These pathways include *AKT* protein kinase B, *JNK* c-Jun N-terminal kinase, *NF-κB* nuclear factor kappa-B, *TLR4* toll-like receptor 4, *MAPK* mitogen-activated protein kinase, *TNF-α* tumor necrosis factor-α, *PI3K* phosphoinositide 3-kinase, and *LPS* lipopolysaccharide. (Modified from Muanprasat and Chatsudthipong 2017)

8.2.3 The Anti-tumor Activity of COS

For several decades, the protection of several kinds of COS against various types of tumors was investigated, and the detailed results were summarized in Table 8.3. The antitumor activity of COS was first reported in the early 1970s, and Muzzarelli believed that this acitvity was due to the positive charge of its amino group (Muzzarelli 1977). The tumor surface contains more negative charges than the typical cell surface (Wang et al. 2010). The positive charge of COS can bind to the negative charges on the cell surface, thereby inhibiting tumor growth (Pae et al. 2001). However, Huang et al. obtained quaternized, methylated, SCOS by chemical modification. The results indicated that COS with a strong negative charge displayed an ability to significantly inhibit tumor growth regardless of whether a negative or positive charge was present. The mechanism driving this action remains unclear (Huang et al. 2006).

In the 1980s, Suzuki et al. (1986) and Tokoro et al. (1988) discovered that the anti-tumor effect of chitohexaose is related to the enhancement of immunity, while chitopentaose does not have the same effect. This result suggests that the anti-tumor effect of COS may be closely related to MW and DP. Qin et al. prepared three different COS mixtures. These include COS with a high polymerization degree of chitosan oligosaccharide, a moderate polymerization degree, and a low polymerization degree. Tumor cell viability experiments established that high polymerization degree of chitosan oligosaccharide exhibited a higher inhibitory effect, indicating that the higher DP is related to stronger anti-tumor activity (Qin et al. 2002). Park et al. and Wu et al. prepared a mixture of different DP and degree of deacetylation (DD), and conducted a tumor cell viability experiment. The results indicated that both the DP and the DD of the COS affected its anti-tumor capabilities (Park et al. 2011; Wu et al. 2012).

A tumor is produced by malignant mutative hyperplasia which is different from healthy tissue. Studies have shown that some immune responses prompted by tumor antigens can limit tumor growth (Wang and Gong 2001). Ferrades et al. found that COS significantly inhibited tumor growth in a mouse model of bladder cancer, presumably by enhancing immunity (Fernandes et al. 2012). Subsequent studies have found that COS stimulate the production of cytokines such as IFN-γ, IL-12, and TNF-α. Furthermore, it increased AMPK activity, and inhibited NF-κB-mediated inflammatory responses, thereby reducing tumor volume (Masuda et al. 2014; Muanprasat et al. 2015; Zou et al. 2016b).

Tumor angiogenesis refers to the process by which tumor cells induce microvascular growth and establish blood circulation in tumors (Hanahan and Weinberg 2011). In 2005, Harish found that COS inhibit angiogenesis in the chicken chorioallantoic membrane (CAM) (Harish Prashanth and Tharanathan 2005). In the years that followed, researchers focused on the role of COS in preventing tumor blood vessel survival and migration. Inhibiting tumor angiogenesis can effectively impede tumor feeding and migration. Van Ta et al. found that COS can reduce MMP-9 expression at the gene and protein levels, thereby hindering cancer cell migration (Van Ta et al. 2006). Nam et al. found that MMP-9 is regulated by COS to reduce

Table 8.3 A summary of anti-tumor activities of COS, ChOS and its derivatives

No.	Ave molecular weight	Degree of deacetylation	Dose	Model	Ending point	Proposed mechanism	References
1	COS DP 6	/	100 mg/mL	180 solid tumor inpanted in ddY male mice;	180 solid tumor growth-inhibitio n 93%	Immunopotentiating effects	Suzuki et al. (1986)
				MM 46 solid tumor inplanted in C3H/He male mice	MM 46 solid tumor inhibition 55%		
2	COS DP 5	/	10 mg/mL	Male BALB/c mice;	Tumor growth-inhibition DP 6–41%, DP5 – no effect	High degree of deacetation shows higher inhibition ability;	Tokoro et al. (1988)
	COS DP 6			Meth-A tumor cells	Delayed type hypersensitivity↑	Acceleration of the production of and response to IL-1 and IL-2	
					IL-1↑, IL-2↑		
3	COS DP 3–10	/	50–1000 mg/mL	HL-60 cell line	Cell growth-inhibition 56%	Positively charged molecules target to cancer cells	Pae et al. (2001)
					Morphological changes		
					G0/G1↑, CD11b↑		
					Apoptosis↑		
4	17.7 kDa	86.0%	50, 200 mg/kg	S180 in Kunming mice	Tumor growth-inhibition 36.8–56.9%	High degree of deacetation shows higher inhibition ability	Qin et al. (2002)
	10.1 kDa	71.5%					
	1.57 kDa	57.9%					
5	COS, DP 5, 6 and higher oligomers mixture	84%	50 mg	Ehrlich ascites tumor (EAT) bearing mice	EAT cells number↓	Activation of apoptosis;	Harish and Tharanathan (2005)
					Apoptosis morphological and biochemical features	Decrease VEGF activity;	
					Chick embryo chorioallantoic membrane (CAM)↓	Inhibit glycolysis	

6	6000–7000 Da (carboxylated, quaternized and sulfated)	75.08%	0.1–1.0 mg/mL	Human colon cancer cell line (SW 480);	IC_{50} HeLa 0.45, 0.20 mg/mL	Highly positive or negative charged molecules target to cancer cells and due to necrosis	Huang et al. (2006)
				Human cervical cancer cell line (HeLa);	IC_{50} SW480 0.52, 0.50 mg/mL		
				Human liver cancer cell line (Hep3B);			
				Normal lung fibroblast cell line (MRC-5)			
7	1–3 kDa	/	100 mg/mL	Human fibrosarcoma cell (HT1080)	No cytotoxic effect	Prevent metastasis in vivo through inhibition of MMP-9 expression;	Van Ta et al. (2006)
	3–5 kDa				MMP-9↓	The molecular masses and concentrations affect the effect	
	5–10 kDa						
8	1–3 kDa	84%	50 and 100 mg	Human colorectal adenocarcinoma HT-29 cell	iNOS↓, MMP-2 ↓	Inhibition of invasiveness	Nam et al. (2007)
9	1 kDa–3 kDa	/	5 mg/mL	Human breast carcinoma cell line MDA-MB-231	Inhibit cell growth 20%	Anti-metastasis	Nam and Shon (2009)
					MMP-9↓		
					Wound healing cell migration↓		

(continued)

Table 8.3 (continued)

No.	Ave molecular weight	Degree of deacetylation	Dose	Model	Ending point	Proposed mechanism	References
10	COS DP 2–18	95%	5–200 mg/mL	Human umbilical vein endothelial cells (HUVECs);	Chick embryo chorioallantoic membrane (CAM)↓	Anti-angiogenic	Wu et al. (2008)
				Human hepatoma carcinoma cells Bel-7402			
11	COS DP 3–9	95%	0–1.6 mg/mL	Human hepatocellular carcinoma cells (SMMC-7721 cells)	Bax↑	Activation of apoptosis	Xu et al. (2008)
12	COS DP 3–8	/	50–1000 mg/mL	Human stomach adenocarcinoma cell line (AGS);	Slightly inhibit AGS and COLO 205	COS-incuced suppression of S-phase DNA synthesis in HepG2 cells;	Shen et al. (2009)
				Human colon adenocarcinoma cell lines (COLO 205);	Inhibition of HepG2 62%	Anti-metastatic	
				Human hepatocellular carcinoma cell lines (HepG2);	S-phase↓, p21↑, PCNA, cyclin A and cdk-2↓		
				HepG2 cells into SCID mice;	Mice tumor growth↓, lung metastasis↓		
				Lewis lung carcinoma (LLC) cell;	MMP-2, MMP-9↓		
				LLC cells in C57BL/6 mice			

13	COS DP 2–8	/	50 mg/mL	Human breast cancer cells MCF-7 and ECV304	Proliferation inhibition of DP 2–8 and mixture 8.2%, 9.8%, 9.1%, 13.7%, 12.7% and 4.6% 2–8 and mixture 8.2%, 9.8%, 9.1%, 13.7%, 12.7% and 4.6%	Anti-angiogenesis	Xiong et al. (2009)
	COS DP 6				Migration inhibition of DP 2–8 and mixture 36.63%, 30.06%, 52.75%, 63.96%, 78.5% and 37.32%		
					DP 6 CAM ↓,mRNA VEGF ↓, uPA ↓		
					MMP-9 and TIMP-2 –, TIMP-1↑		
14	COS DP 2–8	95%	5–200 mg/ mL	Human umbilical vein endothelial cells (HUVECs);	Change the actin polymerization pattern at the cell-cell interface	Inhibit NO-induced angiogenic	Wu et al. (2010)
				Human breast cancer cells (MCF-7);	Wound healing migration↓		
				Subcutaneous xenograft model in mice	NO concentration↓		
					Microvessel density↓		
					SOD activity↑		

(continued)

Table 8.3 (continued)

No.	Ave molecular weight	Degree of deacetylation	Dose	Model	Ending point	Proposed mechanism	References
15	COS DP 2–8	98.5%	25, 50 mg/mL	HepG2, A549, and PC3 cell	IC_{50} of PC3, A549 and HepG2:	Mw and DD might be important factors for the exhibition of antitumor activity in vitro	Park et al. (2011)
					DP 2–8: 25, 25, 25 μg/ml		
	COS DP 3–5	100%			DP 3–5: 25, 5, 12.5 μg/ml		
	COS DP > 5	85.5%			DP > 5: 50, 50, 50 μg/ml		
16	1.763 ± 0.7 kDa	64.14 ± 1.96%	50–500 mg/kg	A rat model of urinary bladder carcinogenesis induced with N-butyl-N-(4-hydroxybutyl) nitrosamine	The size of mice tumor↓	Anti-inflammatory action;	Fernandes et al. (2012)
						Inhibit cellular proliferation control;	
						Improve antioxidant profile	
17	2480 Da	/	125–1000 mg/mL	Human monocytic leukaemia cell line (THP-1);	THP-1 IC_{50} 1 mg/mL	Low molecular weight and the short backbone of COS allow them to get into the intestines and exert biological activities	Salah et al. (2013)
				Normal human foetal lung fibroblastic cell line (MRC-5)	MRC-5 no cytotoxic effects		

18	1–3 kDa	/	50–1000 mg/mL	Human gastric cancer cell line SGC-7901, AGS and NCI-N87	Inhibit SGC-7901 cells growth	Inhibit metastatic through CD147/MMP-2 pathway	Luo et al. (2014b)
					Wound healing cell migration↓		
					MMP-2↓		
19	COS DP 2–8	/	1–4%	MyD-88 knockout mouse	The tumor volume↓	Induction of apoptosis;	Masuda et al. (2014)
						Stimulate the immune system	
20	COS DP 2	/	100–400 mg/mL	Neural cell PC12	Bax/Bcl-2↓	Inhibit glutamate-induced cell death by preventing apoptosis	Hao et al. (2015)
	COS DP 3				Caspase-3↓		
	COS DP 4						
21	5000, 8000 and 14,000 Da	90%	10–500 mg/kg	T84, HT-29 and Caco-2 cells	Statistically significant increase of AMPK phosphorylation ↑	Activation of AMPK;	Muanprasat et al. (2015)
						Suppression of CFTR mediated intestinal fluid secretion and chemoprevention of CRC in vivo	
22	5000 Da	90%	20–500 mg/kg	Mouse model of colitis-associated colorectal cancer	COS at 100 mg/kg/day and 500 mg/kg/day reduced the tumor size in CRC mouse models by ~60%	Reduced rates of cell apoptosis	Mattaveewong et al. (2016)
23	/	/	0.5–4.0 mg/mL	Human hepatoma cells SMMC-7721	Bcl-2↓	Induction of apoptosis	Liu et al. (2017)
					Caspase-3↑		
24	≤1000 Da	95%	200 mg/mL	Human immortalized breast epithelial cell line MCF-10A	Cell migration↓	Inhibit breast epithelial cell migration through down-regulation of GnT-V and its products, branched N-glycans	Xu et al. (2017b)
					GnT-V overexpression↑		

(continued)

Table 8.3 (continued)

No.	Ave molecular weight	Degree of deacetylation	Dose	Model	Ending point	Proposed mechanism	References
25	COS DP 3–7	85%	200 mg/mL	Human umbilical vein endothelial cells (ECV304);	CAM↓	High degree of deacetation and high molecular weight show higher angiogenic ability	Wu et al. (2012)
	COS DP 8–12	85%		Human breast cancer cell MCF-7	Wound healing migration↓		
	COS DP 8–12	70%					
	COS DP 4–7	40%					
	COS DP 8–12	40%					
26	1000 Da	93%	1.25–20 mg/mL	BGC-823, SGC-7901, KCC-853, 786-O, A549, NCI-H460, MCF-7, Bcap-37, HCT-116, HT-29 cell;	IC_{50} 48.6 ± 7.0 to 1329.9 ± 93.4 μg/mL against 10 different tumor cell lines and up to 58.5%	Enhance immune system	Zou et al. (2016b)
	1500 Da	90%		S180-bearing mice	TNF- α↑		
	2000 Da	90%			Tumor weight ↓		

expression (Nam and Shon 2009), and Wu et al. suggested that COS inhibit angiogenesis by reducing NO concentration (Wu et al. 2010). In liver cancer HCC cells, lung cancer LLC cells, breast cancer MDA-MB-231, MCF-7 cells and gastric cancer SGC-7901 cell models, COS were found to obstruct the migration of these cells by down-regulating vascular endothelial growth factor (VEGF) and MMP-2 (Luo et al. 2014b; Shen et al. 2009; Wu et al. 2008; Xiong et al. 2009). Further studies by Xu et al. indicated that COS inhibits the migration of mammary epithelial MCF-10A by down-regulating N-acetylglucosaminyltransferase V (GnT-V) and its products (Xu et al. 2017b).

In recent years, researchers have begun to pay attention to the role of COS in promoting tumor cell apoptosis. Apoptosis is a process of death that maintains an environmentally stable, cell-independent order, involving the activation, expression, and regulation of a range of genes (Fadeel and Orrenius 2005). In 2005, Harish and Tharanathan first discovered that when treated with COS, Ehrlich ascites tumor (EAT) cells showed distinct apoptotic characteristics, including nuclear condensation and membrane destruction. This fact suggested that COS may inhibit tumor growth by promoting tumor cell apoptosis (Harish Prashanth and Tharanathan 2005). Later, Xu et al. found that COS encouraged apoptosis by up-regulating Bax (Xu et al. 2008). Masuda et al. established a MyD-88 knockout mouse model and found that the effect of COS is associated with both Myd-88-dependent and Myd-88-independent pathways (Masuda et al. 2014). Hao et al. and Liu et al. indicate that COS regulate Bax, Bcl-2, and Caspase-3 in the apoptotic pathway (Hao et al. 2015; Liu et al. 2017).

In conclusion, these findings are indicative of the suppression effects due to the ability of COS to promote the inhibition of cancer proliferation, invasion, and metastasis by regulating numerous signal pathways. Figure 8.3 illustrates the possible mechanisms responsible for the anti-cancer ability of COS.

8.2.4 The Anti-microbiocal Activity of COS

COS are recognized to have excellent potential as natural antimicrobial preservatives. The details of this mediation capability against various types microbiota was investigated and is listed in Table 8.4.

The prior study conducted in 1979 by Allan C.R. et al. indicated that chitosan displayed a higher fungicidal activity than chitin (Allan and Hadwiger 1979). Several positive intervention effects against various bacteria and fungi have been reported since then. During the 1990s, researchers focused on evaluating the different levels of activity of the various MW of chitin/chitosan on bacteriostat. They pointed out that the positive charge density of chitin/chitosan was the main limiting factor for their bacteriostat (Hirano and Nagao 1989). It is possible that the underlay mechanism may be connected to the high-affinity binding to an acidic group on the surface of the bacteria (Hirano and Nagao 1989; Zopf and Roth 1996).

In the beginning of the twenty-first century, an increasing number of researchers stated paying attention to anti-bacterial activity with the development and manufacturing of chitin/chitosan and its oligosaccharides (Choi et al. 2001; Jumaa et al.

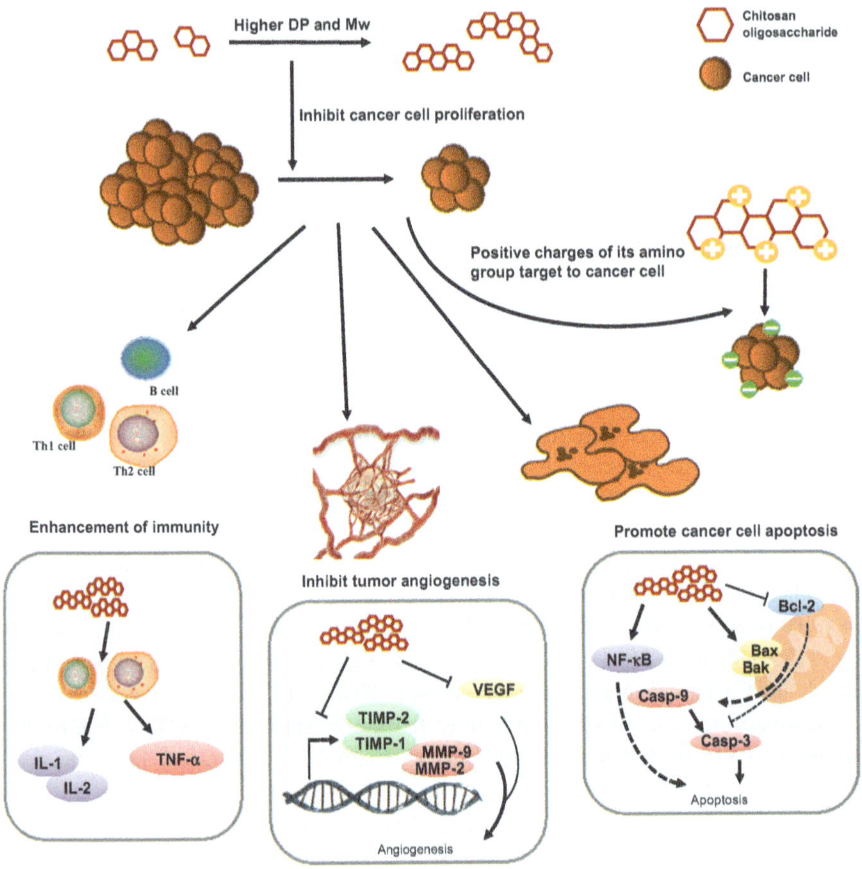

Fig. 8.3 The potential mechanisms that are responsible for the anti-cancer ability of COS. Particularly through the enhancement of immunity, the inhibition of tumor angiogenesis, the promotion of cancer cell apoptosis, and positive charges of its amino group target to cancer cells. *IL* Interleukin, *Bak* Bcl-2 antagonist/killer 1, *Bcl-2* B-cell lymphoma 2, *TIMP* tissue inhibitor of metalloproteinase, *MMP* matrix metalloproteinase, *Bax* Bcl-2 associated X protein, *Casp* Caspase, and *VEGF* vascular endothelial growth factor

2002; Kim et al. 2003; Kittur et al. 2005; No et al. 2002; Tsai et al. 2002; Vishu Kumar et al. 2005; Yang et al. 2005). Both Gram-negative and Gram-positive bacteria were inhibited by the bactericidal properties displayed by COS. Several of the bacteria affected by various concentrations of COS include *Yersinia enterocolitica* (Vishu Kumar et al. 2005), *S. aureus* (Fernandes et al. 2008; Jumaa et al. 2002; Li et al. 2014a; Yang et al. 2005), *Bacillus cereus* (Kittur et al. 2005; Tsai et al. 2002; Vishu Kumar et al. 2005, 2007), *Bacillus licheniformis* (Vishu Kumar et al. 2007), and *Listeria monocytogenes* (Tsai et al. 2002). Furthermore, COS and its derivatives displayed excellent inhibition of eucaryon like mold (Eweis et al. 2006; Li et al. 2012c; Oliveira et al. 2008; Palma-Guerrero et al. 2010; Tsai et al. 2002; Wang et al. 2007) and yeast (Seyfarth et al. 2008; Tikhonov et al. 2006). However, many

Table 8.4 A summary of anti-microbial activities of COS, ChOS and its derivatives

No.	Average molecular weight	Degree of deacetylation	Dose	Model	Ending point	Main conclusion	References
1	Chitin (/)	/	75–1000 μg/mL	Fungi	Chitosan had more fungicidal activity than chitin	Chitin, a polymer of N-acetylglucosamine, is a common constituent of fungal cell walls	Allan and Hadwiger (1979)
	Chitosan (/)					Polymers with consecutive positive charges, such as poly-L-lysine, protamine, and histones have been observed to inhibit the growth of certain fungal pathogens	
						At the concentrations used, chitosan had no effeet on the isolates containing chitin and chitosan in their cell walls	
2	High molecular-weight (HMW)-chitosan: MW 400,000	/	1.0 mg/mL	Fungi	Strong growth inhibition was oserved with LMW chitosan and chitosan oligosaccharides;	The functional groups for the growth inhibition are the cationized amino group (at pH 5.6) of chitosan and the carboxyl group of pectic acid	Hirano and Nagao (1989)
	Low-molecular weight (LMW)-chitosan:/				Decreases in the degree of polymerization of chitosan resulted in decreases in the number of inhibited fungus species	Chitosan may make polyelectrolyte complexes with the acidic and basic groups of the cell surface to disorder it	
	Chitosan-oligosaccharides: DP2–8						

(continued)

Table 8.4 (continued)

No.	Average molecular weight	Degree of deacetylation	Dose	Model	Ending point	Main conclusion	References
3	COS and chitosan mixture (MW 2000–30,000 Da)	91.50%	0.10%	*Actinobacillus actinomycetemcomitans*	0.1% chitooligosaccharides for 30 min, the cells were markedly degraded from a spherical shape to irregularly condensed masses with bleb-like structures (membrane disruption)	Exposure to the chitooligosaccharides either rapidly killed the cells or rendered them non-culturable by directly attacking their membranes	Choi et al. (2001)
				Streptococcus mutans		The site of chitooligosaccharide action is probably the bacterial envelope and that the killing of the organism may be the result of membrane disruption	
4	Chitosan		0.01–1.05% (w/v)	*Pseudomonas aeruginosa*	The antibacterial activity of chitosan is affected by the pH value;	The binding of chitosan to charged molecules in the bacterial cell wall is supported by the observations that a slight shift of the pH resulted in a sudden change in the active concentration of chitosan and in a noticeable difference in the antimicrobial activity of chitosan	Jumaa et al. (2002)
	Type 1: 8.7 × 10^4 g/mol	Type 1: 1.92		*Staphylococcus aureus*	Chitin derivatives inhibit hemagglutination by bacteria and the microbial adherence to epithelial cells	The selective binding of chitin derivatives to sugar receptors on the cell surfaces	
	Type 2: 5.32 × 10^5 g/mol	Type 2: 2.73		*Candida albicans*			
				Aspergillus niger			

5	Chitosan	50%	1000, 2500, 5000 ppm	The gram-negative bacteria	The higher deacetylation degree and the higher concentration of chitosan generally have higher antibacterial activity	The higher deacetylation degree chitosan results in higher amount of free amino groups, which cause more positive charges in the acid condition	Chen et al. (2002)
		70%		*Pseudomonas aeruginosa*, *Salmonella typhimurium*, *Escherichia coli*		The negative surface charges will react with chitosan, and therefore *Escherichia coli* will be destroyed. The antibacterial activity is currently hypothesized as surface interference or permeable inhibition	
		90%		The gram-positive bacteria		It is also found that the deacetylation degree of chitosan is more significant than the molecular weight of chitosan on antibacterial activity	
				Staphylococcus aureus, *Clostridium perfringens*, *Streptococcus faecalis*		This proves that the chitosan with higher molecular weight (shorter persistence length) can easily pass through the cell wall of gram-positive species, but it is blocked outside of the cell wall of the gram-negative species	

(continued)

Table 8.4 (continued)

No.	Average molecular weight	Degree of deacetylation	Dose	Model	Ending point	Main conclusion	References
6	Chitosan and chitosan oligomers	/	1%	Four gram-negative	Chitosan oligomers of 1 kDa showed higher antibacterial activity against gram-negative bacteria while 4 and 2 kDa against gram-positive bacteria	This is probably due to differences in experimental methods, chitosan characteristics, or medium pH	No et al. (2002)
	7. 22 kDa			(*Escherichia coli, Pseudomonas fluorescens, Salmonella typhimurium, and Vibrio parahaemolyticus*)			
	8. 10 kDa			Seven gram-positive bacteria (*Listeria monocytogenes, Bacillus megaterium, B. cereus, Staphylococcus aureus, Lactobacillus plantarum, L. brevis, and L. bulgaricus*)			
	9. 7 kDa						
	10. 4 kDa						
	11. 2 kDa						
	12. 1 kDa						

7	Hetero-COS	50%	0.3125–5 mg/mL	The gram-negative bacteria	Hetero-COS also effectively blocked the growth of most gram-negative and gram-positive bacteria	The inhibitory actibities were different from the factors, the degree of polymerization, and the type of bacterium	Park et al. (2004)
	10,000–5000 Da	75%		*Pseudomonas aeruginosa, Salmonella typhimurium, Escherichia coli*		Chelating is one of the most important actions of chitosan: deprivation of metals, trace elements, or essential nutrients by chelation limits the growth of microorganisms	
	5000–1000 Da	90%		The gram-positive bacteria		The highly reactive amino groups in chitosan have the ability to interact with anionic groups on the cell surface and to form polyelectrolyte complexes with bacterial surface compounds, thereby forming an impermeable layer around the cell, which prevents the transport of essential solutes into the cell	
	<1000 Da			*Staphylococcus aureus,*			
				Staphylococcus epidermidis			
				Micrococcus luteus,			
				Bacillus subtilis,			
				Bacillus cereus			

(continued)

Table 8.4 (continued)

No.	Average molecular weight	Degree of deacetylation	Dose	Model	Ending point	Main conclusion	References
8	COS DP 2–6	35–38%	0.1–0.5 mg% (w/v) levels of chito-oligomeric–monomeric mixture	*B. cereus*, *E. coli*,	Higher the DP of oligomers rich in free GlcN residues, the higher is their growth inhibitory activity	Antibacterial effect of chitosan/chito-oligomers is essentially due to free -NH3 $^{+}$groups	Vishu Kumar et al. (2005)
			0.05–0.1 mg% (w/v) of individual chito-oligomers	*Listeria monocytogenes*,		In gram-positive bacteria, positively charged amino groups of chito-oligomers/GlcN can bind peptidoglycans resulting in cell-wall distortion-disruption, exposure of cell membrane to osmotic shock and exudation of the cytoplasmic contents	
				Yersinia enterocolitica, *Staphylococcus aureus*,		Gram-negative bacteria, the negatively charged O-specific antigenic oligosaccharide-repeating units of the *E. coli* lipopolysaccharide form ionic-type of binding with the amino groups of chito-oligomers. Thus blocking the nutrient flow with concomitant bacterial death due to depletion of the nutrients	
				B. licheniformis			

9	COS DP 1–6	/	0.05–0.1(mg.%)	*Bacillus cereus*	Increase in the oligomer DP resulted in increased growth inhibitory activity	At lower concentrations, entry of COS into the bacterial cells facilitates their action as activator of various physiological functions, whereas at higher concentrations, in addition their binding to microbial cell surface provides a net positive charge, thus segregating the individual cells and hence, without any inhibitory activity	Kittur et al. (2005)
			Inhibition (%)				
			1. GlcN: B. cereus 10–14; E. coli 5–8	*Escherichia coli*			
			2. Dimer: B. cereus 15–19; E. coli 8–12				
			3. Trimer: B. cereus 20–26; E. coli 14–17				
			4. Tetramer: B. cereus 28–32; E. coli 20–25				
			5. Pentamer: B. cereus 28–30; E. coli 19–23				
			6. Hexamer: B. cereus 40–48; E. coli 32–38				

(continued)

Table 8.4 (continued)

No.	Average molecular weight	Degree of deacetylation	Dose	Model	Ending point	Main conclusion	References
10	Chitosan	/	12–18 μg/mL	Fungi	Effect of chitosan on in vitro fungal development	The length of the polymer chain enhances its antifungal activity	Bautista-Baños et al. (2006)
			1–6 mg/mL				
			0.5–10%				
11	COS DP~4	97%	2.5 mg/mL	*E. coli* O157:H7	COS did not result in any reduction in adhesion of the three strains of VTEC, but the three EPEC strains, were inhibited;	At certain levels, COS are known to be toxic to certain bacteria including some strains of *E. coli*	Rhoades et al. (2006)
				Enteropathogenic *E. coli* *Dsv. desulfuricans*	No inhibition of *L. pentosus*, *L. casei* or *L. gasseri* was found;	The reason for the strain selectivity of COS in this assay is not currently known	
				Lactobaccilli	Adhesion of *L. acidophilus* was reduced to a level of 75% of that of the control	In particular, *E. coli* strains carrying G fimbriae containing the GafD adhesin recognise N-acetylglucosamine and are associated with diarrhoeal disease	

12	COS produced by chitosanase from *Pseudomonas* CUY8	50%	Bacteria (0.08–0.12%)	*E. coli*	Antimicrobial effects of chitooligosaccharides against bacteria were higher than fungi;	Low degrees of polymerization are very easy to penetrate through the cell membrane of microorganisms, interact with DNA in cytoplasm and karyon, lead to DNA replication mistake and result in suppression of the growth of microorganisms	Wang et al. (2007)
		60%		*S. aureus*	Antimicrobial effect of COS increased with increase of DD, but decreased with increase of DP	Antimicrobial activity of chitooligosaccharides increased with increase of DD, because chitooligosaccharides with high DD possess a lot of free amines. The free amines interact with the negatively charged residues at cell membrane. The reaction is to adsorb, congregate and precipitate the cells, subsequently result in death of cells	
		70%		*S. lactis*			
		80%		*B. subtilis*			
		90%	Fungi (0.13–0.15%)	*Rhodotorula bacarum*			
				Sac. cerevisiae			
				Mucor circinelloides			
				Rhizopus apiculatus			
				P. charlesii			
				A. niger			
13	Chitohexose	/	1–6 mg/mL	*Candida albicans*	COS-6 could inhibit the hyphae growth of *C. albicans* and induced the yeast proliferation of *C. albicans*	COS-6 remains unclear, they would like to analyze the expression of COS-6 receptor and sugar metabolism in *C. albicans*	Kadowaki et al. (2007)

(continued)

Table 8.4 (continued)

No.	Average molecular weight	Degree of deacetylation	Dose	Model	Ending point	Main conclusion	References
14	Pronase incubated for different periods	1 h 19	*B. cereus* 0.01% (w/v)	*Bacillus cereus*	Increase in the concentration of LMWC showed a linear inhibitory effect on bacterial growth	LMWC and high DD form more amino group and bond easily with the electronegative atoms of the *E. coli* cell membrane that are present on the surface	Vishu Kumar et al. (2007)
	1 h 9.5 kDa	2 h 16	*E. coli* 0.03% (w/v)	*Escherichia coli*		Unlike with gram-positive bacteria, wherein LMWC has to penetrate the cell-wall in search of electronegative groups	
	2 h 9.1 kDa	3 h 14					
	3 h 8.5 kDa						
15	Q1 (DP 5–8)	39%	100, 200, 400, 800,1000 μg/mL	*Alternaria alternata*	Q1 had a growth stimulating effect on *Alternaria alternata* and *Rhizopus stolonifer*	Reversal of inhibition or even stimulation of growth is probably caused by degradation, resulting in detoxification of the oligosaccharides, as well as in the production of dimers and trimers which apparently act as nutrients for the fungi	Oliveira et al. (2008)
	Q2 (DP 2–12) 1133 Da	37%		*Rhizopus stolonifer*	Growth of *Botrytis cinerea* was inhibited by Q3 and Q2	Fungistatic activity is higher at lower pH, to the cationic charge of the oligosaccharides	
	Q3 (DP 2–10) 696 Da			*Botrytis cinerea*	Growth of *Penicillium expansum* was only slightly inhibited by higher concentrations of sample Q3	Enzymatic uptake of simple carbohydrates by permeases could temporally be blocked by the presence of the large oligosaccharides	
				Penicillium expansum			

16	Low molecular weight chitosan hydrochloride (/)	84.90%	0.1%–1%	*Candida albicans,*	Growth inhibition	1. Chitosan can bind calcium and iron and cause a lack of nutrients for fungi;	Seyfarth et al. (2008)
				Candida krusei,		2. Polycationic character of chitosan as its most important factor of antifungal activity;	
				Candida glabrata		3. Anionic cell wall sphingolipids are "attacked" by chitosan	
	High molecular weight chitosan hydrochloride (120 kDa)	85%	0.5%–1%			4. Cell surface perturbing action	
17	COS mixtures<5 kDa and <3 kDa	80–85%	***E. coli*** below 0.10% (COS)-0.25% (Chitosan)	*E. coli*	Chitosans exhibit a stronger bactericidal effect upon gram-positive than gram negative bacteria	It is easier for oligomers to penetrate the gram-negative cell wall, whereas a mechanical barrier is formed by higher MW chitosans in their gram-positive counterparts, which prevents nutrient absorption	Fernandes et al. (2008)
	Chitosans 628, 591 and 107 kDa	80–85%	***S. aureus*** 0.1% (Chitosan)-0.25% (COS)	*S. aureus*			

(continued)

Table 8.4 (continued)

No.	Average molecular weight	Degree of deacetylation	Dose	Model	Ending point	Main conclusion	References
18	LMWS-chitosan 1, 3, 5, 10 kDa	/	0.04–2.5 mg/mL	*Aspergillus fumigatus*	The antifungal activity of LMWS-chitosan (10 kDa) was stronger than that of the other LMWS-chitosan (1, 3 and 5 kDa) against all tested fungal cells	LMWS-chitosna have an affinity for plasma membrane lipids, and their specificities for microbial membranes in many cases have been shown to be related to the positive charges of the chitosan, which favors interaction with the exposed anionic component of the microbial plasma membrane	Park et al. (2008)
				Aspergillus parasiticus		(Cell treated with LMWS-chitosan for 4 h showed cell surface disruption)	
				Fusarium oxysporum			
				Fusarium solani			
				Penicillium verrucosum var, verrucosum			
				Botrytis cinerea			
19	10,000 Da	90–95%	10,000 Da: 1–10 mg/mL	*Vibrio vulnificus*	The inhibition of *V. vulnificus*	The growth of *V. vulnificus* was more effectively suppressed by treatment with 10,000 Da than with 1000 Da	Lee et al. (2009a)
	1000 Da		1000 Da: >10 mg/mL				

20	Chitosan	79.60%	0.1 mg/mL	*Neurospora crassa*	Chitosan permeabilizes the fungal plasma membrane and is internalized by fungal cells;	Chitosan requires ATP to enter conidia of the plant-pathogenic fungus	Palma-Guerrero et al. (2009)
	70 kDa					ATP-dependent uptake is required to kill fungal cells	
					Different cell types (conidia, germ tubes and vegetative hyphae) exhibit differential sensitivity to chitosan	It is now clear that endocytosis is an important part of the vesicle trafficking network in filamentous fungi where it serves a number of important roles	
21	Chemically prepared chitin (CH-chitin): 4.91 × 10^4–1.08 × 10^6 Da	Low (47–53%)	Minimal lethal concentrations 100–2000 ppm	*Bacteria, Gram negative*	The antimicrobial activities of all chitin/chitosan were similar, and the activity increased with increasing DD	Chitosan with a higher DD, which has a higher positive charge, would be expected to have a stronger antimicrobial activity	Tsai et al. (2002)
	Microbiologically prepared chitin (MO-chitin) 5 × 10^4–1 × 10^6 Da	Medium (74–76%)		*Bacteria, Gram positive*		As well as DD, the size and conformation of the chitosan molecule might also affect its antimicrobial activity	
		High (95–98%)		*Fungi*			

(continued)

Table 8.4 (continued)

No.	Average molecular weight	Degree of deacetylation	Dose	Model	Ending point	Main conclusion	References
22	Chitosan 70 kDa	79.60%	0.1 mg/mL	*Beauveria bassiana*	Plasma membrane composition of chitosan-sensitive fungi exhibit increased fluidity	The plasma membranes of chitosan-sensitive fungi were shown to have more polyunsaturated fatty acids than chitosan-resistant fungi, suggesting that their permeabilization by chitosan may be dependent on membrane fluidity	Palma-Guerrero et al. (2010)
				P. chlamydosporia		Lipid components of the plasma membrane that confer chitosan sensitivity	
				F. oxysporum f.sp. radicis-lycopersici		In fungi, they constitute 20–30% of total lipids, and have been previously proposed as a possible target for chitosan binding	
				Neurospora crassa			

23	COS 2, 3, 6 kDa	66%	0.5–4.5%	*Clostridium paraputrificum,*	All bacterial strains were more sensitive to CS than to COS/LMWC	The inhibition effects differed depending on the DP of used polyaminoglycoside products more than with the type of bacterium	Šimůnek et al. (2010)
	LMWC 10, 16 kDa	76%	0.1–1%	*Clostridium beijerinckii,*			
				Roseburia intestinalis,			
				Bacteroides vulgatus,			
				Bacteriodes thetaiotaomicron, Faecalibacterium prausnitzii, Blautia coccoides			
24	Chemical hydrolysis of Chitosan	80%	0.003% (w/v)	Gram-negative bacteria	The oligomers exhibited a bactericidal effect on all bacteria tested	Chitosan possesses a number of polycationic amines which can interact with the negatively charged residues of carbohydrates, lipids and proteins located on the cell surface of bacteria, which subsequently inhibit the growth of bacteria	Benhabiles et al. (2012)
				Escherichia coli, Vibrio cholerae, Shigella dysenteriae, and *Bacteroides fragilis*		The oligomers molecules are assumed to be able to pass through the bacterial cell wall, composed of multilayers of cross-linked murien, and reach the plasma membrane	

(continued)

Table 8.4 (continued)

No.	Average molecular weight	Degree of deacetylation	Dose	Model	Ending point	Main conclusion	References
25	Chitosan 41.2, 18.8, 9, 5.5 kDa	82%	100 mg/L	Crop-threatening pathogenic	Growth inhibition	The four LWCSs exhibited higher antifungal activity against *P. asparagi*, *F. oxysoporum*, and *S. solani* than the original chitosan	Li et al. (2012c)
			200 mg/L	*Phomopsis asparagi*			
			400 mg/L	*Fusarium oxysoporum f. sp. Vasinfectum*			
				Stemphylium solani			
26	COS DP 3–12	85% 70% 35%	16 mg/mL	Enteropathogenic *Escherichia coli* (EPEC)	Low fraction of acetylation (FA) has the most effective adherence inhibition activities	The antiadherence property of oligosaccharides has been attributed to the similarity between the oligosaccharide structure and cell surface receptor to which bacteria attach prior to colonization	Quintero-Villegas et al. (2013)
						Growth of EPEC was not impaired by the COS used in this study, indicating that reduced adherence of EPEC was not due to growth inhibition or cell killing	
27	Irradiated oligochitosan	90%	0.05–1%	*Colletotrichum Gloeosporioides*	Potato Dextrose Agar (PDA) mixed with 0.05–1% irradiated oligochitosan were inhibited or retarded the growth of *C.gloeosporioides*	The inhibition activity was dependent upon the irradiated oligochitosan concentrations, and higher concentration resulted in higher antifungal activity	Kewsuwan et al. (2014)

28	Chitosan (viscosity, 20–200 cps)	Chitosan 75–85%	1–2 mg/mL	*Rhizopus stolonifer*	Both polymers reduced the mycelial growth of *R. stolonifer* and the inhibitory effect was dependent on the concentration;	Both chitosan and oligochitosan affect the development of *R. stolonifer* and might be implicated in the mitochondrial dysfunction	Robles-Martínez et al. (2014)
	COS (viscosity, 6 cps)	COS 90%			The spore germination was also affected by chitosan and oligochitosan		
29	COS2/3/4/5/6, 6–7, 7–8, 9–10, 10–12, >12	100%	62.5–8000 μg/mL	*S. aureus*	COS with high DP, lower pH value showed enhanced antibacterial effect on *S.aureus;*	COS generally show stronger bactericidal effects on gram-positive bacteria than gram-negative bacteria	Li et al. (2014a)
					DP > 12 have stronger antibacterial activity than chitohexaose	Cationic charges have been reported to help antimicrobial agents to bind with the anionic lipid components of bacterial membrane	
						Thecationic oligomers could “glue” negatively charged cells and led to the formation of large bacteria cluster, which might block the nutrition transport of bacterial cell and resulted in impairment of vital bacterial activity	
						The “glue” interaction seemed to requirean enough size of COS	

(continued)

Table 8.4 (continued)

No.	Average molecular weight	Degree of deacetylation	Dose	Model	Ending point	Main conclusion	References
30	$(GlcNAc)_2$-$(GlcNAc)_5$	/	0.1% (w/v)	*Aeromonas hydrophila*	Dietary oligosaccharides improve animal growth performance and enhance the health status of the host	Perhaps the amounts used here were not sufficient to enhance growth performance or such supplementation indeed does not affect the growth of *tilapia*	Qin et al. (2014)
	<1500 Da					Dietary oligosaccharides improve animal growth performance and enhance the health status of the host by modulating the types of intestinal microbiota	
						Chito-oligosaccharides/ chitosan-oligosaccharides may alter the population or structure of the microbial ecosystem in the gastrointestinal tract of tilapia	
31	COS 1, 2 and 3 kDa	70%	10–50 mg/L	Beer-spoilage bacteria	Growth inhibition	COS with an average 2 kDa molecular weight was the best at inhibiting all tested beer-spoilage bacteria	Zhao et al. (2016)
				Lactic acid bacteria			

32	Chitosan/COS 0.7–20 kDa	78–99%	100–1000 μg/mL	*Klebsiella pneumoniae*	Antibacterial activity of chitosan against *K.lebsiella pneumoniae* was associated with its polycationic nature, and depended on the degree of protonation of the chitosan amino groups	The fraction of the charged amino sugar residues will increase with decreasing the polymerization degree of the sample	Kulikov et al. (2015)
						The higher molecular weight samples with the increased polymerization degree (>50) will contain the slightly lower proportion of the protonated amino groups	
						Higher molecular weight samples lose completely their antibacterial properties	
33	COS 1760 Da	91.08%	0.25 and 0.50%	*Trichophyton rubrum*	0.5% and 1% COS significantly suppressed *T. rubrum* cell growth	The anti-dermatophytic activity of COS appears to involve a defensive response	Mei et al. (2015)
					No inflammation or tissue destruction in the groups treated with 5% COS		

(continued)

Table 8.4 (continued)

No.	Average molecular weight	Degree of deacetylation	Dose	Model	Ending point	Main conclusion	References
34	COS		1–10 mg/mL	*E. coli*	The gram negative bacteria *E. coli* was inhibited more strongly than the gram positive *L. monocytogenes*	Antimicrobial activities are clearly closely dependent on the average molecular weight of COS and decreased with declining molecular weight as shown by the reduction of growth inhibition	Sánchez et al. (2017)
	P1: 1–17.2 kDa	P1: 10–14%		*L. monocytogenes*	Acetylated sequences and its balance with fully deacetylated are essential for using COS antimicrobial	The fact that gram negative bacteria have greater negative charges on the cell surface than gram positive bacteria could explain why *E. coli* was more sensitive to the presence of different COS than gram positive *L. monocytogenes*	
	P2: 1–8.1 kDa	P2: 9–10%				The high number of free amino groups present in these molecules could generate repulsive forces that reduce the interaction with the bacterial cell wall, and consequently decrease the antibacterial effects against *E. coli* and *L. monocytogenes* when compared to COS obtained using the one-step process	

35	Highmolecular weight chitosan HMW 624 kDa	HMW ≥ 75%	0.5–2 mg/mL	*Acinetobacter baumannii*	The HMW MBC value registered was two folds superior to the one of obtained for LMW chitosan	LMW chitosan antibacterial activity has been attributed to its interaction with electronegative compounds in the interior of the bacterial cell	Costa et al. (2017b)
	Low molecular weight chitosan LMW 107 kDa	LMW 75–85%				Chitosan may be more effective in preventing the subsequent development of the biofilm structure than in preventing the establishment and metabolic activity of the micro-colonies	
						A. baumannii's quorum signaling could be similarly affected by chitosan resulting in a critical inhibition of the later stages of biofilm formation	
						LMW chitosan has been described as being able to penetrate the bacterial cell wall, combine with DNA and inhibit the synthesis of mRNA and DNA transcription	

factors, such as the microbe characteristics (origins of strain, and cell age), COS (MW distribution, DD, and DP), environmental factors (pH, temperature, light, metal ion concentrations, and oxygen levels), and more have an impact on the activity of COS and its derivatives (Li et al. 2014a 2016; Zou et al. 2016a). Therefore, these factors may contribute to the various antimicrobial activities of COS/Chitin oligosaccharide (ChOS) and could provide more comprehensive interpretations when compared with existing studies.

On the basis of these reports, the proposed mechanisms of antimicrobial activities were summarized as follows:

(a) Leakage and distortion of the bacterial cell wall is caused by the interaction of negatively charged proteoglycans and positively charged COS on its surface. This process subsequently leads to the destruction of the Gram-positive bacteria due to osmotic shock (Vishu Kumar et al. 2005, 2007; Fernandes et al. 2008);
(b) The death of the Gram-negative bacteria occurs when its nutrient flow is obstructed. This lack of nutrition is prompted when the negatively charged O-antigen of the LPS situated on the bacterial cell wall binds with COS (Vishu Kumar et al. 2005, 2007; Fernandes et al. 2008);
(c) Low molecular weight chitosan (LMCS) could be absorbed and formed an adduct with the DNA, inhibiting both the mRNA transcription and protein synthesis (Costa et al. 2017b);
(d) COS could form a chelate with metal ions and other nutrients and decrease the support for bacterial growth (Park et al. 2004).
(e) Regarding the anti-fungal mechanism, cationicity is a vital factor relating to the anti-fungal activity of COS. Besides microbial surface reaction, COS might also be responsible for adjustments to the framework of the phospholipid bilayer within the cell membrane. Therefore, the membranal absorbance capacity is modified leading to the liberation of individual cellular elements. Several studies indicated that higher levels of polyunsaturated fatty acids were discovered in fungal material sensitive to chitosan and COS, resulting in a higher fluidity of the cell membrane and simplified permeability (Palma-Guerrero et al. 2010; Park et al. 2004).

In some species of bacteria, their resistance and pathogenicity will be greatly improved following biofilm formation (Danhorn and Fuqua 2007; Johnson 2008). Chitosan (including its derivatives)/COS had an excellent inhibition effect on bacterial biofilm. However, the mechanism with which this was probably differed from its antibacterial activity (Gonçalves et al. 2017; Ilk et al. 2017). For example, *Cronobacter spp.* and its biofilm flora is opportunistic foodborne pathogens found in milk powder that can prove fatal to consumers. These bacteria can cause serious infections in infants such as neonatal bacteraemia, meningitis and necrotizing enterocolitis (Farmer et al. 1980; Hunter and Bean 2013). In our more recent study, COS combined with 2000 Da achieved excellent results in inhibiting the *Cronobacter* biofilm (up to 80%), with a concentration of 20 μg/mL in reconstituted powdered milk. In addition, the COS was able to reduce the number of extracellular polymeric substances at the initial stage of biofilm development. According to qRCR, the inhibition may be related to the down-regulation of the *bcsA* expression, and the up-regulation of the *flhD*, *flgJ*, *luxR*, *ompA*, and *wcaJ* expressions, subsequently altering the primary attachment phase of *Cronobacter* (Chen et al. 2019; Lu et al. 2019).

8.2.5 The Activity in Gut Microbiota

With the development of food science and nutrition, a new concept of "microbiome nutrition" is gaining acceptance by most nutritionists. The term means "feed me and my microbes." Humans are super organisms consisting of both human cells and microbial cells, particularly the gut microbiota (Flint 2012; Kau et al. 2011). Gut microbes participate in the regulation of many activities within the human body. Research regarding the health of the human intestinal microbes by employing prebiotics, diet, probiotics, drugs and small functional molecules, are increasingly recognized by scientists (Biesalski 2016; Kau et al. 2011; Yadav et al. 2018).

Typically prebiotics is non-digestible food compounds consisting mostly of fiber that encourages the growth of microorganisms in the intestines that are advantageous to human health (Verbeke et al. 2015). In 2010, Roberfroid et al. updated this concept to read: "A component that can be selectively fermented can cause a specific change in the composition and/or the function of the entire gastrointestinal microbe, thereby producing a beneficial effect on the health of the host" (Roberfroid et al. 2010). From those definitions, whether it is "selectively" fermented, and then "specifically" enriched, the beneficial bacteria become the primary criterion for determining whether a substrate is prebiotic. Most of the non-digestible carbohydrates that meet the prebiotic evaluation requirements include polysaccharides and oligosaccharides (polysaccharides refer to more than ten carbohydrates in sugar units, and oligosaccharides are between two and ten sugar units). The results obtained relating to the activity of chitosan and COS in gut microbiota are illustrated in Table 8.5.

COS stimulate the growth of probiotics and produce beneficial metabolites via in vitro fermentation in human intestinal microbes. Lee et al. (2002) studied the influence of COS on the growth of probiotics by using the enzymatic method. COS display promise as a possible health food with beneficial prebiotic qualities since it encourages the growth of enteric bacteria especially *B. bifidium KCTC 3440* and *Lactobacillus sp. B.* bifidium in particular, exhibited significant growth in the presence of elevated COS levels. Vernazza et al. (2005) studied the in vitro fermentation of chitosan and COS with varying commercial MW measurements in human fecal cultures. In the amount of *bifidobacteria* showed no significant change, while the bacterial groups that were potentially harmful to human health inceased. By measuring absorbance, Simunek et al. (2010) discovered that chitosan displayed significant antibacterial activity in the presence of a variety of bacteria. These species included *Eubacterium* lentum, *B.* thetaiotaomicron, *Roseburia* intestinalis, *Bacteroidesvulgatus, F.* praunsnitzii, and *Clostridium* paraputrificum. Furthermore, as a study in 2012 established that the growth of six types of *bifidobacteria* are

Table 8.5 The activity of COS, ChOS and its derivatives in gut microbiota

No.	Average molecular weight	Degree of deacetylation	Model	Dose	Main conclusion	References
1	COS (Degree of polymerization 2–8)	99.90%	Specific microbial culture in vitro	0.2–0.4%	COS stimulates the growth of *Lactobacillus sp.* and *B. bifidium KCTC 3440*. *B. bifidium* has a concentration dependence on COS	Lee et al. (2002)
2	Low molecular (LM) weight chitosan (90–190 kDa);	/	A three-stage fermentation model	10 g/L	COS reduced *Clostridium* and increased *Lactobacillus* in the area representing the proximal colon. COS increases *Bacteroides* and *Clostridium* in areas representing the transverse colon. COS increases the butyrate content of intestinal contents	Vernazza et al. (2005)
	Medium molecular weight chitosan (190–300 kDa);					
	High molecular (HM) weight chitosan (310–375 kDa);					
	COS (<5 kDa)					
4	Chitosan (>60 kDa)	/	In vitro culture of colonic microorganisms	0.025–0.5%	The antimicrobial activity increased with the degree of polymerization (DP). MIC ranged from 0.25 to 4.5% in dependence on bacterial strain and DP of CS/Low molecular weight chitosan. MBC also decreased with DP. The most effective antimicrobial action was detected in LMWC with 16 kDa and CS. Weak antimicrobial activity was found in COS with small molecules (2 and 3 kDa)	Šimůnek et al. (2012)
	LMCS (10–16 kDa)					
	COS (2, 3, 6 kDa)					

5	COS (609 Da)	/	In vitro culture	0.05–0.5%	The minimum inhibitory concentration of COS against *Escherichia coli* and *Staphylococcus aureus* was 0.3%, and the minimum inhibitory concentration against *Salmonella paratyphi A.* and *Shigella sonnei* was 0.1%. As the mass concentration of COS increases, the bacteriostatic effect gradually becomes stronger	Cao et al. (2008)
7	Chitin oligosaccharide (ChOS), Chitosan, LMCS, COS	(ChitoClear® 90%)	In vitro batch fermentation of human microorganisms	1% w/v	Chitosan reduces *Bifidobacterium spp.*, *Eubacterium rectale/Clostridium coccoides*, *C. Histolyticum* and *Bacteroides/Prevotella*;	Mateos-Aparicio et al. (2016)
					COS keeps the flora stable and increases *Lactobacillus/Enterococcus*;	
					The amount of short-chain fatty acids produced by COS is comparable to that of FOS	
8	COS	/	Piglet intervention	200, 400, or 600 mg/kg	The present results indicate that dietary supplementation of COS at 400 or 600 mg/kg promotes growth performance and improves gut barrier function, increases the population of *Bifidobacteria* and *Lactobacilli*, and Decreases *S. aureus* in the cecum of weanling pigs	Yang et al. (2012)
9	COS	/	Weanling pigs	0.1–0.3%	COS could improve feed efficiency in youngpigs and reduce the growth of harmful bacteria, such as *Clostridium spp.*	Han et al. (2007)
					However, further experiments are needed to assess the effect of COS on feed intake in pigs	
10	COS	/	Fragrance pig	Add 0.5% COS to the base diet	COS significantly increased the number of *Lactobacillus* and decreased the number of *Streptococcus* in the ileum	Lian et al. (2012)
					COS significantly increased the number of *Bifidobacteria* and *Lactobacilli* and significantly decreased the number of *E. coli* and *Streptococci* in the cecum	

(continued)

Table 8.5 (continued)

No.	Average molecular weight	Degree of deacetylation	Model	Dose	Main conclusion	References
11	COS (DP 2–6)	95%	Weaning piglets	0.5 g/kg diet added	Dietary supplementation with COS modifies the composition of ileal and colonic microbiota, and increased the amounts of some presumably beneficial intestinal bacteria and concentrations of SCFAs in the intestinal luminal content, while suppressed the growth of potential bacterial pathogens and the amount of several protein-derived catabolites	Kong et al. (2014)
12	COS<5000 Da	91.25%	Weaning piglets	200–400 mg/kg	The dominant bands of the CS group were in the genus *Lactobacillus*, while the relative numbers of *Staphylococcus*, *Escherichia* and *Salmonella* were significantly reduced	Xiao (2011)
					In the CS group, the number of *Lactobacilli* and *Bifidobacteria* in the small intestine increased, *Escherichia coli* and *Streptococcus* decreased, and it was found that CS was better in the jejunum than in the duodenum and ileum	
13	COS	90%	Growing pigs	5 g/kg diet added	The COS administration also decreased the number of fecal *Escherichia coli*, whereas the number of fecal *Lactobacilli* was not influenced by either COS or tylosin administration	Wang et al. (2009a)
14	COS 1500 Da	/	Nursery pigs	100, 200, and 400 mg/kg	COS can be an effective alternative to the use of antibiotic growth promoters to increase growth through enhancing small intestinal structure, preventing diarrhea, and modifying fecal shedding of *E. coli* and *Lactobacillus*	Liu et al. (2008)
					On the basis of the current study, COS would be most effective in improving the growth of nursery pigs when supplemented in the diet at a concentration of 158.8 mg/kg	

15	COS	/	Tiger puffer	0–2 g/kg	The Shannon index of intestnal bacterial flora was signifcantly higher in the control group than in other groups	Su et al. (2017)
					However, dietary COS supplementaton did not alter the phyla species of intestnal bacterial flora	
					There were signifcantly more *Aliivibrio* and *Bacillus* in LC3 than in other groups, and there were signifcantly fewer *Faecalibacterium*, *Escherichia*,and *Bacteroides* than in other groups at the genus level	
16	The COS (GlycoBio Company, Dalian, China) used in this study contained 40%	/	Quail chicks	75–150 mg/kg	The number of bacteria and yeast in the intestine were lower in the COS groups than in the control group	Tufan et al. (2015)
	COS and 60% cyclodextrine as a carrier				In conclusion, the addition of 75 mg/kg of COS had no adverse effect on the tested parameters, and it increased the crypt depth, villus length, and beneficially on intestinal microflora	
17	COS	/	Quail	30, 50 mg/kg	Adding 30, 50 mg/kg COS group can significantly increase the average daily gain of sputum, immune organ index, Newcastle disease antibody level, T lymphocyte transformation rate and the number of intestinal lactobacilli, significantly reducing intestinal *E. coli* and *Salmonella* quantity	Li and Liu (2013)
18	COS 1500 Da	/	Broiler chickens	50–100 mg/kg	Compared with the birds in thecontrol or chlortetracycline treatments, the birds receiving 100 mg/kg of COS had better nutrient digestibility of energy, calcium and phosphorus; higher ($P < 0.05$) concentrations of cecal *Lactobacillus*	Li et al. (2007)
19	COS	/	Broilers	5 g/kg diet added	The COS administration also decreased the number of fecal *Escherichia coli*, whereas the number of fecal *Lactobacilli* was not influenced by either COS or tylosin administration	Wang et al. (2003)

(continued)

Table 8.5 (continued)

No.	Average molecular weight	Degree of deacetylation	Model	Dose	Main conclusion	References
20	Chitosan and COS	/	Healthy mouse intervention model	10 g/kg feed	COS have, in most cases, an opposite effect compared with CS;	Koppová et al. (2012)
					Only the *Bacteroides–Prevotella bacterial* group and *Enterobacteriaceae* were influenced in the same way	
					The *Bifidobacteria* group was not influenced by the administration CS and COS	
21	COS (DP 2–6)	88%	T2DM model mice	200 mg/kg	COS treatment inhibited the reduction of occludin ($P < 0.01$) and relieved the gut dysbiosis in diabetic mice by promoting *Akkermansia* ($P < 0.01$) and suppressing *Helicobacter* ($P < 0.05$)	Zheng et al. (2018b)
22	ChOS (DP2–6)	88%	High-fat diet (HFD) model mice	200 mg/kg	16S ribosome RNA sequencing of fecal samples demonstrates that ChOS promoted the growth of beneficial intestinal bacteria remarkably and decreased the abundance of inflammogenic taxa	Zheng et al. (2018a)
23	LCS (83.9 kD)	83.9%	High-fat-diet-fed model mice	5%	Both low molecular weight chitosan and chitosan oligosaccharide significantly decreased the fecal microflora mucinase and β-glucuronidase activities	Chiu et al. (2017)
	COS (709 kD)	100%				
24	COS (1.5 kD)	90%	Colitis model mice	500 mg/kg	The COS could function as prebiotics by increasing the levels of *Bacteroidetes* and *Actinobacteria phyla*, the relative ratio of *Bacteroidetes* to *Firmicutes*, as well as common probiotics such as *Lactobacillus* and *Bifidobacterium* and inhibiting the growth of *Firmicutes* and *Proteobacteria phyla*, as well as potential pathogens such as *Enterococcus*	Long et al. (2018)
					Oral intake of COS were found to enhance the colonic concentrations of short-chain fatty acids (SCFAs)	

effectively impeded by both LMCS and standard chitosan. However, the preventative effect of COS was not noteworthy (Šimůnek 2012). Li et al. (2012d) showed that COS have apparent proliferative effects on intestinal probiotics *Lactobacillus bulgaricus* and *Streptococcus thermophilus* in certain concentrations (0.1–0.8 g/100 mL). With an increase in the COS concentration in the culture medium, its inhibitory effect on *Staphylococcus aureus* and *Salmonella* was enhanced. Furthermore, the MW of 1 kDa of COS is more significant than the 3 kDa effect on microbial growth. Cao et al. (2008) found that the minimum inhibitory concentration of COS (average molecular mass 609 Da), prepared by using hydrogen peroxide, on *Escherichia coli* and *Staphylococcus aureus* was 0.3%, and that the minimum inhibitory concentration against *Salmonella paratyphi A.* and that *Shigella sonnei* was 0.1%. Moreover, as the mass concentration of COS increased, the bacteriostatic effect gradually strengthened. Mateos-Aparicio et al. (2016) found that chitosan can maintain the stability of intestinal flora, inhibit the growth of *Bifidobacterium spp.*, *Eubacterium rectale/Clostridium coccoides*, *C.Histolyticum,* and *Bacteroides/ Prevotella,* and promote the growth of *Lactobacillus/Enterococcus.*

The addition of LMCS and COS to animal feed can provide practical application possibilities and economic value to the animal nutritional industry. Yang et al. (2012) studied dietary supplementation of 400–600 mg/kg of COS and found that it promoted growth performance and improved gut barrier function. The population of *Bifidobacteria* and *Lactobacilli* increased, while the concentration of *S. aureus* was reduced in the cecum of weanling pigs. A study conducted by Han et al. (2007) examined the extent to which nutrient digestibility, pH and growth performance responded to COS in the gastro-intestinal tracts (GI) of weanling pigs. According the results, the feeding ability of young pigs was successfully enhanced by COS, while bacteria detrimental to their health were effectively eliminated. Lian et al. (2012) showed that 0.5% of COS could significantly increase the number of *Lactobacilli* in the ileum of model pigs, and greatly reduce the amount of *Streptococcus*. Furthermore, the number of *Bifidobacteria* and *Lactobacilli* in the cecum displayed dramatically elevated levels, while *E. coli* and *streptococci* were significantly reduced. COS improved the protein metabolism and intestinal flora balance of pigs, thereby enhancing its intestinal health. Kong et al. (2014) showed that COS increased bacterial species that are considered to have beneficial properties, such as *Bifidobacterium spp.*, *Bifidobacterium breve*, *Faecalibacterium prausnitzii*, *Lactobacillus spp.*, *Prevotella*, *Fusobacterium prausnitzii*, *Methanobrevibacter smithii* and *Roseburia,* and short-chain fatty acids (SCFAs) concentrations. Furthermore, COS decreases the amount of several potentially harmful pathogens, ammonia concentrations, and branched chain fatty acids in the intestinal luminal content of weaning piglets. Xiao (2011) found that in the sample pigs the dominant jejunal flora with LMCS (average MW <5000 Da) intervention was *Lactobacillus*, while the relative numbers of *Staphylococcus*, *Escherichia,* and *Salmonella* were significantly reduced. Further real-time PCR confirmed that the numbers of *Lactobacilli* and *Bifidobacteria,* as well as *Escherichia coli* and *Streptococcus* in the small intestine of the chitosan group were reduced. Examination by Wang (2009a) of fecal samples taken from growing pigs and subsequently cultured, indicated that the application of COS failed to have any significant influence on the visual counts of *Lactobacillus*. Research conducted by Liu et al. (2008) indicates that the industries utilizing antibiotic growth enhancers can consider COS as a potential replacement application. The efficacy of COS has been proven in diarrhea prevention, the structural enhancement of the small intestine, growth stimulation, and encouraging the purging of *Lactobacillus* and *E. coli* through fecal matter. The current research suggests that the addition of a supplemental concentration of 158.8 mg/kg to the diet of nursery pigs is optimal for growth enhancement.

It was also found that chitosan and COS have similar effects when added to fish and poultry feed. Su et al. (2017) found that 0.2% COS increased the levels of *Alivivibrio* and *Bacillus* in fish, while the numbers of *Faecalibacterium*, *Escherichia* and *Bacteroides* were significantly reduced. It was established that the supplemental addition of COS to the diet of the tiger puffer could modify the intestinal bacterial flora. Furthermore, COS can enhance enzyme activity relating to digestive functions in the intestines and encourage growth. Research conducted by Tufan et al. (2015) established that compared to the control group, the intestinal yeast and bacteria levels were significantly reduced in the experimental COS groups. Besides enhancing intestinal microflora, and increasing both the length of the villus and the depth of the crypt, adding 75 mg/kg COS had no unfavorable effect on the measured limits. Li and Liu (2013) found that 30 and 50 mg/kg of low-polymerization COS could significantly increase the number of *Lactobacillus*, while reducing the numbers of *E. coli* and *Salmonella* in the gut of quail. Li et al. (2007) found an increase of *Lactobacillus* in visual counts of cultured cecum contents and fecal samples respectively, in broiler chickens that were fed with COS. Wang et al. (2003) found that the number of *E.coli*, *Bifidobacteria,* and *Lactobacilli* in the cecal contents of broilers decreased after COS intervention and that the density of ileum microvilli increased. Therefore, COS can inhibit the intestinal bacteria of broilers and promote the growth and development of intestinal microvilli.

COS affect intestinal microbes and produce metabolites such as SCFAs to indirectly regulate metabolic diseases such as HFD, type 2 diabetes, and obesity. Koppová et al. (2012) studied the effect of the administration of chitosan and COS on rat fecal microbiota. The administration of both chitosan and COS, changed the profile and structure of the microbial ecosystem of the GI of healthy rats. Compared with chitosan, COS had the opposite effect in most cases. Only the *Bacteroides Prevotella* bacterial group and *Enterobacteriaceae* were similarly influenced. Results obtained from a study conducted by Zheng et al. (2018a) indicated that the gut dysbiosis in mice afflicted with diabetes was successfully alleviated following ChOS application. This treatment prevented occludin from decreasing while facilitating the production of *Akkermansia* and impeding *Helicobacter*. The possible regulation of the metabolic pathways of intestinal microbiota by ChOS is suggested by the functional profiling supported by the particular composition of the microbiota. The presence of COS caused a significant reduction in the abundance of inflammogenic taxa, as well as an exceptional increase in the growth of beneficial bacteria in the intestines according to research by Zheng et al. (2018b). Furthermore, COS was responsible for reconstructing the microbial population, providing relief for HFD-fed mice afflicted with the metabolic syndrome. Chiu et al. (2017) found that LMCS and COS significantly decreased the fecal microflora mucinase and β-glucuronidase activities in HFD-fed rats. LMCS promoted a more significant positive improvement than COS in lipid metabolism and intestinal disaccharidase activity in HFD-induced obese rats. A study by Long et al. (2018) indicated that COS exhibited prebiotic characteristics since it was responsible for elevating the relative ratio of *Firmicutes* and *Bacteriodetes*, as well as the *Actinobaceria* and *Bacteroidetes* levels. Furthermore, COS restricted the augmentation of *Proteobacteria* and *Firmicutes* while encouraging the production of standard probiotics, such as *Bifidobacterium* and *Lactobacillus* in normal mice, as well as mice affected by colitis. Additionally, the intestinal SCFA levels were improved by the oral ingestion of COS.

In summary, it is evident that LMCS and COS can improve intestinal micro-ecology to a certain extent, promote the growth of beneficial bacteria, and inhibit harmful bacterial flora in the intestine as a micro-ecology. Moreover, as a fermentable oligosaccharide, the ability of COS to produce SCFAs is recognized (Gentile and Weir 2018). For example, acetate provides an energy source for the synthesis for fatty acids, while gluconeogenesis uses propionate as a substrate in the liver, which is the primary metabolic site. Butyrate is the primary components in epithelial cells and is used as a main source of energy. The regulation of gene expression results from the inhibition of histone deacetylases by butyrate, and to a lesser degree, propionate. Furthermore, immune responses, energy metabolism, and intestinal homeostasis are regulated by SCFAs that bind in varying degrees to intestinal G protein receptors and other targets (Fig. 8.4).

Future studies into the probiotic effect of amino oligosaccharides on intestinal microbes, or the regulation of the flora should focus on the following three aspects.

(a) A comprehensive macroscopic view of the role of amino oligosaccharides in intestinal probiotics. The study of the probiotic effects of amino oligosaccharides on intestinal microbes cannot rely solely on the positive effects of probiotics in the traditional sense, and the adverse effects on conditional pathogens. Studying the migration and overall regulation of intestinal micro-ecology by amino oligosaccharides should be at the core of this research. Given the intervention of amino oligosaccharides, then the study of intestinal microflora migration, the expression of related transcription factors and vital metabolic enzymes, and the production of metabolites on host health based on multi-omics technology (metagenomics, transcriptomics, proteomics, and metabolomics) should become the focus of the research work.
(b) Under the conditions created by COS intervention, the systematical studies based on the combination of in vitro models, animal models, and clinical trials should be considered. Considering the characteristics of prebiotics, it was evident that they are not easily digested in the intestines and can be metabolized by the flora in the gut. Regulating the composition and function of the gut flora is

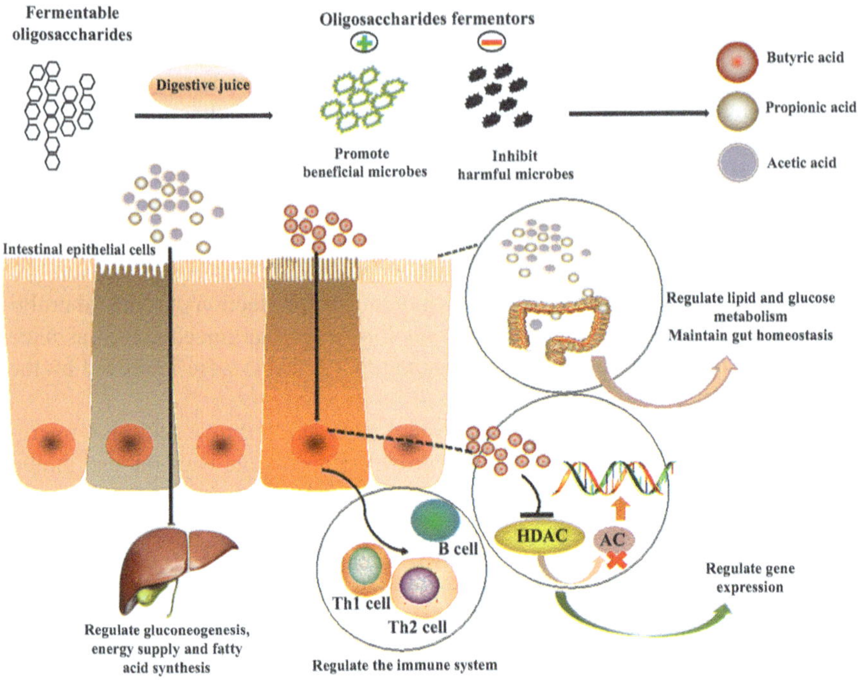

Fig. 8.4 SCFAs are generated by oligosaccharides, and can interact with host tissues on various levels. (Modified from Gentile and Weir 2018)

a fundamental feature of prebiotics. The in vitro fermentation model is mainly used to reveal the regulation of the intestinal flora structure and metabolite composition by prebiotics. The cell model can be used to evaluate the possible effects of the flora or metabolites produced during the in vitro fermentation of the prebiotics in the intestinal epithelial cells. Furthermore, it provides a tool for the direct evaluation of prebiotic function to regulate the flora. Combining animal models with clinical trials can provide a comprehensive understanding of the overall effects of prebiotic-regulated flora.

(c) Further studies should be conducted on the probiotic effect of additional amino oligosaccharide derivatives. Marine-derived amino oligosaccharides occur in a wide variety and have different structures. Therefore, the effects of various structures of amino oligosaccharides on intestinal microflora, as well as the effects of metabolites produced in hosts are also very different. Current researches are limited to the evaluation of the function of the natural structure of amino oligosaccharides. Future research should be conducted into the effects of amino oligosaccharide derivatives on intestinal microbes to develop amino oligosaccharide prebiotic products with improved health benefits.

8.2.6 Alternative Therapy for Obesity with COS

Obesity refers to the extreme amassment of body fat due to an energy imbalance (Manulu and Sutanegara 2006). It is a chronic metabolic condition related to dyslipidemia, inclusive of both hypercholesterolemia and hypertriglyceridemia (Zhang et al. 2009). Factors such as improper dietary habits and modern inactive lifestyles are the primary reasons for obesity. These elements lead to considerable energy storage without adequate exercise to expend energy (Khan 2012; Monteiro and Azevedo 2010). Moreover, severe health conditions including hyperlipidemia, cancer, coronary heart disease, hypertension, and diabetes can potentially result from obesity (Khan 2012; Mellitus and Amanti 1998). However, treating obesity is not without its challenges. In some cases it is necessary for medication to be taken for extended periods of time, potentially leading to dependency and various other side effects (Zhang et al. 2014). Therefore, the prevention obesity largely depends on the successful formulation of safe, efficient and high-quality treatment for the condition. Recently, various herbal medications and food components were subjected to several trials in an attempt to construct a new anti-obesity treatment (Riccardi et al. 2005).

COS displays exceptional potential for application as an anti-obesity and anti-dyslipidemia resource. After the report, which body weight reduction induced by COS was higher than that by orlistat (Huang et al. 2015b), a drug currently used for obesity intervention, the anti-obesity effect of COS is currently extensively investigated (Table 8.6). Studies have shown that COS significantly decreased lipid accumulation, a marker of adipogenesis, in a dose-dependent manner (Cho et al. 2008). In the phenotype, COS was demonstrated to significantly reduce the serum levels of triglyceride (TG), total cholesterol (TC), and low-density lipoprotein-cholesterol (LDLC), and alleviating lipid accumulation in the liver and adipose tissues in rodent models of obesity (Huang et al. 2015b; Kang et al. 2012; Sumiyoshi and Kimura 2006). In addition, COS increased the serum level of high-density lipoprotein-cholesterol (HDLC), responsible for removing cholesterol from cells and protecting against atherosclerosis Therefore, the potential risks of cardiovascular diseases were decreased by lowering values of the atherogenic index and cardiac risk factor (Kang et al. 2012). Among these studies, Huang et al. compared the effects of COS with varying molecular distribution (glucosamine, GLC) and COS on HFD-induced obese rats. They selected GLC, COS1 (COS, number-average MW $\leq$1000), and COS2 (COS, number-average MW $\leq$3000) for processing during the research. According to the results, the manifestation of peroxisome proliferator-activated receptor-γ (PPARγ) and liver X receptor alpha (LXRα) mRNA in the white fatty tissue was significantly lowered by COS1 and COS2, as well as GLC. Furthermore, the effects of these elements were examined in obese rats that were fed an HFD. The adipocyte distinction was impeded, leading to reduced body weight and improved dyslipidemia (Huang et al. 2015b).

Furthermore, several researchers attempted to combine functional components to assist the lipid-lowering effects. Kang et al. (2012) proposed a new diet therapy that employs a mixture of protamine and COS to treat overweight and obesity. In animal experiments, Sprague Dawley rats that were fed a HFD showed lipid

Table 8.6 Alternative therapy for obesity with COS, ChOS and its derivatives

No.	Average molecular weight	Degree of deacetylation	Dose	Model	Ending point	Proposed mechanism	References
1	46 kDa	/	300 mg/kg	C57BL/6 J mice	Increases in bodyweight, tissue weights	The binding of bile acids;	Sumiyoshi and Kimura (2006)
					Liver lipids↓	The inhibition of pancreatic lipase activity	
					The faecal bile acid and fat↑		
2	1–3 kDa	/	/	3T3-L1 cell line	Lipid accumulation ↓	The downregulated expression of adipogenic transcription factors and other specific genes	Cho et al. (2008)
					C/EBP and PPAR↓		
					Adipogenic marker proteins ↓		
3	/	/	4 mg/mL	3T3-L1 cell line	FABP, GPDH, and MDH↑	Mediated by C/EBPa and PPARγ pathway through significant downregulations of important adipogenic molecules	Rahman et al. (2008)
					NADH dehydrogenase ↓		
					ED↑		
					Residual basal triacylglycerol lipase activity↓		
4	/	/	200 mg/kg	ob/ob mice	Diet intake and body weight gain↓	Act as a potent down-regulator of obesity-related gene expression that may normalize altered plasma proteins	Kumar et al. (2009)
					Blood glucose levels ↓		
					Glucose tolerance ↑		

5	<1000 Da	>70%	100 mg/mL	Mouse 3T3-L1 pre-adipocytes	GAPDH, CEBPa, PPARg and ADIPOQ↓	Regulate the expression of IL-6 and PTGS2 genes that may inhibit adipogenesis	Bahar et al. 2011
	1000–3000 Da				FABP4, FABP5 ↓		
	3000–5000 Da						
	5000–10,000 Da						
6	3.5 kDa	/	3% COS	Mouse model of HF diet-induced obesity	Body weight gain and adipocyte size↓	Alteration of adipose tissue-specific gene expression	Choi et al. (2012b)
					Serum and hepatic lipid abnormalities↑		
7	/	98% and 88%	0, 5, 25, 50, 75, 100, 200, and 300 μg/mL	SD-rats	TG, T-CHO and LDLC↓	Interrupt the digestion and absorption of dietary lipids in the body by inhibiting pancreatic lipase activity	Kang et al. (2012)
					Fecal TG and T-CHO contents↑		
8	COS1 ≤ 1000 Da	/	COS1-H:1000 mg/kg	Obese rats	CHO↓	Inhibiting the adipocyte differentiation	Huang et al. (2015b)
	COS2 ≤ 3000 Da		COS1-M: 500 mg/kg		LDLC levels↓		
			COS1-L: 250 mg/kg		PPARγ and LXRα mRNA↓		
			COS2-H: 1000 mg/kg				
			COS2-M: 500 mg/kg				
			COS2-L: 250 mg/kg				

(continued)

Table 8.6 (continued)

No.	Average molecular weight	Degree of deacetylation	Dose	Model	Ending point	Proposed mechanism	References
9	<1000 Da	/	1 mg/mL	Male C57BL/6 J mice	Hyper-triglyceride ↓	Rebuilding the structure of the gut microbiota community	Zheng et al. (2018b)
	DP 2–6				HDL↓		
10	≤1000 Da	95.60%	COSC-L: 150 mg/kg	The obese model	Weight gain ↓	Activate the JAK2-STAT3 signaling pathway to alleviate leptin resistance and suppress adipogenesis to reduce lipid accumulation	Pan et al. (2018)
			COSC-M: 300 mg/kg		The serum total cholesterol CHO ↓		
			COSC-H: 600 mg/kg		The HDL-C levels↓		
					The expression levels of PPARγ and LXRα mRNA↓		
11	≤1000 Da	/	1 mg/mL	SD rat obesity model	The differentiation rate of fat cells ↓	Inhibiting the increase of fat cells and leptin resistance	Ding (2015)
					Triglyceride accumulation ↓		
					PPARγ and CEB/Pα mRNA ↓		
					Serum leptin levels ↓		
					Brain leptin receptor-related protein expression ↑		
12	≤1000 Da	90%	COS capsules:	SD rat obesity model	PPARγ↓T-CHO↓TG↓LDLC↓	Inhibits the expression of LXRα in adipose tissue, and thus PPARγ↓	Huang (2016)
			600 mg/kg		HDL-C↑		
			300 mg/kg				
			150 mg/kg				

13	/	/	25 mg/100 g	SD rat obesity model	Body weight↓	Relying on the characteristics of positive charges	Wang et al. (2009b)
			50 mg/100 g		CHO, LDLC↓		
			150 mg/100 g		HDLC↑		
14	/	/	0.3 g	SD rat obesity model	Inhibition of weight gain	With a positive charge on the surface, which can adsorb negatively charged fats and fatty acids, which helps to control body weight and lipid-lowering	Zhou et al. (2017)
					Liver index, TC, TG, LDLC ↓		
15	COS1 ≤ 1000 Da COS2 ≤ 3000 Da	/	COS1/COS2	SD rat obesity model	Accumulation of liver fat	Lipid profile regulation	Chen (2014)
			250 mg/kg		Improvement of hepatocyte necrosis		
			500 mg/kg		Blood glucose level, TG, CHO, LDLC↓		
			1000 mg/kg		HDL-C↑		
16	≤1000 Da	/	300 mg/kg	C57BL/6 J induced by high fat diet and ob/ob mice	T-CHO, TC, LDLC, Glu ↓	In the promotion of the JAK2/STAT3 pathway	Kong (2017)
					HDL-C↑	Mitigate the occurrence of endoplasmic reticulum stress	
					Accumulation of fat droplets in the liver↓	Improve the leptin resistance	
					ALT, AST ↓, Leptin↓		

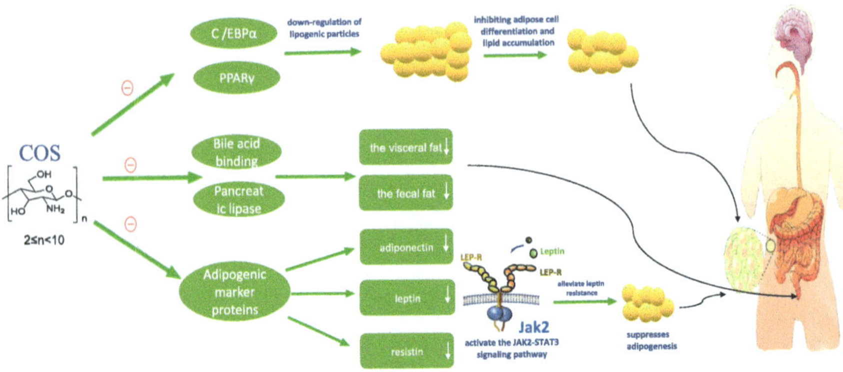

Fig. 8.5 The undelay mechanism responsible for the obesity intervention with COS. *C/EBP* CCAAT-enhancer-binding proteins, *PPAR* peroxisome proliferator-activated receptor, *Jak* Janus tyrosine kinase, *LEP-R* leptin receptor. (Modified from Muanprasat and Chatsudthipong 2017)

accumulation in the liver tissue and increased levels of TG and T-CHO in the feces. The mixture significantly reduced serum TG, T-CHO, and LDLC levels. Therefore, the results suggest that protamine and COS mixtures can be utilized as an essential dietary therapy technique to prevent overweight, obesity, and cardiovascular disease associated with hyperlipidemia.

Multiple mechanisms have been proposed to mediate the anti-obesity and lipid-lowering effects of COS (Fig. 8.5). Firstly, COS can inhibit adipogenesis in adipose tissue. COS may promote weight loss by inhibiting adipose cell differentiation and lipid accumulation, therefore reducing the amount of adipose tissue and the fat percentage in the body. For example, Huang et al. (2016) demonstrated that COS improved dyslipidemia and prevented weight gain by inhibiting the adipocyte differentiation in obese rats that were fed an HFD. Secondly, the PPAR family plays an important role in obesity intervention. According to the results obtained from some study, the brain and stomach exhibited PPAR activity, particularly PPAR-α and PPAR-γ (Kao et al. 2012). Meanwhile, the up-stream C/EBPα and the down-stream PPARγ pathways were responsible for the arbitration of the preventative effect of COS on the adipocyte distinction (Rahman et al. 2008). Thirdly, the COS is able to inhibit the activity of pancreatic lipase and bile acid binding, thereby reducing the absorption of visceral fat and increasing the excretion of fecal fat (Kang et al. 2012). Fourthly, COS significantly down-regulated adipogenic marker proteins, such as leptin, adiponectin, and resistin. For example, Cho et al. (2008) reported that compared to other types of COS, LMCS (1–3 kDa) were more effective in inhibiting adipocyte differentiation in 3T3-L1 cells. The undelay mechanism of the intervention may be related to the down-regulation of the adipogenic marker proteins. Kumar et al. (2009) found the anti-obesity and anti-diabetic properties of COS on *ob/ob* mice by conducting a plasma proteomic difference analysis. The results suggest that COS may be a potent downstream regulator of adipogenic marker proteins' gene expression, prompting the modification of plasma proteins in *ob/ob* mice to overcome metabolic disorders in obesity. Moreover, leptin resistance was recovered by COS with multiple signaling. COS can improve the sensitivity of leptin receptors in the rat brain and activate the JAK2-STAT3 signaling pathway to alleviate leptin resistance. This process ultimately suppresses adipogenesis to reduce lipid accumulation (Pan et al. 2018).

8.2.7 The Liver Protection Ability of COS

Liver diseases, such as liver cirrhosis, alcoholic liver, fatty liver, hepatitis B, hepatitis A, hepatitis C, liver cancer, and more are currently among the most detrimental challenges affecting human health (Powell et al. 2005, 2010).

In 1998, Kim and his colleagues showed that COS has the ability to prevent liver disease by partially lowering serum cholesterol levels (Kim et al. 1998). Since then, many studies have demonstrated the protective effects of COS on chemical or drug induced liver damage, fatty acid liver disease, and liver cancer. The liver protection characteristics of COS are summarized in Table 8.7. The most common liver damage is caused by chemicals or drugs. Yang Yan et al. found that the pretreatment with COS, $GlcNH_2$ and GlcNAc could effectively protect mice from the toxic effects of CCl_4. COS, $GlcNH_2$, and GlcNAc could reduce the activity of AST and ALT in serum, alleviating liver lipid peroxidation. Furthermore, these components can significantly reduce serum creatinine and uric acid levels and inhibit lipid peroxidation in the kidneys (Yang et al. 2006). Similarly, the protective effect of COS on hydrogen peroxide (H_2O_2)-induced oxidative stress in human embryonic hepatocytes (L02) was studied by Xu et al. The results showed that the survival rate of cells induced by H_2O_2 was significantly restored after a 24 h treatment with COS. This restorative effect may be related to the antioxidative properties of COS. Moreover, COS prevented apoptosis that was caused by H_2O_2. This result was accomplished by inhibiting the cleavage of polyadenosine diphosphate and increasing the expression of the anti-apoptotic protein B-cell lymphoma-extra large (Bcl-XL). The results indicated that COS can effectively protect L02 cells from oxidative stress, suggesting COS may hold clinical significance in oxidative stress-related liver injury (Xu et al. 2010). Furthermore, COS display protective effects on ERT-butyl hydroperoxide (T-BHP) induced hepatocyte injury in a dose-dependent manner. COS can increase the cell viability of hepatocytes that are exposed to T-BHP and significantly reduce ROS production and lipid peroxidation. In addition, COS help increase moisture content and antioxidant enzyme activity in the liver. Therefore, COS protect liver cells from T-BHP induced oxidative damage by inhibiting the production of ROS and over-activated lipids (Senevirathne et al. 2011).

Regarding preliminary cancer intervention, Xu et al. reported on the inhibitory effect of COS on human hepatoma cells (SMMC-7721) and its mechanism. The results showed that COS could induce the apoptosis of SMMC-7721 cells in a dose-dependent manner and increase the cleavage of the polymerase. The mechanism was likely linked to the regulation of apoptosis-promoting protein (Bax) and the initiation of the apoptosis process (Xu et al. 2008).

Table 8.7 Liver protection effects of COS, ChOS and its derivatives

No.	Function	Average molecular weight	Degree of deacetylation	Dose	Model	Ending point	Proposed mechanism	References
1	Liver protection	Water-soluble	/	1%	Ethanol induced liver injury	Plasma cholesterol level (↓)	Partly decrease serum cholesterol levels;	Kim et al. (1998)
						Prevent alcoholic liver disease	Prevent alcohol from producing a fatty liver	
2		COS(MW = 3100)	/	1.5 g/kg	Carbon tetrachloride induced liver injury in mice	Induction of liver MT expression;	Avoid oxidative damage	Yang et al. (2006)
		D-glucosamine ($GlcNH_2$)				Inhibition of lipid peroxidation;		
		N-acetyl-d-glucosamine (GlcNAc)				Restoring of total antioxidant capability		
3		/	Below 5%	0.8 mg/mL	Human hepatoma cells (SMMC-7721)	Induce apoptosis of SMMC-7721 cells	Up-regulate pro-apoptotic protein Bax;	Xu et al. (2008)
							Trigger the cells a start-up of the apoptosis program	
4		300–500 kDa	90%	0.1–0.4 mg/mL	Human embryonic hepatocytes (L02 cells)	Protect L02 cells against oxidative stress	The inhibition of the cleavage of poly (adenosine diphosphate-ribose) polymerase;	Xu et al. (2010)
							Increase expression of the anti-apoptotic protein Bcl-XL	
5		COS I (5–10 kDa)	90%	100, 200, 500 and 1000 μg/mL	Chang liver cells	ROS generation and lipid peroxidation (↓)	Protect Chang liver cells against oxidative damage	Senevirathne et al. (2011)
		COS II (1–5 kDa)				GSH content and antioxidant enzyme activity (↑)		
		COS III (below 1 kDa)						

6		100 kDa	90%	80 mg/kg per day	Blast injury in male Kunming mice	Reactive oxygen species (↓)	Inhibition of oxidative damage	Liu et al. (2018)
						MDA5 and IREα (↓)	Anti-apoptosis activity	
						Cell apoptosis (↓)		
						Bax and Caspase-3 (↓)		
						ADMA and p38 (↓)		
						Superoxide dismutase-1 (↑)		
						Bcl-2 and Caspase-8 (↑)		
						DDAH1 (↑)		
1	Lipid lowering effect	/	/	4 mg/mL	Adipocyte 3T3-L1	Cell differentiation was significantly inhibited by COS at the concentration of 4 mg/mL	C/EBPα and PPARγ pathway (downregulations of fatty acid binding protein and glucose transporter 4)	Rahman et al. (2008)
2		<4 kDa	/	1% COS diet	Male C57BL/6 N mice	Help ameliorate HF diet-induced weight gain	COS reversed the changes in the expression of these genes	Choi et al. (2012b)
				3% COS diet		Improve serum and liver lipid profile abnormalities	(Mmps, Timp1, and Ctsk) in the obese mice	
3		<1000 Da	90%	500 mg	Healthy men (11 smokers and 8 non-smokers)	Total plasma cholesterol and LDL cholesterol (↓)	Low MW chitooligos-accharides would be an effective dietary supplement for lowering cholesterol level	Choi et al. (2012a)
4		High molecular weight chitosan (712.6 kDa)	HMWC 90.4%	5% (w/w) chitosan	Male Sprague-Dawley rats fed with high fat diets	The liver and serum LPL activities of rats in LMWC group were higher than those of HMWC group	The increase in activities of enzymes regulating lipid metabolism might be an important way contributing to the higher hypolipidemic activity of LMWC	Zhang et al. (2012)
		Low molecular weight chitosan (39.8 kDa)	LMWC 90.2%					

(continued)

Table 8.7 (continued)

No.	Function	Average molecular weight	Degree of deacetylation	Dose	Model	Ending point	Proposed mechanism	References
5		5–10 kDa	70%	0, 600, 1200, 2400, 4800 mg/mL	Mouse 3T3-L1 preadipocytes	COS inhibited the de-methylation of three CpG sites (at 252, 262 and 295) in the LEP promoter	The action of COS in three CpG sites is considered adequate for alteration in LEP gene expression	Bahar et al. (2013)
						Blocking of de-methylation of CpG-52 by COS most likely inhibited leptin expression.		
						Inhibits LEP gene expression in differentiating pre-adipocytes		
6		Chitosan (CTS) 350 kDa	CTS 96.2%	WSC (225 mg/kg/day)	Hyperlipidemic rats	CTS and WSC microspheres were more effective in improvinghyperlipidemia in rats than common CTS and WSC	CTS and WSC microspheres were found to be more effective in improving hyperlipidemia in rats than common CTS and WSC	Tao et al. (2013)
		Water-soluble chitosan (WSC) 200 kDa	WSC 85%	CTS (450 mg/kg/day)		Serum lipid levels and plasma viscosity (↓)		
						SOD levels (↑)		
7		/	/	5% (w/w) COS	Male Sprague-Dawley rats fed with high fat diets	Inhibit the increase of body weight	Regulate lipid levels	Xia et al. (2013)
						Plasma TC, TG and LDLC levels (↓)		
						HDL-C level (↑)		
8		1000 Da	/	1 μg/mL, 10 μg/mL,	3T3-L1 preadipocytes	Inhibit the differentiation of 3T3-L1 preadipocytes;	Be related to the down-regulation expression of PPARc and C/EBPα	Kong et al. (2017)
				100 μg/mL		PPARc and C/EBPa (↓)		

9		/	/	0.5 mg/mL	The L02 cells (Fatty acid-induced)	Inhibit the expression of SREBP-1c and downstream FAS	TG contents and related genes were decreased in the hepatocyte model of nonalcoholic fatty liver disease and in the liver in rats fed with high-fat diet	Wang et al. (2016)
						Reduce the synthesis of TG		
						Increase the expression of CPT1A		
						Accelerate the breakdown of TG		
10		/	/	20 and 40 mg/kg	Hyperlipidemia model	TG and LDLC (↓)	Appropriate supplementation of COS may improve lipid metabolism, promote immune organ development, and inhibit lymphocyte apoptosis in broilers	Li et al. (2017a)
						Periarterial lymphatic sheath area of the spleens (↑)		
						Average number of bursa of Fabricius nodes (↑)		
11		Water-soluble chitosan (WSC)	82%	60 mg	A biopharmaceutical model of the digestive tract	WSC samples exhibit great fat- and cholesterol-binding capacities	WSC as an excellent resource in the development of functional foods against obesity for its adsorption, electrostatic binding and entrapment of cholesterol, fat, sterols and triglycerides in the diet	Jin et al. (2017)
		Mw (1000, 3000, 5000, 7000 and 9000 Da)				There is a significant correlation between the binding capacity of WSC and its Mw		

(continued)

Table 8.7 (continued)

No.	Function	Average molecular weight	Degree of deacetylation	Dose	Model	Ending point	Proposed mechanism	References
12		Mw ≤ 1000 Da	>95%	500 mg/kg	ApoE−/− mice (HFD) (21% fat and 0.15% cholesterol)	PCSK9(↓)	FOXO3a levels are up-regulated,leading to a reduction in the PCSK9 promoter binding capacity of HNF-1α and thus suppressing PCSK9 gene expression;	Yang et al. (2018b)
				50–200 μg/mL	HepG2 cell line	LDLR(↑)	Up-regulating LDLR levels, and enhancing the lipid droplets in HepG2 cells	
						Lipid intake by liver cells (↑)		
						The concentration of lipid in the serum (↓)	Decreased expression of the PCSK9 gene was also contributed to down-regulation of SREBP-2 by COS	
13		1000 Da	96%	Low 150 mg/kg	Sprague-Dawley rats (Hyperlipidemia model)	CYP7A1, LXRA, and PPARα (↓)	Improve lipid metabolism;	Jiang et al. (2018)
				Medium 300 mg/kg		HMG-CoA reductase and SREBP2 (↓)	Promote the conversion of cholesterol into bile acid;	
				High 600 mg/kg		LDLR (↑)	Reduce the de novo synthesis of cholesterol	
14		~1000 Da	≥95%	4 mg/mL	The human hepatocellular carcinoma cell line (HepG2) (OA-induced)	Hepatic TG, FFAs and LDLC content (↓)	$(GlcN)_2$ provides the best active effect on anti-hyperlipidemia and steatosis regulation via decreasing fatty acid uptake and TG synthesis in HepG2 cells	Li et al. (2018)
						DGAT2, LXRα, PPARγ, PXR and CD36 (↓)		

Studies showed that chitosan increases fat excretion in feces and inhibits pancreatic lipase activity, leading to the prevention of fatty liver formation (Guerciolini et al. 2001). Compared to chitosan, Because COS was easily dissolvable in water and possessed a lower density, the intestines could assimilate it more readily. Data accumulated from a cultural cell model, as well as animal and human clinical trials, indicate that COS and their derivatives are effective in reducing blood lipid levels and providing protection from excessive fatty acid-induced liver damage (Cho et al. 2008; Choi et al. 2012a, b; Huang et al. 2015b; Kong et al. 2017).

Oleic acid or palmitic acid was used for inducing the simulation of a high-fat environment in vitro. Wang et al. (2016) indicated that the TG content in L02 hepatocytes was elevated and the steatosis was aggravated by oleic acid treatment. Following the intervention of COS, the expression levels of SREBP-lc and fas cell surface death receptor (FAS) were significantly lower than those in the OA-model group. The level of the CPT1A gene was elevated, while the TG content was reduced. It is speculated that COS may inhibit TG synthesis by obstructing the expression of the lipid synthesis-related genes SREBP-lc and FAS. Furthermore, COS accelarated TG decomposition by up-regulating the gene CPT1A, eventually alleviating TG accumulation in the fatty liver cell model.

Employing a broiler model, Li et al. (2017a). studied the effects of COS on lipid metabolism, immune organ development, and lymphocyte apoptosis. Following treatment with COS, serum TG and LDLC levels were significantly reduced, as were the lymphatic sheaths around the artery. The spleen area and the average number of bursa dramatically increased, and the levels of serum total protein (TP) and HDLC in the bursa were higher. Moreover, the number of caspase-3 positive cells in both the bursa and spleen were considerably reduced. These results indicated that the addition of COS improved lipid metabolism, promoted immune organ development, and inhibited lymphocyte apoptosis in broilers (Li et al. 2017b).

Due to suffering from obesity and non-alcoholic fatty liver disease (NAFLD), Wang et al. studied the effects of COS on hepatic TG metabolism by using HFD mice. It is possible that COS inhibits the expression of SREBP-lc and its downstream FAS gene. COS reduces the synthesis of TG, increases the expression of CPT1A gene, and accelerates the decomposition of TG (Wang et al. 2016). NAFLD is closely related to cholesterol homeostasis (Caballero et al. 2009). Jiang et al. studied the cholesterol-lowering effect of COS and its possible mechanism. The study concluded that the role COS plays in lowering cholesterol may be related to the following two pathways. The first refers to the process when the manifestation of the LXRα, the receptor-α as activated by the peroxisome proliferator, and cholesterol 7α-hydroxylase (CYP7A1), is up-regulated when cholesterol is transformed into bile acid. The second pathway followed the method down-regulating both the sterol response element binding protein-2 and the gene presentation of 3-hydroxy-3-methylglutaryl coenzyme A reductase (HMGCR) to reduce the generation of additional cholesterol. Furthermore, the process required the up-regulation of the low-density lipoprotein receptor (LDLR) (Jiang et al. 2018). Similarly, Yang et al. explored the molecular mechanism of cholesterol lowering effect of COS by using COS in ApoE $^{-/-}$ male mice. Both lipid deposition in the aorta, and hepatic proprotein convertase subtilisin/kexin type 9 (PCSK9) protein levels were significantly reduced. For in vitro mechanism confirmation, COS could regulate lipid transfer by up-regulating forkhead box O3 (FOXO3a), and it decreased the binding capacity of hepatocyte nuclear factor-1α (HNF-1α) (PCSK9 promoter). Therefore, COS inhibited PCSK9 gene expression, up-regulated LDLR levels, and enhanced lipid droplets in HepG2 cells. Furthermore, COS reduced the expression of the PCSK9 gene by down-regulating sterol regulatory element-binding protein 2 (SREBP-2). By combining with HNF-1α and SREBP-2 silencing, the results demonstrated that COS treatment synergistically altered PCSK9 expression and LDLR activity through a combination of HNF-1α and SREBP-2 (Yang et al. 2018b).

In clinical trials, Choi et al. investigated the lipid-lowering effect of COS and whether oral COS can reduce the plasma level of lipid in smokers and non-smokers. They selected 19 people to participate in the experiment, and the results suggest that LMCS would be an useful dietary supplement for lowering the cholesterol level (Choi et al. 2012a).

Most current studies used COS mixtures with varying MW and DD. The MW, DP, and DD are closely related to the biological activity of COS. The role of singular DP COS on lowering the effect of lipids remains unclear. Screening the active ingredients of oligosaccharides and its lipid-lowering effect, and exploring the mechanism by which this is accomplished are crucial. Both animal and cell culture models were employed by Li et al. to examine the remedial influence exerted on lipid accumulation by unrefined COS, as well as an additional five specific COS ($(GlcN)_{2\text{–}6}$). The aggregation of fatty deposits in the liver were considerably reduced by $(GlcN)_2$. Furthermore, significantly lower levels were evident in the protein and mRNA manifestations of PPARγ, diacylglycerol acyltransferase 2 (DGAT2), LXRα, the pregnenolone X receptor (PXR), and the cluster of differentiation 36 (CD36) in the lipid uptake pathway and TG synthesis (Li et al. 2018).

In a summary of the lipid-lowering effects (Fig. 8.6), COS mainly influence the metabolism of TG and TC including the specific pathway related to uptake, synthesis, metabolism, decomposition, and more.

8.2.8 The Utilization of COS as an Alternative Therapy for Diabetes Mellitus

Diabetes has become the third largest chronic diseases that seriously endangers human health. Diabetes is a metabolic disorder characterized by hyperglycemia, including type 1 diabetes, which is mainly caused by absolute insulin deficiency, and non-insulin-dependent diabetes mellitus (type 2 diabetes). Type 2 diabetes is mainly caused by insulin deficiency and insulin resistance (Pandey et al. 2011). About 200 million people worldwide suffer from this disease, and more than 90% of them have type 2 diabetes. Insulin resistance and islet β-cell function defects are the basic pathological features of type 2 diabetes (Bellamy et al. 2009; Hussain et al. 2007). The marine pharmaceutical industry is one of the critical areas for future development. It is vital to actively explore the possible ways to prevent type

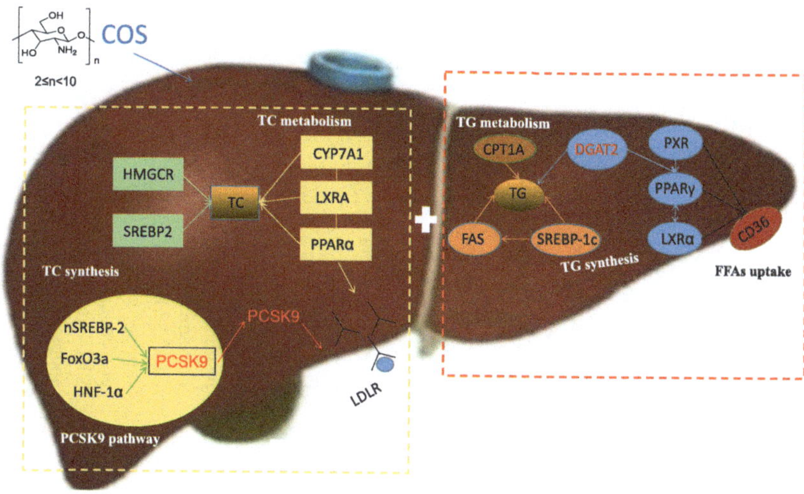

Fig. 8.6 The related signaling pathway relating to the lipid-lowering effects of COS. *CD36* cluster of differentiation 36, *DGAT2* diacylglycerol acyltransferase 2, *HMGCR* 3-hydroxy-3-methyl-glutaryl-CoA reductase, *PCSK9* proprotein convertase subtilisin/kexin type 9, *FoxO3a* forkhead box O3, *HNF* hepatocyte nuclear factor, *CYP7A1* cholesterol 7 alpha-hydroxylase, *LXRα* liver X receptor, *PPARγ* peroxisome proliferator-activated receptor-γ, *CPT1A* carnitine palmitoyltransferase I, *PXR* pregnenolone X receptor, *SREBP2* sterol regulatory element binding proteins, *TC* total cholesterol, *TG* total triglyceride, *FAS* Fas cell surface death receptor

2 diabetes. COS possesses a similar structure to acarbose which is commonly used in treating diabetes (Kim et al. 2014a, b). Results regarding the anti-diabetic effect of COS are summarized in Table 8.8 (Huang et al. 2015b; Jo et al. 2013, 2014; Ju et al. 2010; Kang et al. 2016; Karadeniz et al. 2010; Katiyar et al. 2011; Kim et al. 2014a, b; Lee et al. 2003; Wu et al. 2017b; Yu et al. 2017; Zheng et al. 2018a, b).

The hypoglycemic effect of COS was first investigated in 2003. Treatment with COS increased glucose tolerance, and reduced TG in Sprague-Dawley rats with diabetes induced by a streptozocin (STZ) injection (i.p.) (Lee et al. 2003). Moreover, it was possible for COS to increase the insulin sensitivity index and oral glucose tolerance, which contributed to promoting the proliferation of β cells. Regarding the mechanism, stimulated insulin release, up-regulation of glucose transporter-2 (GLUT-2) mRNA expression, and the prevention of STZ-induced apoptosis play a vital role in COS protected INS-1 cells (Ju et al. 2010). Carbohydrate hydrolase is a key target in diabetes intervention. COS could inhibit various types of carbohydrate hydrolase, subsequently improving the control of fasting blood glucose levels and reducing postprandial glucose levels in both mice models and clinical trials (Jo et al. 2014, 2013; Kim et al. 2014a, b). A Korean study discovered that COS could inhibit intestinal α-glucosidase and regulate the glucose uptake transporters sodium-glucose co-transporters (SGLT1) and GLUT2. The same study found an additional related signaling pathway responding to the anti-diabetes effects. COS supplements could enhance adipocyte differentiation, the active PPARγ pathway and its

Table 8.8 Alternative therapy for diabetes mellitus with COS, ChOS and its derivatives

No.	Average molecular weight	Degree of deacetylation	Dose	Model	Ending point	Proposed mechanism	References
1	20 kDa	/	0.3% COS	Sprague-Dawley mice (High-energy diet and streptozotocin induced)	Glucose levels ↓	COS has a TG-lowering effect in diabetic rats, and COS reduces signs of diabetic cardiomyopathy, such as vacuolization of mitochondria and separation and denaturation of myofibrils	Lee et al. (2003)
					Glucose tolerance↑		
					Glucose-induced insulinsecretion↑		
					Triglyceride (TG) levels ↓		
2	1.5 kDa	86.5%	500 mg/kg	Sprague-Dawley mice (High-energy diet and streptozotocin induced)	Fasting blood glucose (FBG) ↓	COS plays an important role in INS-1 cells by promoting proliferation, increasing glucose-stimulated insulin release, up-regulating GLUT-2 mRNA expression, and preventing STZ-induced apoptosis	Ju et al. (2010)
					Fasting insulin (FINS) ↓		
					Insulin sensitivity index (ISI) ↑		
					Oral glucose tolerance ↑		

3	<1 kDa	/	/	HIT-T15 β-cell	Cell viability ↑	Free radical scavenging activity	Karadeniz et al. (2010)
	1–3 kDa						
	3–5 kDa						
	5–10 kDa						
4	/	/	10 mg/kg	Healthy male Swiss albino mice	Serum total cholesterol ↓	Free radical scavenging activity and lipid profiles regulation	Katiyar et al. (2011)
			5 mg/kg		Triglyceride ↓		
					Low density lipoprotein cholesterol ↓		
					Very low density lipoprotein cholesterol ↓		
					High density lipoprotein cholesterol ↓		
5	<1 kDa	/	/	Sprague-Dawley mice	Postprandial blood glucose level ↓	Inhibition of carbohydrate hydrolase α-glucosidase	Jo et al. (2013)
	1–10 kDa						
	>10 kDa						
6	1000 Da	/	40 g/kg	db/db mice model	Fasting blood glucose level ↑	Inhibition of carbohydrate hydrolase α-glucosidase	Kim et al. (2014b)
					HbA1c level ↓		
					Total cholesterol ↓		
					Cecal weight ↑		
7	1000 Da	/	500 mg	13 healthy volunteers	Blood sugar level ↓	Inhibition of carbohydrate hydrolase α-glucosidase	Jo et al. (2014)

(continued)

Table 8.8 (continued)

No.	Average molecular weight	Degree of deacetylation	Dose	Model	Ending point	Proposed mechanism	References
8	/	/	1500 mg/day	Healthy volunteers	Blood pressure ↓	Improve serum glucose levels, HbA1c, pro-inflammatory cytokines such as IL-6 and TNF-α, and plasma adiponectin	Kim et al. (2014a)
					Blood lipid level ↓		
					Plasma HbA1c level ↓		
9	1000 Da	/	1 mg/mL	Caco-2 cells	Intestinal α-glucosidase ↓	Inhibiting intestinal alpha-glucosidase and different glucose transporters SGLT1 and GLUT2;	Yu et al. (2017)
			10 mg/mL	3T3-L1 cells	Glucose uptake ↑		
					Glucose transporter activity and gene expression ↓	Enhance adipocyte differentiation, PPARγ expression and its target gene	
10	1000 Da	/	/	db/db mice	Sucrose ↓	Inhibition of carbohydrate hydrolase α-glucosidase	Kang et al. 2016
					Glucoamylase activity ↓		

11	/	/	/	CD-1 mice	Lower blood sugar ↑ Total cholesterol ↑ Triglycerides ↑ HDL-cholesterol ↑	The MAPK signaling pathway is attenuated by the IL-1β-Erk/p38-histone H3 pathway in T2D model mice	Wu et al. (2017b)
12	/	/	200 mg/kg	C57BL/6 J mice	Plasma insulin levels ↑ Triglycerides ↓ High-density lipoprotein cholesterol concentration ↓	Reduce LPS concentration in the circulatory system	Zheng et al. (2018a)
13	<1 kDa	88%	200 mg/kg	C57BL/6 J mice or diabetic db/db mice	Blood glucose ↓ Fat production ↓ Adipocyte differentiation↓	Inhibits reduction of occludin and promotes intestinal loss control in diabetic mice by promoting *Akkermansia* and inhibiting *Helicobacter pylori*	Zheng et al. (2018b)

downstream target genes like fatty acid binding protein 4 (FABP4), adiponectin, and GLUT4, subsequently increasing the glucose tolerance and maintain the blood glucose levels (Yu et al. 2017).

With developments in COS bio-manufacturing, additional types of mono-DP COS were obtained. Wu et al. (2017b) reported that *N*-acetyl-chitobiose ameliorated metabolism dysfunction through the Erk/p38 MAPK signaling pathway in mice suffering from type 2 diabetes. Furthermore, *N*-acetyl-chitobiose sufficiently regulated lipid metabolism by decreasing the T-CHO (~26.5%) and TG (~16.1%), and increasing the HDL-cholesterol (~107.2%), as well as lowering inflammation factors including TNF-α, IL-1β, and NF-κB. Further studies should test a greater variety of mono-DPs COS, as well as investigate an enhanced structure-function relationship.

The derivatives of COS were reported to have anti-diabetic effects as well. COS chelating vanadium (COS-V) and COS chelating chromium (COS-Cr) can alter diabetic symptoms such as weight loss, polydipsia, polyphagia, the decrease of both blood and urine glucose levels, as well as increase glucose tolerance. Moreover, COS-Cr and COS-V provide protection to the structures of the liver, kidneys, and spleen (Liu 2007; Qu 2012; Wei 2012). Research successfully demonstrated that the anti-diabetic effect of the COS chelated complex was more evident than that of only COS. However, the subsequent mechanism remains unclear.

Results indicated that the treatment and prevention of diabetes mellitus could be beneficially impacted by the pharmaceutical properties of both COS and chitosan. Figure 8.7 shows the potential mechanisms responsible for the anti-diabetic effect of COS.

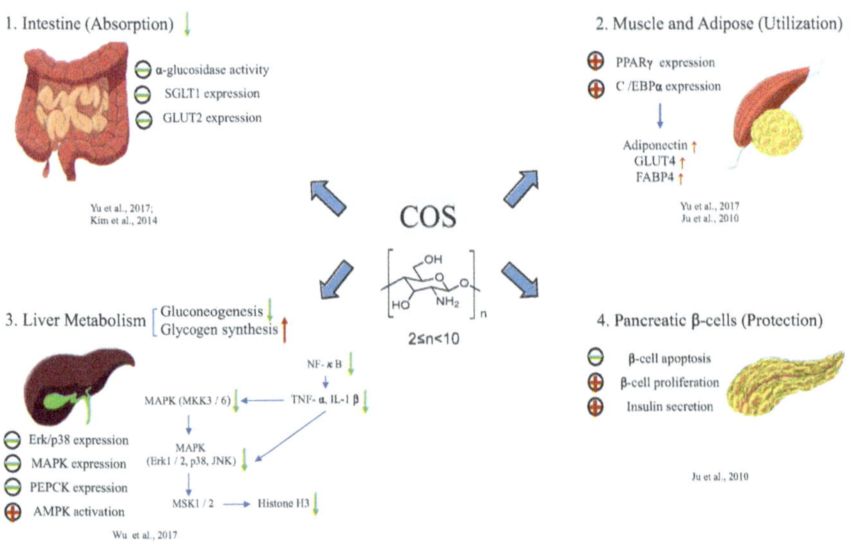

Fig. 8.7 Possible mechanisms of alternative therapy for diabetes mellitus with COS. COS regulated the blood glucose by impeding glucose absorption, promoting glucose utilization, altering liver metabolism and protecting pancreatic β-cells. The activation and suppression by COS are denoted by negative (−) and positive (+) symbols. *FABP4* fatty acid binding protein 4, *GLUT* glucose transporter, *PEPCK* phosphoenolpyruvate carboxykinase, *SGLT1* sodium-glucose co-transporters. (Modified from Muanprasat and Chatsudthipong 2017)

8.2.9 The Neuroprotective Effects of COS

Neuronal disorder is the primary cause of cognitive and motor diseases. This disorder is commonly observed in Alzheimer's disease (AD), Parkinson's disease, Huntington's disease, neurotrophic infections, neoplastic disorders, prion diseases, multiple sclerosis, amyotrophic lateral sclerosis, stroke, and traumatic brain and spinal cord injuries (Hao et al. 2017). As neuronal disorder is mainly caused by apoptosis, oxidative stress, and *β*-amyloid (A*β*) aggregation, the current therapeutic strategy is to alleviate these disorders, thus minimizing disturbance to neuronal function (Kostrzewa and Segura-Aguilar 2003). However, despite the synthetic neuronal protection agents displaying specific side effects, it is necessary for patients to continue using them for the rest of their lives (Tucci and Bagetta 2008). Therefore, developing new anti-neuronal disorder agents that are effective with a low toxicity level is crucial. Evidence obtained from numerous studies indicate that COS possess neuronal protective properties, leading to the possibility that it can provide new solutions for the treatment of anti-neuronal disorder (Table 8.9).

A crucial factor in the progression of AD is known as A*β*. Aβ refers to an accumulating piece of protein that is capable of self-association. This process induces neuronal injury and ultimately death, through the production of neurotoxic fibrils that stimulates a sequence of detrimental cellular incidents (Hamley 2012). Dai et al. (2015) investigated the effect of COS (MW < 1000, DD 91.3%) in suppressing A*β*'s aggregation. The results indicated that a COS volume of 5 mg/mL was successful in prohibiting transformation of A*β* into a β-sheet structure. Furthermore, the application of COS at specific dosage levels was able to substantially impede the production of fibrils. For the mechanism, it was assumed that COS molecules altered the environment of the surrounding water molecules, rendering it more energetically favorable for A*β* aggregation. β-secretase is an aspartic protease also known as BACE. This enzyme cleaves an easily accessible site at the luminal side of the A*β* precursor protein, and its activity is the rate-limiting step in A*β* peptide production in vivo (Vassar et al. 2014). Lee et al. (2009b) found that COS could significantly inhibit β-secretase activity, while COS with MW ranging from 3 to 5 kDa exhibited the optimal inhibitory effect. Furthermore, it was evident that COS could inhibit β-secretase in a DD-dependent manner. The obstruction pattern of the inhibitor was found to be noncompetitive by a Dixon plot, and Ki was 3.87–6.47 mM.

AD is attributed to various contributory effects and the functionality of pathogens. Oxidative stress is a dominatingly responsible for the pathogenesis of neuron death (Markesbery 1997). Huang et al. (2015a) found that COS with a DP below ten could attenuate Cu^{2+}-induced cellular oxidative stress damage. An increase in cell viability, a decrease in lactate dehydrogenase (LHD) release, and a reduction in the level of H_2O_2 and intracellular ROS were observed following the administration of COS. Furthermore, activation of the Nrf2-ARE pathway and a substantial increase in the expression of the anti-oxidant gene HO-1 was apparent. Nidheesh et al. (2016) used monocrotophos affected *C.Elegans* to examine the anti-oxidative and neuronal protective effect of COS. Results showed that with the administration of

Table 8.9 Neuro-protection effect of COS, ChOS and its derivatives

No.	Average molecular weight	Degree of deacetylation	Dose	Model	Ending point	Proposed mechanism	References
1	1–10 kDa, and sulfated	90,75,50%	0.6–6 μM	β-secretase	β-secretase activity↓	High degree of deacetation shows higher inhibition ability;	Byun et al. (2005)
						For the molecular weight,3–5 kDa is better, IC_{50} = 3.87–6.47 μm	
2	800 Da	92.30%	0.5–2 mg/mL	Glutamate-induced rat primary hippocampal neurons apoptosis	Apoptosis↓	Decrease intracellular calcuim levels;	Zhou et al. (2008)
						Decrease caspase-3 activity	
3	<1000 Da	90%, 50%	0.5–4 mg/mL	AChE, $A\beta_{23-25}$ induced PC12 cell lines	IC_{50} 2.59, 1.67, 3.52, 1.98, 2.93, >4 mg/mL	Not mentioned	Lee et al. (2009b)
	1000–5000 Da				AChE expression and activity↓		
	5000–10,000 Da						
4	800 Da	92.30%	1.5, 3 mg/kg	Regenerate crushed common peroneal nerve	Levels of electrophysiological properity of animals↑	Regulate N-cadherin expression;	Gong et al. (2009)
					Reconstruction of injured common peroneal nerve	Improve neuronal differentiation and neurite growth	
					Nerve regeneration		

5	800–3000 Da	90%	/	AChE	IC_{50} 56.5 ± 0.26, 24.1 ± 0.39 and 9.2 ± 0.33 μg/mL	AE-COS, non-competetive inhibitor against AChE;	Yoon et al. (2009)
						DMAE-COS, DEAE-COS, competetive inhibitor against AChE	
6	800 Da	92.30%	3–6 mg/kg	Sciatic nerve crushed rats	Withdrawal reflex lantency↓	Regenerate nerve	Jiang et al. (2009)
					Sciatic function index↓		
					Recovery index of CMAPS↑		
					Muscle wet weight ratio↑		
					CSA of gastrocnemius muscle fibers↑		
7	800 Da	92.30%	500 μg/mL	Glucose deprivation-induced primary cortical neuron cells apoptosis	Cell viability↑	Activation of PI3K/AKT and MEK/ERK	Xu et al. (2011)
					Apoptosis↓		
					TUNEL positive cells↓		
					Bcl2/Bax↑		
					Cleavage of caspase3↓		

(continued)

Table 8.9 (continued)

No.	Average molecular weight	Degree of deacetylation	Dose	Model	Ending point	Proposed mechanism	References
8	<1400 Da	>90%	200 μg/mL	A$\beta_{1\text{-}42}$ induced primary neuronal cell death	Cell viability↑	Decrease caspase3 activity;	Dai et al. (2013)
					ROS↓	Decrease JNK phospholation	
					Lipid preoxidation↓		
					LHD release↓		
9	<2000 Da	/	100, 200, 400, 800 mg/L	$CuCl_2$ treated SH-SY5Y cells	Cell viability↑	Activation Nrf2-ARE pathway;	Huang et al. (2015a)
					Cellular oxidative stress↓	Increase antioxidant gene HO-1;	
					Apoptosis↓	Attenuate oxidative stress	
					LHD release↓		
10	<900 Da	0%	100, 200, 400 μg/mL	Glutamate influenced PC12 cell	Cell vialibity↑	Down-regulate caspase3 activity;	Hao et al. (2015)
		100% peracetylation			LHD leakage↓	Attenuate cell death	
					ROS production↓		
					MMP↑		
11	<1000 Da	/	0.2 mM	Monocrotophos affected *C.elegans*	Life span↑	Attenuate reduction of CaE activity;	Nidheesh et al. (2016)
					Egg laying↑	Reduce oxidative stress;	
					Brood size↑	Increase dopamine content;	
						Regenerate neurons	

12	/	/	200, 400, 800 mg/kg	Aggregrated $A\beta_{1-42}$ affected mice	Performance in water maze test and passive avoidance test↑	Reduce apoptosis;	Jia et al. (2016)
					GSH-Px↑	Reduce inflammation	
					SOD↑		
					MDA↓		
					Hippocampal neuronal morphological change↓		
13	/	/	0.5–5 mg/mL	$A\beta42$ affected cortical neuronal primary cell	Cell viability↑	Inhibite aggregration of Aβ42 into B-sheet structure;	Dai et al. (2015)
					Apoptosis↓	Inhibit Aβ42 fibril formation;	
						Disrupt performed fibrils	
14	/	/	5–10 mg/kg	Elevation of ocular tension induced retinal ischemia/ reperfusion injury rats	Reduction of B-wave decrement in the ERG	Inhibit activation of NF-κB;	Fang et al. (2015)
					Preserve retinal morphology and neurons	Down-regulate p-JNK and p-ERK;	
					Retinal oxidative damage↓	Up-regulate p-p38	
					Apoptosis↓		

(continued)

Table 8.9 (continued)

No.	Average molecular weight	Degree of deacetylation	Dose	Model	Ending point	Proposed mechanism	References
15	/	/	25, 100, 400 mg/kg	HIBD induced brain damaged mouse	Early neurological reflex↑	Upregulate energy metabolism level (ATP, lactic acid);	Wu et al. (2017a)
					Brain infarction volume↓	Increase antioxidant enzyme activity (GSH-px, SOD, T-AOC);	
					Hecrosis cell loss and degeneration↓	Alleviate inflammatory	

COS, a considerable increase was apparent in the life span of insects, number of eggs and brood size as compared to monocrotophos group. Moreover, COS caused a significant decrease in ROS and increased both the GSH level and SOD and catalase activity. Wu et al. (2017a) explored the neuronal protective effect on hypoxic-ischemic brain damage in neonatal rats. They found that post-treatment with COS could improve early neurological reflex, reduce brain infarction volume, and prevent cell loss by necrosis, and degeneration. The proposed mechanism was up-regulated energy metabolism level (ATP, lactic acid) as well as an increase in antioxidant enzyme activity (GSH-Px, SOD, T-AOC).

Apoptosis, also known as programmed cell death, is a series of biochemical events leading to morphological cell change and death (Green 2012). Studies indicate that cell apoptosis is responsible, either directly or indirectly, for numerous diseases found in humans including neurodegenerative disorders (Fadeel and Orrenius 2005). Several researchers have reported that COS could inhibit neuron apoptosis, thus alleviating neuronal disorders. Zhou et al. (2008) reported that COS (0.5–2 mg/mL) could increase cell viability, and attenuate LHD release, as well as apoptosis induced by glutamate. Furthermore, a decrease in intracellular calcium levels and caspase-3 activity was observed. Similar results were obtained by Dai et al. (2013). Results show that COS (200 μg/mL) attenuates cell death, ROS level, lipid peroxidation level, LDH release, and apoptosis induced by $A\beta_{1-42}$. The proposed mechanism was a decrease in caspase-3 activity and JNK phosphorylation. Xu et al. (2011) used a glucose deprivation-induced primary cortical neuron cell apoptosis model to estimate the anti-apoptosis properties of COS (500 μg/mL). A decrease in apoptotic cells, a decrease in TUNEL positive cells, an increase in Bcl-2/Bax ratio, and a decrease in the cleavage of caspase-3 were observed. The proposed mechanism was the activation of the PI3K/AKT and the MEK/ERK pathways. Jia et al. (2016) reported that COS attenuates cognitive deficits in an $A\beta_{1-42}$-induced rat model via anti-apoptosis. They observed an increase in performance in a water maze test and a passive avoidance test, improving the activity of GSH-Px, and SOD, as well as decreasing hippocampus neuronal morphological change. Hao et al. (2015) synthesized peracetylated COS and N-acetylated COS and investigated their anti-apoptosis properties. Results indicated that peracetylated COS possessed the optimal anti-apoptosis properties for treatment against glutamate-induced PC12 cell death, indicating that acetyl might be vital to anti-apoptosis.

The neurotransmitter acetylcholine (ACh) is located in the nervous system and is delivered to acetate and choline by ADhE. Impulse transmission is profoundly dependent on this oligomeric enzyme that affixes to the neuromuscular junction to facilitate the degradation of ACh at this location, as well as the brain cholinergic synapses (Siow et al. 2002). Currently, cognitive inadequacy in AD patients is mitigated by the application of acetylcholinesterase (AChE) suppressors. These medications impede the ACh reaction leading to the activation of the central cholinergic system. Lee et al. (2009b) used six kinds of COS to explore their suppressing activity against AChE and found that COS with MW 1–5 kDa and DD 90% provide the optimal properties in repressing AChE activity, with IC_{50} 1.67 mg/mL. Furthermore, results show that COS could suppress the protein expression of AChE. Yoon et al.

(2009) examined the preventative effect of three types of COS against AChE when integrated with various replacement groups including diethylaminoethyl (DEAE), aminoethyl (AE), and dimethylaminoethyl (DMEM). The respective IC_{50} levels were measured at 56.5 ± 0.26, 24.1 ± 0.39, and 9.2 ± 0.33 μg/mL. By employing the Lineweaver-Burk plots technique, it was revealed that the restrictive effect of AE-COS on AChE appeared to be non-competitive, while both DEAE-COS and DMAE-COS displayed competitive suppression behavior toward AChE.

Various factors such as acute compression, accidental trauma and deliberate surgery influence injuries to the peripheral nerve that is frequently encountered in medical professionals. Jiang et al. (2009) explored the regeneration properties of COS in a sciatic crushed nerve rat model. A decrease in withdrawal reflex latency and the sciatic function index were observed, as well as an increase in the recovery index of compound muscle action potentials (CMAPS), muscle wet weight ratio, and the cross-sectional area (CSA) of gastrocnemius muscle fibers. Further investigation was conducted into the regeneration properties of COS in a common crushed peroneal nerve rabbit model. The recovery of electrophysiological prosperity in animals, the reconstruction of an injured common peroneal nerve, and nerve regeneration were studied (Gong et al. 2009).

In summary, several studies indicate that COS possess multiple bio-properties and related mechanisms (Fig. 8.8) that are capable of suppressing A*β* aggregation, and inhibiting β-secretase. Moreover, COS can inhibit ADhE, while also displaying anti-oxidative, anti-apoptotic, and regenerative qualities. Therefore, COS might be a promising therapeutic agent for neuronal disorders including AD, Parkinson's disease, as well as crushed peripheral nerve.

8.2.10 Other Biological activities

Studies indicate additional biological functions of COS. The acquired immune deficiency syndrome is the result of the human immunodeficiency virus type-1 (HIV-1). Some SCOS displayed anti-HIV activities. However, due to poor water solubility, their potential application as anti-HIV therapy is not clear (Jayakumar et al. 2007). Compared to chitosan, COS display better water solubility. The anti-HIV qualities of sulfated SCOS were investigated following integration by Artan et al. (2010). The production of the p24 antigen (7.76 μg/mL and EC50 4.33 μg/mL for HIV-1Ba-L and HIV-1RF, respectively), HIV-1-induced syncytia formation (EC_{50} 2.19 μg/mL), and the lytic effect (EC_{50} 1.43 μg/mL) all responded favorably to the repressive influence of SCOS (MW 3–5 kDa). The connection of the CD4 cell surface receptor to HIV-1 gp120 was successfully interrupted by the suggested method. The result suggested that SCOS might be a potential treatment for HIV infection. Furthermore, due to its capacity to encourage the formation of fibroblasts by influencing its growth rate, COS displayed wound-healing qualities (Howling et al. 2001). Additionally, we are investigating the effect of COS and ChOS against insect pest and parasite such as snail and *schistosome.*

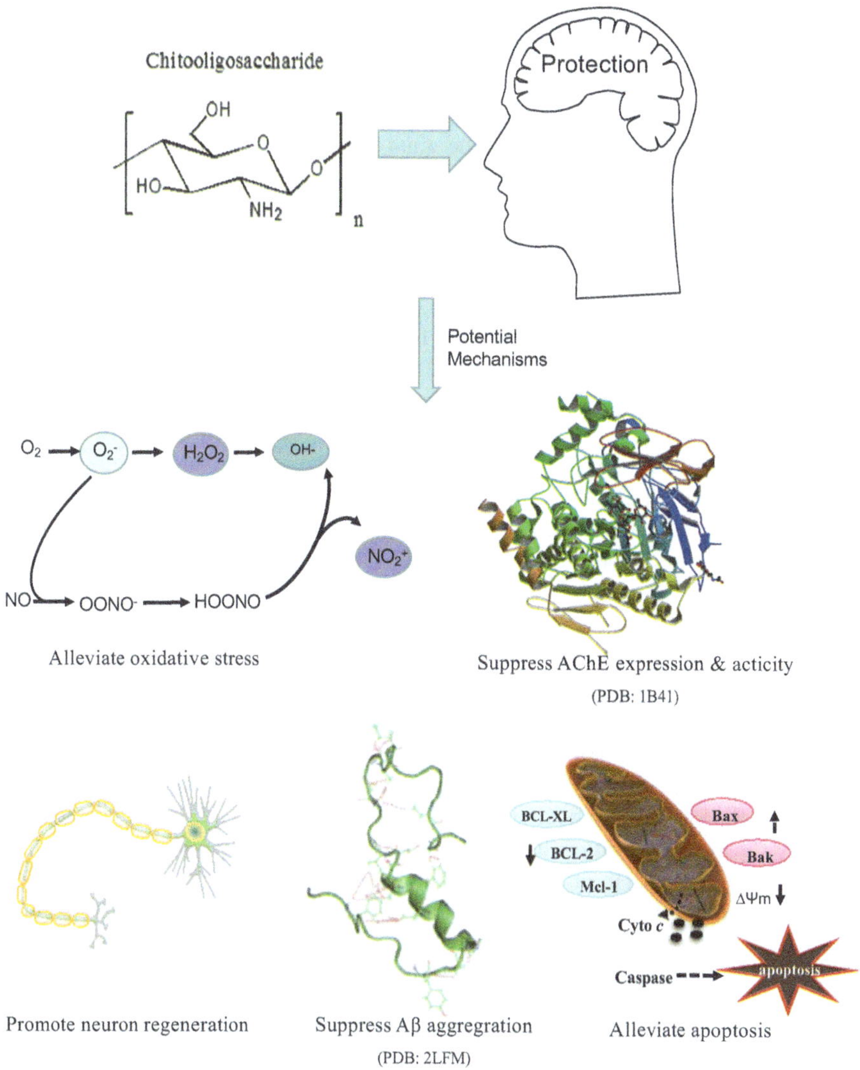

Fig. 8.8 The multiple bio-properties and related mechanisms underlying the neuro-protective effects of COS. *AChE* acetylcholinesterase, *Bcl-XL* B-cell lymphoma-extra large, *Mcl-1* Myeloid cell leukemia-1, *Cyto c* cytochrome C, $\Delta\Psi_m$ mitochondrial membrane potential

8.3 The Potential Applications of COS in the Food Industry

8.3.1 COS as Food Additives

The inhibitory effects of COS against different microbes and their excellent water-soluble ability encourage their application in the food industry (Hamed et al. 2016; Xia et al. 2011). Due to their excellent anti-microbial activity, COS have always been used as food additive in dairy products and beverages to retain product quality and improve the storage shelf-life. A study from Liu et al. (2005) reported that 50 g/L COS displayed significant inhibitory effects against the total bacterial count and psychrophiles in fresh milk. In yoghurt, a 0.1% w/w COS addition diminishes bacterial acidification and renders the final product sensory acceptable (Vela Gurovic et al. 2015). Similar results were obtained by Ji et al. (2016), who found that limiting amounts of COS (w/w $\leq 0.1\%$) could slow down the post-acidification progress and maintain the good qualities of yogurt. Yang et al. indicated that COS inhibited the formation of staling compounds and scavenged the radicals in beer (Yang et al. 2017).

COS also exhibit significant promise for utilization in the coatings of edible packaging (Dutta et al. 2011; Hamed et al. 2016). Foods of animal, aquatic origin (meat, fish, and seafood) and fresh fruit (strawberries, peaches, apples, and pears) are easily contaminated with a wide range of microorganisms during transportation processing, and storage. The effects of COS on the processing and storage quality of beef and aquatic products were reported by Guo et al. (2016) and Yuan et al. (2018), respectively. Guo et al. (2016) reported that the addition of 0.06 g/100 mL COS is ideal to extend the shelf-life (10–20 days) and retain the color and nutrients of fresh beef. Calculations by Yuan et al. (2018) indicated that COS, mixed with tea polyphenol, caffeine, algin, and acetic acid in a water solution could be used as a seafood preservative, especially in shrimp, shellfish, and fish (CN 108633987 A). For improvement of the post-harvest quality of strawberries, Nie et al. (2010) found a 1.5% w/w of COS coating could maintain the sensory, flavor, and nutritional quality during a storage period of 5 days at room temperature. Chen et al. (2012) reported that 2% w/w succeeded in retaining the firmness of peaches for 10 days, as well as decreasing the relative conductivity and MDA, polyphenoloxidase activity (PPO) activities during refrigeration. The effect of COS on the storage quality of Nanguo pears was studied by He et al. (2016). Immersed treatment with 100 mg/L COS significantly reduced the bad fruit rate and delayed fruit softening, as well as preserved the content of soluble solids and titratable acid during the 15 days' storage period at room temperature. For fresh-cut fruit, Sa et al. (2017) reported that coating treatment with COS with the addition of 0.25% thyme oil and 0.06% cinnamon oil, could significantly maintain the color, ascorbic acid content, and decrease the PPO and peroxidase activity in fresh-cut apples during storage.

Additionally, researchers also developed new food formulas with COS in snacks due to the popularity of this food resource in China. For example, Bu et al. (2018) developed a oinner roll formula by adding 2% w/w COS and 6–9% w/w yam/soybean fiber. This product, enriched with dietary fiber, will be beneficial to human health (CN 108669146 A).

For decades, chitin and chitosan were widely used as food additives, especially as preservatives in the food industry. With the improvement of the manufacture and purification methods of COS and specific regulatory conditions in the food industry, COS can satisfy the limitations exhibited by chitin and chitosan. Figure 8.9 illustrates the potential application of COS as food ingredients.

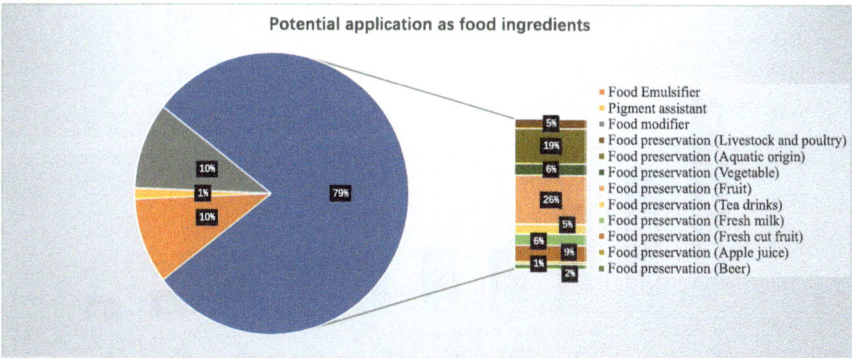

Fig. 8.9 Potential application of COS as food ingredients

Moreover, Singh (2016) examined the role of chitosan in the quality of processed and stored food products derived from marine and other animal sources, while its applications as a biopreservative, an immuno-stimulator, and a prebiotic were reviewed by Liaqat and Eltem (2018) as well.

8.3.2 Functional Daily Supplement

Diet and age-related diseases might be inhibited by functional food containing COS as active ingredients. Various biological activities of COS such as anti-oxidative, anti-microbial, hypocholesterolemic, anti-tumor, immune-enhancing, anti-obesity activities, and lower lipid effects show considerable promise for the production of useful products that will add value to the functional food industry. Figure 8.10 shows the potential application of COS in daily supplements. Glucose control, liver protection, and anti-tumor activity are the areas that can benefit the most from COS.

To illustrate the anti-tumor effects, Cai et al. (2015) combined the functional ingredients from *Antrodia cinnamomea* and COS to create a health food recipe (CN 104523877 A, 2015). Similarly, many COS-based food supplements which were mixed with oligopeptides of marine origin (CN 102228181 A, 2011), *haematococcaceae* polysaccharides and oligosaccharides (CN 107873933 A, 2018), and more, indicated that these products display excellent anti-tumor properties. To show the immune-enhancing effects, Du et al. (2003) optimized a formula containing a combination of different ratios of COS and *ganoderma lucidum* spore powder. These products exhibit immune-boosting effects and improve recovery following illness (CN1463744 A, 2003). Due to the shared targets of obesity, hyperlipidemia, and hypercholesterolemia (Monteiro and Azevedo 2010; Takahashi et al. 2012), the amount of COS-enriched functional supplemental recipes were illustrated. For example, Lei and Zhang (2015) reported that combining L-carnitine with COS at a ratio of (0.1–10):1, displayed beneficial properties regarding the intervention in obesity, NAFLD, and protecting against cardiovascular disease (CN 105326053 A,

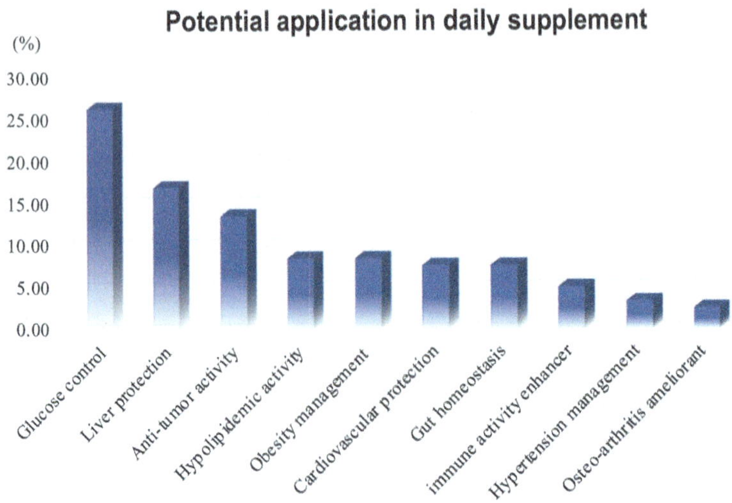

Fig. 8.10 Potential application of COS in daily supplement

2015). Huang (2017) invented an oral liquid which is mainly consists of black garlic, COS, *Lycium barbarum*, and Chinese dates are shown to enhance immunity and modulate the lipid profile in serum and weight control (CN 106617024 A, 2017). Furthermore, the free radical and scavenging abilities of COS are well documented in inhibiting oxidative damage and number of functional formulas were reported. Cao et al. reported that hesperidin mixed with COS at a ratio of 1:1–10 improved the water solubility of hesperidin and enhanced its antioxidant activity (CN 107568737 A, 2018). Wang et al. indicated that a functional food formula containing ascorbic acid, vitamin C, COS, and collagen peptide could enhance immunity and improve antioxidant activity (CN 103005255 A, 2013).

With the development of food science and technology, this functional food trend is pursued and propelled by food marketing entities and companies aiming to provide a satisfactory answer to pressing health concerns by providing access to food resources and recipes. On the other hand, consumers, especially those that are highly educated, think that a wholesome diet is fundamental for the maintenance of satisfactory health (Fogliano and Vitaglione 2005). Therefore, the health product industry provides an accessible environment for the development of thousands of products and formulas. The beneficial qualities of COS and its derivatives have mostly been established, and the raw products are commercially widely available. A multitude of parameters should be considered when adding functional ingredients of food products such as the bioavailability of the ultimate product, interaction with the food matrix, the stability of the production process, the safety assessment of each formula, and the possible side-effects on particular population groups (such as adolescents, as well as progestational and post-menopausal individuals).

8.3.3 Foods for Special Medical Purpose (FSMP)

FSMP is an emerging topic in clinical supervision. FSMP are manufactured according to a distinct formula to accommodate individuals with specific dietary requirements such as metabolic disorders, limited feeding problems, digestions absorption issues or other particular disease conditions. Therefore, FSMPs are attracting increased attention from informed patients seeking natural products. Additionally, patient associations regularly recommend FSMP since their utilization is considered a viable alternative to conventional tablets and pills, and are also better tolerated by patients (Wang 2016, 2017). Essentially, COS-enriched FSMP contain biologically active oligosaccharides with various ratios of DP. Furthermore, additional functional factors like dietary fiber, small functional peptides, and polysaccharides were added to the recipe. For example, COS and water-soluble fiber enriched FSMP reduced the content of TC, LDLC, increased the content of glucagon-like peptide-1, and finally modulated the postprandial blood glucose levels (Ding and Zhang 2015). You (2018) invented an FSMP for elderly patients that included functional factors such as COS, γ-aminobutyric acid, L-carnitine, resveratrol, and β-glucan from oats to enhance the effects on immunity and modulate digestive absorption disorders (CN 108260802 A, 2018). Similarly, COS were added to FSMP to address issues such as abnormal lipid metabolism (CN 108703364 A, 2018a), lactose intolerance (CN108013453 A, 2018c), postpartum abdominal flatulence (CN108013452 A, 2018b), cardiovascular protection (CN108552320 A, 2018), and sarcopenia (CN108618129 A, 2018d).

8.4 Conclusion

8.4.1 Recent Limitation

Interests in COS-related activities have increased substantially during the past several decades. Much remains to be clarified before they can be optimally applied to improve human health. The utilization of COS is subject to several limitations. Firstly, further extensive research is necessary regarding the procedural mechanisms, since the majority of the studies dealing with the biological actions of COS fail to provide clear information on its molecular operation. Therefore, it becomes challenging to clarify which types of COS are responsible for specific biological activities. Secondly, since the COS that are employed throughout the various studies display significant physicochemical diversity, controversy persists regarding these biological activities. This severely restricts unmitigated comprehension of their operational processes. Therefore, it is essential to establish bioactive COS products with distinct physicochemical characteristics to facilitate successful mass production and commercialization. Particularly, the bulk production for singular COS containing a DP level higher than 6, is currently unavailable. Therefore, the biological activities of these COS with higher DP could not be thoroughly investigated. Thirdly, although the customary way to understand a functional food is by adding COS to a traditional food matrix and not the contemporary food formula, it is difficult to infer the safety parameters of these products. Additional safety assessments are necessary to evaluate the role of COS in FSMPs and require a case-by-case study. Additionally, the current manufacture and separation technology of COS are not adequate to meet the demands for preparing the products with uniform DP or MW, moreover, the processing technology of COS in food matrix is always absent in the research.

8.4.2 Additional Perspectives

Considering the various limitations described in this study, further extensive research is required in both the food industry and academic community.

(a) In functional evaluation studies, the variety of appropriate doses for multiple biological conditions should be suggested in previous research. Studies involving the biological activities of COS need to be substantiated with further research on well-defined singular COS, to clarify complex molecular mechanisms. The availability of well-characterized COS favors the creation of a knowledge base for a structure-function relationship determination. What's most important, clinical trials should be considered in future to correct the in vivo or in vitro experimental results. Conversely, the remarkable biological importance of singular COS ensures continued interest in new technologies for their preparation, purification, and analysis.
(b) Manufacturers should take into account the different compositions of COS when determining their biological effects. Moreover, the development of a unique method to mass-produce purified and well characterized COS remains the ultimate challenge and requires further research to find an adequate solution.
(c) It is necessary to enhance the risk assessment database regarding the presence of COS in FSMPs and food supplements. Specific areas that require improvement include data regarding product consumption and the related COS that is current and reliable, COS concentration measurement specifically in plasma and tissue, as well as identifying possible additional risk factors.

Acknowledgements The authors thank to PhD student Xiaoguo Ji, Shumin Chen, and MS student Liming Gu, Xin Shen, Chunfeng Zhu, An-na Zheng, Jiangshan You, Jun Lu, Yang Liu for their assistant of materials and references collection. Especially thanks to PhD student Shumin Chen for her assistant of proof reading this chapter.

References

Adefegha SA (2018) Functional foods and nutraceuticals as dietary intervention in chronic diseases; novel perspectives for health promotion and disease prevention. J Diet Suppl 15(6):977–1009. https://doi.org/10.1080/19390211.2017.1401573

Allan CR, Hadwiger LA (1979) The fungicidal effect of chitosan on fungi of varying cell wall composition. Exp Mycol 3(3):285–287. https://doi.org/10.1016/S0147-5975(79)80054-7

Artan M, Karadeniz F, Karagozlu MZ, Kim MM, Kim SK (2010) Anti-HIV-1 activity of low molecular weight sulfated chitooligosaccharides. Carbohydr Res 345(5):656–662. https://doi.org/10.1016/j.carres.2009.12.017

Azuma K, Osaki T, Kurozumi S, Kiyose M, Tsuka T, Murahata Y et al (2015) Anti-inflammatory effects of orally administered glucosamine oligomer in an experimental model of inflammatory bowel disease. Carbohydr Polym 115:448–456. https://doi.org/10.1016/j.carbpol.2014.09.012

Bahar B, O'Doherty JV, Sweeney T (2011) A potential role of IL-6 in the chito-oligosaccharide-mediated inhibition of adipogenesis. Br J Nutr 106(8):1142–1153. https://doi.org/10.1017/S0007114511001486

Bahar B, O'Doherty JV, O'Doherty AM, Sweeney T (2013) Chito-oligosaccharide inhibits the de-methylation of a 'CpG' island within the leptin (LEP) promoter during Adipogenesis of 3T3-L1 cells. PLoS One 8(3):e60011. https://doi.org/10.1371/journal.pone.0060011

Bautista-Baños S, Hernández-Lauzardo AN, Velázquez-Del Valle MG, Hernández-López M, Ait Barka E, Bosquez-Molina E, Wilson CL (2006) Chitosan as a potential natural compound to control pre and postharvest diseases of horticultural commodities. Crop Prot 25(2):108–118. https://doi.org/10.1016/j.cropro.2005.03.010

Bellamy L, Casas JP, Hingorani AD, Williams D (2009) Type 2 diabetes mellitus after gestational diabetes: a systematic review and meta-analysis. Lancet 373(9677):1773–1779. https://doi.org/10.1016/S0140-6736(09)60731-5

Benhabiles MS, Salah R, Lounici H, Drouiche N, Goosen MFA, Mameri N (2012) Antibacterial activity of chitin, chitosan and its oligomers prepared from shrimp shell waste. Food Hydrocoll 29(1):48–56. https://doi.org/10.1016/j.foodhyd.2012.02.013

Biesalski HK (2016) Nutrition meets the microbiome: micronutrients and the microbiota. Ann N Y Acad Sci 1372(1):53–64. https://doi.org/10.1111/nyas.13145

Bo H, Xiao D, Bie T, Hao X, Jing W, Jie Y (2016) Chitosan oligosaccharide reduces intestinal inflammation that involves calcium-sensing receptor (casr) activation in lipopolysaccharide (LPS)-challenged piglets. J Agric Food Chem 64(1):245–252

Bu X, Xie Y, Lu D, Chang J, Xiong W, Wu K (2018) CN 108669146 A. China

Byun HG, Kim YT, Park PJ, Lin X, Kim SK (2005) Chitooligosaccharides as a novel β-secretase inhibitor. Carbohydr Polym 61(2):198–202. https://doi.org/10.1016/j.carbpol.2005.05.003

Caballero F, Fernández A, De Lacy AM, Fernández-Checa JC, Caballería J, García-Ruiz C (2009) Enhanced free cholesterol, SREBP-2 and StAR expression in human NASH. J Hepatol 50(4):789–796. https://doi.org/10.1016/j.jhep.2008.12.016

Cai Q, Chen J, Xu L (2015) CN 104523877 A. China

Cao W, Wang J, She Y, Zhao H, Zhang Y, Yang X, Zhang H (2008) Study on inhibitive effect of chitooligosaccharides on intestinal causative microorganism. Food Sci 29:114–116

Cao R, Ma Q, Zhou Z, Zhao X (2018) CN 107568737 A. China

Chang SH, Wu CH, Tsai GJ (2018) Effects of chitosan molecular weight on its antioxidant and antimutagenic properties. Carbohydr Polym 181(November 2017):1026–1032. https://doi.org/10.1016/j.carbpol.2017.11.047

Chen J (2014) Study on the weight-lossing effect of chitooligosacchrides, gucosamine and preparation of capsaicin-chitosan microspheres enteric coated tablets. Guangdong Pharmaceutical University

Chen Y, Chung Y, Wang LW, Chen K, Li S-Y (2002) Antibacterial properties of chitosan in waterborne pathogen. J Environ Sci Health A37(7):1379–1390. https://doi.org/10.1081/ESE-12005993

Chen AS, Taguchi T, Sakai K, Kikuchi K, Wang MW, Miwa I (2003) Antioxidant activities of chitobiose and chitotriose. Biol Pharm Bull 26(9):1326–1330. https://doi.org/10.1248/bpb.26.1326

Chen A-S, Taguchi T, Sakai K, Matahira Y, Wang M-W, Miwa I (2005) Effect of chitobiose and chitotriose on carbon tetrachloride-induced acute hepatotoxicity in rats. Biol Pharm Bull 28(10):1971–1973. https://doi.org/10.1248/bpb.28.1971

Chen Q, Jun L, Qiu Y, Zhao L (2019) Short communication: bioinformatics-based mining of novel gene targets for identification of Cronobacter turicensis using PCR. J Dairy Sci 102(7):6023–6026

Chen Y, Wang Y, Gang C, Li J, Gao J, Qian W et al (2012) Study on preservation effects of chitosan ologosaccharide on peaches during refrigeration. Food Ferment Ind 38(4):208–211. https://doi.org/10.13995/j.cnki.11-1802/ts.2013.05.028

Chiu CY, Feng SA, Liu SH, Chiang MT (2017) Functional comparison for lipid metabolism and intestinal and fecal microflora enzyme activities between low molecular weight chitosan and chitosan oligosaccharide in high-fat-diet-fed rats. Mar Drugs 15(7). https://doi.org/10.3390/md15070234

Cho JE, Rahman A, Kim SW, Baek Y, Hwang HJ et al (2008) Chitosan oligosaccharides inhibit adipogenesis in 3T3-L1 adipocytes. J Microbiol Biotechnol 18:80–87

Choi BK, Kim KY, Yoo YJ, Oh SJ, Choi JH, Kim CY (2001) In vitro antimicrobial activity of a chitooligosaccharide mixture against Actinobacillus actinomycetemcomitans and Streptococcus mutans. Int J Antimicrob Agents 18(6):553–557. https://doi.org/10.1016/S0924-8579(01)00434-4

Choi CR, Kim EK, Kim YS, Je JY, An SH, Lee JD et al (2012a) Chitooligosaccharides decreases plasma lipid levels in healthy men. Int J Food Sci Nutr 63(1):103–106. https://doi.org/10.3109/09637486.2011.602051

Choi EH, Yang HP, Chun HS (2012b) Chitooligosaccharide ameliorates diet-induced obesity in mice and affects adipose gene expression involved in adipogenesis and inflammation. Nutr Res 32(3):218–228. https://doi.org/10.1016/j.nutres.2012.02.004

Chung MJ, Park JK, Park YI (2012) Anti-inflammatory effects of low-molecular weight chitosan oligosaccharides in IgE-antigen complex-stimulated RBL-2H3 cells and asthma model mice. Int Immunopharmacol 12(2):453–459. https://doi.org/10.1016/j.intimp.2011.12.027

Costa C, Tsatsakis A, Mamoulakis C, Teodoro M, Briguglio G, Caruso E et al (2017a) Current evidence on the effect of dietary polyphenols intake on chronic diseases. Food Chem Toxicol 110(August):286–299. https://doi.org/10.1016/j.fct.2017.10.023

Costa EM, Silva S, Vicente S, Veiga M, Tavaria F, Pintado MM (2017b) Chitosan as an effective inhibitor of multidrug resistant Acinetobacter baumannii. Carbohydr Polym 178(September):347–351. https://doi.org/10.1016/j.carbpol.2017.09.055

Dai X, Chang P, Zhu Q, Liu W, Sun Y, Zhu S, Jiang Z (2013) Chitosan oligosaccharides protect rat primary hippocampal neurons from oligomeric β-amyloid 1-42-induced neurotoxicity. Neurosci Lett 554:64–69. https://doi.org/10.1016/j.neulet.2013.08.046

Dai X, Hou W, Sun Y, Gao Z, Zhu S, Jiang Z (2015) Chitosan oligosaccharides inhibit/disaggregate fibrils and attenuate amyloid β-mediated neurotoxicity. Int J Mol Sci 16(5):10526–10536. https://doi.org/10.3390/ijms160510526

Danhorn T, Fuqua C (2007) Biofilm formation by plant-associated bacteria. Annu Rev Microbiol 61(1):401–422. https://doi.org/10.1146/annurev.micro.61.080706.093316

Di Meo S, Reed TT, Venditti P, Victor VM (2016) Role of ROS and RNS sources in physiological and pathological conditions. Oxidative Med Cell Longev 2016. https://doi.org/10.1155/2016/1245049

Ding C (2015) Study on the anti-obesity activity and mechanism of COST particles. Guangdong Pharmaceutical University

Ding W, Zhang Y (2015) Clinical effect of chitosanoligosaccharide diabetic food on type 2 diabetes mellitus. Food and Drug 17(6):407–410

Dou J, Tan C, Du Y, Bai X, Wang K, Ma X (2007) Effects of chitooligosaccharides on rabbit neutrophils in vitro. Carbohydr Polym 69(2):209–213. https://doi.org/10.1016/j.carbpol.2006.09.029

Du YG, Bai X, Li S, Qu T, Zhao X, Lu H (2003) CN1463744A. China

Dutta J, Tripathi S, Dutta PK (2011) Progress in antimicrobial activities of chitin , chitosan and its oligosaccharides : a systematic study needs for food applications. Food Sci Technol Int 18(1):3–34. https://doi.org/10.1177/1082013211399195

Eweis M, Elkholy SS, Elsabee MZ (2006) Antifungal efficacy of chitosan and its thiourea derivatives upon the growth of some sugar-beet pathogens. Int J Biol Macromol 38(1):1–8. https://doi.org/10.1016/j.ijbiomac.2005.12.009

Fadeel B, Orrenius S (2005) Apoptosis: a basic biological phenomenon with wide-ranging implications in human disease. J Intern Med 258(6):479–517. https://doi.org/10.1111/j.1365-2796.2005.01570.x

Fang IM, Yang CH, Yang CM (2014) Chitosan oligosaccharides attenuate ocular inflammation in rats with experimental autoimmune anterior uveitis. Mediat Inflamm 2014. https://doi.org/10.1155/2014/827847

Fang IM, Yang CM, Yang CH (2015) Chitosan oligosaccharides prevented retinal ischemia and reperfusion injury via reduced oxidative stress and inflammation in rats. Exp Eye Res 130:38–50. https://doi.org/10.1016/j.exer.2014.12.001

Farmer JJ, Asbury MA, Hickman FW, Brenner DJ (1980) Enterobacter sakazakii: a new species of "Enterobacteriaceae" isolated from clinical specimens. Int J Syst Bacteriol 30(3):569–584. https://doi.org/10.1099/00207713-30-3-569

Fernandes JC, Tavaria FK, Soares JC, Ramos ÓS, João Monteiro M, Pintado ME, Xavier Malcata F (2008) Antimicrobial effects of chitosans and chitooligosaccharides, upon Staphylococcus aureus and Escherichia coli, in food model systems. Food Microbiol 25(7):922–928. https://doi.org/10.1016/j.fm.2008.05.003

Fernandes JC, Eaton P, Nascimento H, Gião MS, Ramos ÓS, Belo L et al (2010a) Antioxidant activity of chitooligosaccharides upon two biological systems: erythrocytes and bacteriophages. Carbohydr Polym 79(4):1101–1106. https://doi.org/10.1016/j.carbpol.2009.10.050

Fernandes JC, Spindola H, De Sousa V, Santos-Silva A, Pintado ME, Malcata FX, Carvalho JE (2010b) Anti-inflammatory activity of chitooligosaccharides in vivo. Mar Drugs 8(6):1763–1768. https://doi.org/10.3390/md8061763

Fernandes JC, Sereno J, Garrido P, Parada B, Cunha MFX, Reis F et al (2012) Inhibition of bladder tumor growth by chitooligosaccharides in an experimental carcinogenesis model. Mar Drugs 10(12):2661–2675. https://doi.org/10.3390/md10122661

Flint HJ (2012) The impact of nutrition on the human microbiome. Nutr Rev 70(SUPPL. 1):S10–S13. https://doi.org/10.1111/j.1753-4887.2012.00499.x

Fogliano V, Vitaglione P (2005) Functional foods: planning and development. Mol Nutr Food Res 49(3):256–262. https://doi.org/10.1002/mnfr.200400067

Franken L, Schiwon M, Kurts C (2016) Macrophages: sentinels and regulators of the immune system. Cell Microbiol 18(4):475–487. https://doi.org/10.1111/cmi.12580

Franz MJ (2016) Diabetes nutrition therapy: effectiveness, macronutrients, eating patterns and weight management. Am J Med Sci 351(4):374–379. https://doi.org/10.1016/j.amjms.2016.02.001

Gangemi S, Gofita E, Costa C, Teodoro M, Briguglio G, Nikitovic D et al (2016) Occupational and environmental exposure to pesticides and cytokine pathways in chronic diseases (Review). Int J Mol Med 38(4):1012–1020. https://doi.org/10.3892/ijmm.2016.2728

Gentile CL, Weir TL (2018) The gut microbiota at the intersection of diet and human health. Science 780(November):776–780. https://doi.org/10.1126/science.aau5812

Gonçalves RC, da Silva DP, Signini R, Naves PLF (2017) Inhibition of bacterial biofilms by carboxymethyl chitosan combined with silver, zinc and copper salts. Int J Biol Macromol 105:385–392. https://doi.org/10.1016/j.ijbiomac.2017.07.048

Gong Y, Gong L, Gu X, Ding F (2009) Chitooligosaccharides promote peripheral nerve regeneration in a rabbit common peroneal nerve crush injury model. Microsurgery 29:650–656. https://doi.org/10.1002/micr

Green D (2012) Means to an end: apoptosis and other cell death mechanisms. Q Rev Biol 87:68

Guerciolini R, Radu-Radulescu L, Boldrin M, Dallas J, Moore R, Jianguo Z (2001) Comparative evaluation of fecal fat excretion induced by orlistat and chitosan. Obes Res 9(6):364–367. https://doi.org/10.1038/oby.2001.47

Guo X, Liu C, You X, Dai X, Liang S, Zhong X (2016) Effects of chitosan oligosaccharide on preservation of cold fresh beef. Food Ferment Ind 42(10):204–209

Hamed I, Özogul F, Regenstein JM (2016) Industrial applications of crustacean by-products (chitin, chitosan, and chitooligosaccharides): a review. Trends Food Sci Technol 48:40–50. https://doi.org/10.1016/j.tifs.2015.11.007

Hamley IW (2012) The amyloid beta peptide: a chemist's perspective. Role in Alzheimer's and fibrillization. Chem Rev 112(10):5147–5192. https://doi.org/10.1021/cr3000994

Han KN, Kwon IK, Lohakare JD, Heo S, Chae BJ (2007) Chito-oligosaccharides as an alternative to antimicrobials in improving performance, digestibility and microbial ecology of the gut in weanling pigs. Asian Australas J Anim Sci 20(4):556–562. https://doi.org/10.5713/ajas.2007.556

Hanahan D, Weinberg RA (2011) Hallmarks of cancer: the next generation Douglas. Cell 144:646–674. https://doi.org/10.1016/j.cell.2011.02.013

Hao C, Gao L, Zhang Y, Wang W, Yu G, Guan H et al (2015) Acetylated chitosan oligosaccharides act as antagonists against glutamate-induced PC12 cell death via Bcl-2/Bax signal pathway. Mar Drugs 13(3):1267–1289. https://doi.org/10.3390/md13031267

Hao C, Wang W, Wang S, Zhang L, Guo Y (2017) An overview of the protective effects of chitosan and acetylated chitosan oligosaccharides against neuronal disorders. Mar Drugs 15(4). https://doi.org/10.3390/md15040089

Harish Prashanth KV, Tharanathan RN (2005) Depolymerized products of chitosan as potent inhibitors of tumor-induced angiogenesis. Biochim Biophys Acta Gen Subj 1722(1):22–29. https://doi.org/10.1016/j.bbagen.2004.11.009

He Y, Liu T, Lu H, Wang W, Hu J e, Yin H (2016) Effect of chitosan oligosaccharide on the storage quality of Nanguo pear. Liaoning Agricultural Science 3:1–5

Hirano S, Nagao N (1989) Effects of chitosan, Pectic acid, lysozyme, and chitinase on the growth of several phytopathogens. Agric Biol Chem 53(11):3065–3066. https://doi.org/10.1002/14651858.CD005489.pub2

Howling GI, Dettmar PW, Goddard PA, Hampson FC, Dornish M, Wood EJ (2001) The effect of chitin and chitosan on the proliferation of human skin fibroblasts and keratinocytes in vitro. Biomaterials 22(22):2959–2966. https://doi.org/10.1016/S0142-9612(01)00042-4

Huang L (2016) Preparation of chitosan oligosaccharide capsules, chitosan capsules and assessment of their weight control function. Guangdong Pharmaceutical University

Huang X (2017) CN 106617024 A. China

Huang R, Mendis E, Rajapakse N, Kim SK (2006) Strong electronic charge as an important factor for anticancer activity of chitooligosaccharides (COS). Life Sci 78(20):2399–2408. https://doi.org/10.1016/j.lfs.2005.09.039

Huang HC, Hong L, Chang P, Zhang J, Lu SY, Zheng BW, Jiang ZF (2015a) Chitooligosaccharides attenuate Cu2+-induced cellular oxidative damage and cell apoptosis involving Nrf2 activation. Neurotox Res 27(4):411–420. https://doi.org/10.1007/s12640-014-9512-x

Huang L, Chen J, Cao P, Pan H, Ding C, Xiao T et al (2015b) Anti-obese effect of glucosamine and chitosan oligosaccharide in high-fat diet-induced obese rats. Mar Drugs 13(5):2732–2756. https://doi.org/10.3390/md13052732

Huang B, Xiao D, Tan B, Xiao H, Wang J, Yin J et al (2016) Chitosan oligosaccharide reduces intestinal inflammation that involves Calcium-Sensing Receptor (CaSR) activation in lipopolysaccharide (LPS)-challenged piglets. J Agric Food Chem 64(1):245–252. https://doi.org/10.1021/acs.jafc.5b05195

Hunter CJ, Bean JF (2013) Cronobacter: an emerging opportunistic pathogen associated with neonatal meningitis, sepsis and necrotizing enterocolitis. J Perinatol 33(8):581–585. https://doi.org/10.1038/jp.2013.26

Hussain A, Claussen B, Ramachandran A, Williams R (2007) Prevention of type 2 diabetes: a review. Diabetes Res Clin Pract 76(3):317–326. https://doi.org/10.1016/j.diabres.2006.09.020

Hyung JH, Ahn CB, IL Kim B, Kim K, Je JY (2016) Involvement of Nrf2-mediated heme oxygenase-1 expression in anti-inflammatory action of chitosan oligosaccharides through MAPK activation in murine macrophages. Eur J Pharmacol 793(November):43–48. https://doi.org/10.1016/j.ejphar.2016.11.002

Ilk S, Sağlam N, Özgen M, Korkusuz F (2017) Chitosan nanoparticles enhances the anti-quorum sensing activity of kaempferol. Int J Biol Macromol 94:653–662. https://doi.org/10.1016/j.ijbiomac.2016.10.068

Jayakumar R, Nwe N, Tokura S, Tamura H (2007) Sulfated chitin and chitosan as novel biomaterials. Int J Biol Macromol 40(3):175–181. https://doi.org/10.1016/j.ijbiomac.2006.06.021

Je J-Y, Park P-J, Kim S-K (2004a) Free radical scavenging properties of hetero-chitooligosaccharides using an ESR spectroscopy. Food Chem Toxicol: Int J Published Br Ind Biol Res Assoc 42(3):381–387. https://doi.org/10.1016/j.fct.2003.10.001

Je JY, Park PJ, Kim SK (2004b) Radical scavenging activity of hetero-chitooligosaccharides. Eur Food Res Technol 219(1):60–65. https://doi.org/10.1007/s00217-004-0881-x

Ji X, Wang T, Wang Z, Li JG, Mao X (2016) Effect of chitooligosaccharide addition on postacidification and storage stability of yogurt. Zhonggue Rupin Gongye 44(6):4–7

Jia S, Lu Z, Gao Z, An J, Wu X, Li X et al (2016) Chitosan oligosaccharides alleviate cognitive deficits in an amyloid-β1-42-induced rat model of Alzheimer's disease. Int J Biol Macromol 83:416–425. https://doi.org/10.1016/j.ijbiomac.2015.11.011

Jiang M, Zhuge X, Yang Y, Gu X, Ding F (2009) The promotion of peripheral nerve regeneration by chitooligosaccharides in the rat nerve crush injury model. Neurosci Lett 454(3):239–243. https://doi.org/10.1016/j.neulet.2009.03.042

Jiang Y, Fu C, Liu G, Guo J, Su Z (2018) Cholesterol-lowering effects and potential mechanisms of chitooligosaccharide capsules in hyperlipidemic rats. Food Nutr Res 62:1–15. https://doi.org/10.29219/fnr.v62.1446

Jin Q, Yu H, Wang X, Li K, Li P (2017) Effect of the molecular weight of water-soluble chitosan on its fat–/cholesterol-binding capacities and inhibitory activities to pancreatic lipase. PeerJ 5:e3279. https://doi.org/10.7717/peerj.3279

Jo SH, Ha KS, Moon KS, Kim JG, Oh CG, Kim YC et al (2013) Molecular weight dependent glucose lowering effect of low molecular weight chitosan oligosaccharide (GO2KA1) on postprandial blood glucose level in SD rats model. Int J Mol Sci 14(7):14214–14224. https://doi.org/10.3390/ijms140714214

Jo SH, Ha KS, Lee JW, Kim YC, Apostolidis E, Kwon YI (2014) The reduction effect of low molecular weight chitosan oligosaccharide (GO2KA1) on postprandial blood glucose levels in healthy individuals. Food Sci Biotechnol 23(3):971–973. https://doi.org/10.1007/s10068-014-0131-3

Johnson LR (2008) Microcolony and biofilm formation as a survival strategy for bacteria. J Theor Biol 251(1):24–34. https://doi.org/10.1016/j.jtbi.2007.10.039

Ju C, Yue W, Yang Z, Zhang Q, Yang X, Liu Z, Zhang F (2010) Antidiabetic effect and mechanism of chitooligosaccharides. Biol Pharm Bull 33(9):1511–1516. https://doi.org/homogenate

Jumaa M, Furkert FH, Müller BW (2002) A new lipid emulsion formulation with high antimicrobial efficacy using chitosan. Eur J Pharm Biopharm 53(1):115–123. https://doi.org/10.1016/S0939-6411(01)00191-6

Kadowaki O, Ogasawara A, Watanabe T, Mikami T, Matsumoto T, Misawa Y et al (2007) Chitohexose induce the yeast proliferation of Candida albicans. Biol Pharm Bull 30(3):583–584. https://doi.org/10.1371/journal.pone.0043789

Kang N-H, Lee WK, Yi B-R, Park M-A, Lee H-R, Park S-K et al (2012) Modulation of lipid metabolism by mixtures of protamine and chitooligosaccharide through pancreatic lipase inhibitory activity in a rat model. Laboratory Animal Res 28(1):31–38. https://doi.org/10.5625/lar.2012.28.1.31

Kang YR, Choi HY, Lee JY, Jang SI, Oh JB, Kim JS et al (2016) Effect of supplementation of low-molecular-weight chitosan oligosaccharide, GO2KA1, on postprandial blood glucose levels in healthy individuals following bread consumption. Food Sci Biotechnol 25(3):911–914. https://doi.org/10.1007/s10068-016-0149-9

Kao CH, Hsiang CY, Ho TY (2012) Assessment of chitosan-affected metabolic response by peroxisome proliferator-activated receptor bioluminescent imaging-guided transcriptomic analysis. PLoS One 7(4):1–10. https://doi.org/10.1371/journal.pone.0034969

Karadeniz F, Artan M, Kong CS, Kim SK (2010) Chitooligosaccharides protect pancreatic β-cells from hydrogen peroxide-induced deterioration. Carbohydr Polym 82(1):143–147. https://doi.org/10.1016/j.carbpol.2010.04.046

Katiyar D, Singh B, Lall AM, Haldar C (2011) Efficacy of chitooligosaccharides for the management of diabetes in alloxan induced mice: a correlative study with antihyperlipidemic and antioxidative activity. Eur J Pharm Sci 44(4):534–543. https://doi.org/10.1016/j.ejps.2011.09.015

Kau AL, Ahern PP, Griffin NW, Goodman AL, Gordon JI (2011) Human nutrition, the gut microbiome and the immune system. Nature 474(7351):327–336. https://doi.org/10.1038/nature10213

Kewsuwan P, Rujitanapanich S, Bhasabutra T, Puripunyavanich V, Busamongkol A, Injarean U, Pichetpong P (2014) Irradiated oligochitosan against Colletotrichum gloeosporioides in chili. Energy Procedia 56(C):274–279. https://doi.org/10.1016/j.egypro.2014.07.158

Khan SR (2012) Is oxidative stress, a link between nephrolithiasis and obesity, hypertension, diabetes, chronic kidney disease, metabolic syndrome? Urol Res 40:95–112. https://doi.org/10.1126/scisignal.274pe36.Insulin

Kim S, Kang S, Park S, Shin T, Ko Y (1998) Effects of chitooligosaccharides on liver function in the mouse. Korean J Food …, (November 2016). Retrieved from http://www.kpbio.com/PDF/eng/Effect_of_Chitoligo_fuction_in_the_mouse.PDF

Kim JY, Lee JK, Lee TS, Park WH (2003) Synthesis of chitooligosaccharide derivative with quaternary ammonium group and its antimicrobial activity against Streptococcus mutans. Int J Biol Macromol 32(1–2):23–27. https://doi.org/10.1016/S0141-8130(03)00021-7

Kim HM, Hong SH, Yoo SJ, Baek KS, Jeon YJ, Choung SY (2006) Differential effects of chitooligosaccharides on serum cytokine levels in aged subjects. J Med Food 9(3):427–430

Kim HJ, Ahn HY, Kwak JH, Shin DY, Kwon YI, Oh CG, Lee JH (2014a) The effects of chitosan oligosaccharide (GO2KA1) supplementation on glucose control in subjects with prediabetes. Food Funct 5(10):2662–2669. https://doi.org/10.1039/c4fo00469h

Kim JG, Jo SH, Ha KS, Kim SC, Kim YC, Apostolidis E, Kwon YI (2014b) Effect of long-term supplementation of low molecular weight chitosan oligosaccharide (GO2KA1) on fasting blood glucose and HbA1c in db/db mice model and elucidation of mechanism of action. BMC Complement Altern Med 14(1):1–7. https://doi.org/10.1186/1472-6882-14-272

Kim JH, Kim YS, Hwang JW, Han YK, Lee JS, Kim SK et al (2014c) Sulfated chitosan oligosaccharides suppress LPS-induced NO production via JNK and NF-κB inactivation. Molecules 19(11):18232–18247. https://doi.org/10.3390/molecules191118232

Kittur FS, Vishu Kumar AB, Varadaraj MC, Tharanathan RN (2005) Chitooligosaccharides – preparation with the aid of pectinase isozyme from Aspergillus niger and their antibacterial activity. Carbohydr Res 340(6):1239–1245. https://doi.org/10.1016/j.carres.2005.02.005

Kong S (2017) Study on the mechanism of COST particles inhibiting the endoplasmic reticulum stress regulating JAK2/STAT3 signaling pathway to improve leptin resistance. Guangdong Pharmaceutical University

Kong XF, Zhou XL, Lian GQ, Blachier F, Liu G, Tan BE et al (2014) Dietary supplementation with chitooligosaccharides alters gut microbiota and modifies intestinal luminal metabolites in weaned Huanjiang mini-piglets. Livest Sci 160(1):97–101. https://doi.org/10.1016/j.livsci.2013.11.023

Kong S, Ding C, Huang L, Bai Y, Xiao T, Guo J, Su Z (2017) The effects of COST on the differentiation of 3T3-L1 preadipocytes and the mechanism of action. Saudi J Biol Sci 24(2):251–255. https://doi.org/10.1016/j.sjbs.2016.09.008

Koppová I, Bureš M, Šimůnek J (2012) Intestinal bacterial population of healthy rats during the administration of chitosan and chitooligosaccharides. Folia Microbiol 57(4):295–299. https://doi.org/10.1007/s12223-012-0129-2

Kostrzewa RM, Segura-Aguilar J (2003) Novel mechanisms and approaches in the study of neurodegeneration and neuroprotection. A review. Neurotox Res 5(6):375–383. https://doi.org/10.1007/BF03033166

Kowaltowski AJ, de Souza-Pinto NC, Castilho RF, Vercesi AE (2009) Mitochondria and reactive oxygen species. Free Radic Biol Med 47(4):333–343. https://doi.org/10.1016/j.freeradbiomed.2009.05.004

Kulikov S, Tikhonov V, Bezrodnykh E, Lopatin S, Varlamov V (2015) Comparative evaluation of antimicrobial activity of oligochitosans against Klebsiella pneumoniae. Bioorg Khim 41(1):67–73

Kumar SG, Md. Rahman A, Lee SH, Hwang HS, Kim HA, Yun JW (2009) Plasma proteome analysis for anti-obesity and anti-diabetic potentials of chitosan oligosaccharides in ob/ob mice. Proteomics 9(8):2149–2162. https://doi.org/10.1002/pmic.200800571

Kunanusornchai W, Witoonpanich B, Pichyangkura R, Chatsudthipong V, Muanprasat C (2016) Chitosan oligosaccharide suppresses synovial inflammation via AMPK activation: an in vitro and in vivo study. Pharmacol Res 113:458–467. https://doi.org/10.1016/j.phrs.2016.09.016

Lee HW, Park YS, Jung JS, Shin WS (2002) Chitosan oligosaccharides, dp 2-8, have prebiotic effect on the Bifidobacterium bifidium and lactobacillus sp. Anaerobe 8(6):319–324. https://doi.org/10.1016/S1075-9964(03)00030-1

Lee H-W, Park Y-S, Choi J-W, Yi S, Shin W-S (2003) Antidiabetic effects of chitosan oligosaccharides in neonatal streptozotocin-induced noninsulin-dependent diabetes Mellitus in rats. Biol Pharm Bull 26(8):1100–1103. https://doi.org/10.1248/bpb.26.1100

Lee BC, Kim MS, Choi SH, Kim KY, Kim TS (2009a) In vitro and in vivo antimicrobial activity of water-soluble chitosan oligosaccharides against Vibrio vulnificus. Int J Mol Med 24:327–333. https://doi.org/10.3892/ijmm

Lee SH, Park JS, Kim SK, Ahn CB, Je JY (2009b) Chitooligosaccharides suppress the level of protein expression and acetylcholinesterase activity induced by Aβ25-35in PC12 cells. Bioorg Med Chem Lett 19(3):860–862. https://doi.org/10.1016/j.bmcl.2008.12.019

Lee SH, Senevirathne M, Ahn CB, Kim SK, Je JY (2009c) Factors affecting anti-inflammatory effect of chitooligosaccharides in lipopolysaccharides-induced RAW264.7 macrophage cells. Bioorg Med Chem Lett 19(23):6655–6658. https://doi.org/10.1016/j.bmcl.2009.10.007

Lei F, Zhang Y (2015) CN 105326053A. China

Li L, Liu Z (2013) Effect of oligochitosan on growth performance, immune function and intestinal microflora of quails. Cereal Feed Ind 7(5):50–52

Li XJ, Piao XS, Kim SW, Liu P, Wang L, Shen YB et al (2007) Effects of Chito-oligosaccharide supplementation on performance, nutrient digestibility , and serum composition in broiler chickens. Poult Sci 86:1107–1114

Li K, Xing R, Liu S, Li R, Qin Y, Meng X, Li P (2012a) Separation of chito-oligomers with several degrees of polymerization and study of their antioxidant activity. Carbohydr Polym 88(3):896–903. https://doi.org/10.1016/j.carbpol.2012.01.033

Li K, Xing R, Liu S, Qin Y, Li B, Wang X, Li P (2012b) Separation and scavenging superoxide radical activity of chitooligomers with degree of polymerization 6-16. Int J Biol Macromol 51(5):826–830. https://doi.org/10.1016/j.ijbiomac.2012.07.031

Li K, Xing R, Liu S, Qin Y, Meng X, Li P (2012c) Microwave-assisted degradation of chitosan for a possible use in inhibiting crop pathogenic fungi. Int J Biol Macromol 51(5):767–773. https://doi.org/10.1016/j.ijbiomac.2012.07.021

Li N, Xiao K, Pang H, Wang Z (2012d) Effect of oligo-chitosan on the growth of intestinal microorganisms as determined by MTT assay. Food Sci 33(03):101–104

Li K, Xing R, Liu S, Qin Y, Yu H, Li s P (2014a) Size and pH effects of chitooligomers on antibacterial activity against Staphylococcus aureus. Int J Biol Macromol 64:302–305. https://doi.org/10.1016/j.ijbiomac.2013.11.037

Li Y, Liu H, Xu Q, Du Y, Xu J (2014b) Chitosan oligosaccharides block LPS-induced O-GlcNAcylation of NF-κB and endothelial inflammatory response. Carbohydr Polym 99:568–578

Li Y, Xu Q, Wei P, Cheng L, Peng Q, Li S et al (2014c) Chitosan oligosaccharides downregulate the expression of E-selectin and ICAM-1 induced by LPS in endothelial cells by inhibiting MAP kinase signaling. Int J Mol Med 33(2):392–400. https://doi.org/10.3892/ijmm.2013.1589

Li K, Xing R, Liu S, Li P (2016) Advances in preparation, analysis and biological activities of single chitooligosaccharides. Carbohydr Polym 139:178–190. https://doi.org/10.1016/j.carbpol.2015.12.016
Li S, Jin E, Qiao E, Wu G, Li K (2017a) Chitooligosaccharide promotes immune organ development in broiler chickens and reduces serum lipid levels. Histol Histopathol 32(9):951–961. https://doi.org/10.14670/HH-11-860
Li X, Ding X, Peng X, Chi X, Cui H, Zuo Z, Fang J (2017b) Effect of chitosan oligosaccharides on antioxidant function, lymphocyte cycle and apoptosis in ileum mucosa of broiler. Kafkas Univ Vet Fak Derg 23(4):571–577. https://doi.org/10.9775/kvfd.2016.17329
Li X, Zhao M, Fan L, Cao X, Chen L, Chen J et al (2018) Chitobiose alleviates oleic acid-induced lipid accumulation by decreasing fatty acid uptake and triglyceride synthesis in HepG2 cells. J Funct Foods 46C:202–211
Lian G, Zhou X, Kong X, Feng Z, Yin Y (2012) Effects of dietary supplementing chitooligosaccharide on biochemical parameters and intestinal Flora in Huanjiang mini-piglets. Nat Prod Res Dev 24:1605–1609. https://doi.org/10.16333/j.1001-6880.2012.11.032
Liaqat F, Eltem R (2018) Chitooligosaccharides and their biological activities: a comprehensive review. Carbohydr Polym 184(December 2017):243–259. https://doi.org/10.1016/j.carbpol.2017.12.067
Liu B (2007) Antidiabetic effects of chitooligosaccharides and it's derivatives. Ocean University of China
Liu F, Du P, Huo G (2005) Study on inhibitant activities of chitosan oligosaccharide against microbes from milk. Zhonggue Rupin Gongye 33(6):28–20
Liu P, Piao XS, Kim SW, Wang L, Shen YB, Lee HS, Li SY (2008) Effects of chito-oligosaccharide supplementation on the growth performance, nutrient digestibility, intestinal morphology, and fecal shedding of Escherichia coli and lactobacillus in weaning pigs. J Anim Sci 86(10):2609–2618. https://doi.org/10.2527/jas.2007-0668
Liu HT, Li WM, Xu G, Li XY, Bai XF, Wei P et al (2009) Chitosan oligosaccharides attenuate hydrogen peroxide-induced stress injury in human umbilical vein endothelial cells. Pharmacol Res 59(3):167–175. https://doi.org/10.1016/j.phrs.2008.12.001
Liu HT, Li WM, Li XY, Xu QS, Liu QS, Bai XF et al (2010) Chitosan oligosaccharides inhibit the expression of interleukin-6 in lipopolysaccharide-induced human umbilical vein endothelial cells through p38 and ERK1/2 protein kinases. Basic Clin Pharmacol Toxicol 106(5):362–371. https://doi.org/10.1111/j.1742-7843.2009.00493.x
Liu X, Wang Z, Jiao N (2011) CN 102228181A. China
Liu L, Xin Y, Liu J, Zhang E, Li W (2017) Inhibitory effect of chitosan oligosaccharide on human hepatoma cells in vitro. Afr J Tradit Complement Altern Med 14(4):272–277. https://doi.org/10.21010/ajtcam.v14i4.30
Liu YE, Tong CC, Zhang YB, Cong PF, Shi XY, Liu Y et al (2018) Chitosan oligosaccharide ameliorates acute lung injury induced by blast injury through the DDAH1/ADMA pathway. PLoS One 13(2):1–14. https://doi.org/10.1371/journal.pone.0192135
Long T, Yu ZJ, Wang J, Liu J, He BS (2018) Orally administered chitooligosaccharides modulate colon microbiota in normal and colitis mice. Int J Pharmacol 14(2):291–300. https://doi.org/10.3923/ijp.2018.291.300
Lu J, Chen Q, Pan B, Qin Z, Fan L, Xia Q, Zhao L (2019) Efficient inhibition of Cronobacter biofilms by chitooligosaccharides of specific molecular weight. World J Microbiol Biotechnol 35(6):87. https://doi.org/10.1007/s11274-019-2662-5
Luo Z, Dong X, Ke Q, Duan Q, Shen L (2014a) Chitooligosaccharides inhibit ethanol-induced oxidative stress via activation of Nrf2 and reduction of MAPK phosphorylation. Oncol Rep 32(5):2215–2222. https://doi.org/10.3892/or.2014.3463
Luo Z, Dong X, Ke Q, Duan Q, Shen L (2014b) Downregulation of CD147 by chitooligosaccharide inhibits MMP-2 expression and suppresses the metastatic potential of human gastric cancer. Oncol Lett 8(1):361–366. https://doi.org/10.3892/ol.2014.2115
Ma P, Liu HT, Wei P, Xu QS, Bai XF, Du YG, Yu C (2011) Chitosan oligosaccharides inhibit LPS-induced over-expression of IL-6 and TNF-α in RAW264.7 macrophage cells through blockade

of mitogen-activated protein kinase (MAPK) and PI3K/Akt signaling pathways. Carbohydr Polym 84(4):1391–1398. https://doi.org/10.1016/j.carbpol.2011.01.045
Manulu MS, Sutanegara IN (2006) Pathogenesis and pharmacologic treatment of obesity: the role of energy regulatory mechanism. Acta Med Indones 38:42–50
Markesbery WR (1997) Oxidative stress hypothesis in Alzheimer's disease. Free Radic Biol Med 23:134–147
Masuda S, Azuma K, Kurozumi S, Kiyose M, Osaki T, Tsuka T et al (2014) Anti-tumor properties of orally administered glucosamine and N-acetyl-d-glucosamine oligomers in a mouse model. Carbohydr Polym 111:783–787. https://doi.org/10.1016/j.carbpol.2014.04.102
Mateos-Aparicio I, Mengíbar M, Heras A (2016) Effect of chito-oligosaccharides over human faecal microbiota during fermentation in batch cultures. Carbohydr Polym 137:617–624. https://doi.org/10.1016/j.carbpol.2015.11.011
Mattaveewong T, Wongkrasant P, Chanchai S, Pichyangkura R, Chatsudthipong V, Muanprasat C (2016) Chitosan oligosaccharide suppresses tumor progression in a mouse model of colitis-associated colorectal cancer through AMPK activation and suppression of NF-κB and mTOR signaling. Carbohydr Polym 145:30–36. https://doi.org/10.1016/j.carbpol.2016.02.077
Mei Y x, Dai X y, Yang W, Xu X w, Liang Y x (2015) Antifungal activity of chitooligosaccharides against the dermatophyte Trichophyton rubrum. Int J Biol Macromol 77:330–335. https://doi.org/10.1016/j.ijbiomac.2015.03.042
Mellitus D, Amanti M (1998) Coronary artery disease : a link between hypertension, diabetes mellitus, hyperlipidemia, and obesity. J Am Acad Nurse Pract 10(2):77–81
Mendis E, Kim MM, Rajapakse N, Kim SK (2007) An in vitro cellular analysis of the radical scavenging efficacy of chitooligosaccharides. Life Sci 80(23):2118–2127. https://doi.org/10.1016/j.lfs.2007.03.016
Monteiro R, Azevedo I (2010) Chronic inflammation in obesity and the metabolic syndrome. Mediators Inflamm 2010(Atp Iii). https://doi.org/10.1155/2010/289645
Muanprasat C, Chatsudthipong V (2017) Chitosan oligosaccharide: biological activities and potential therapeutic applications. Pharmacol Ther 170:80–97. https://doi.org/10.1016/j.pharmthera.2016.10.013
Muanprasat C, Wongkrasant P, Satitsri S, Moonwiriyakit A, Pongkorpsakol P, Mattaveewong T et al (2015) Activation of AMPK by chitosan oligosaccharide in intestinal epithelial cells: mechanism of action and potential applications in intestinal disorders. Biochem Pharmacol 96(3):225–236. https://doi.org/10.1016/j.bcp.2015.05.016
Muzzarelli R (1977) Chitin. Oxford Pergamon Press, London
Nam KS, Shon YH (2009) Suppression of metastasis of human breast cancer cells by chitosan oligosaccharides. J Microbiol Biotechnol 19(6):629–633. https://doi.org/10.4014/jmb.0811.603
Nam KS, Kim M-K, Shon Y-H (2007) Inhibition of proinflammatory cytokine-induced invasiveness of HT-29 cells by chitosan oligosaccharide. J Microbiol Biotechnol 17:2042–2045
Ngo DN, Kim MM, Kim SK (2008) Chitin oligosaccharides inhibit oxidative stress in live cells. Carbohydr Polym 74:228–234
Ngo DN, Lee SH, Kim MM, Kim SK (2009) Production of chitin oligosaccharides with different molecular weights and their antioxidant effect in RAW 264.7 cells. J Funct Foods 1(2):188–198. https://doi.org/10.1016/j.jff.2009.01.008
Nidheesh T, Salim C, Rajini PS, Suresh PV (2016) Antioxidant and neuroprotective potential of chitooligomers in Caenorhabditis elegans exposed to Monocrotophos. Carbohydr Polym 135:138–144. https://doi.org/10.1016/j.carbpol.2015.08.055
Nie Q, Liu D, Wang Y, Zheng M (2010) Effects of chitosan oligosaccharide on the quality of strawberry during storage. Innovational Ed Farm Prod Process 215(7):31–33. https://doi.org/10.3969/jissn.1671
No HK, Park NY, Lee SH, Meyers SP (2002) Antibacterial activities of chitosans and chitosan oligomers with different molecular. Int J Food Microbiol 67(4):65–72
Oliveira EN, El Gueddari NE, Moerschbacher BM, Peter MG, Franco TT (2008) Growth of phytopathogenic fungi in the presence of partially acetylated chitooligosaccharides. Mycopathologia 166(3):163–174. https://doi.org/10.1007/s11046-008-9125-0

Pae HO, Seo WG, Kim NY, Oh GS, Kim GE, Kim YH et al (2001) Induction of granulocytic differentiation in acute promyelocytic leukemia cells (HL-60) by water-soluble chitosan oligomer. Leuk Res 25(4):339–346. https://doi.org/10.1016/S0145-2126(00)00138-7

Palma-Guerrero J, Huang IC, Jansson HB, Salinas J, Lopez-Llorca LV, Read ND (2009) Chitosan permeabilizes the plasma membrane and kills cells of Neurospora crassa in an energy dependent manner. Fungal Genet Biol 46(8):585–594. https://doi.org/10.1016/j.fgb.2009.02.010

Palma-Guerrero J, Lopez-Jimenez JA, Pérez-Berná AJ, Huang IC, Jansson HB, Salinas J et al (2010) Membrane fluidity determines sensitivity of filamentous fungi to chitosan. Mol Microbiol 75(4):1021–1032. https://doi.org/10.1111/j.1365-2958.2009.07039.x

Pan H, Fu C, Huang L, Jiang Y, Deng X, Guo J, Su Z (2018) Anti-obesity effect of chitosan oligosaccharide capsules (coscs) in obese rats by ameliorating leptin resistance and adipogenesis. Mar Drugs 16(6). https://doi.org/10.3390/md16060198

Pandey A, Tripathi P, Pandey R, Goswami S (2011) Alternative therapies useful in the management of diabetes: a systematic review. J Pharm Bioallied Sci 3:504–512

Pangestuti R, Bak SS, Kim SK (2011) Attenuation of pro-inflammatory mediators in LPS-stimulated BV2 microglia by chitooligosaccharides via the MAPK signaling pathway. Int J Biol Macromol 49(4):599–606. https://doi.org/10.1016/j.ijbiomac.2011.06.014

Park PJ, Je JY, Kim SK (2003) Free radical scavenging activity of chitooligosaccharides by electron spin resonance spectrometry. J Agric Food Chem 51(16):4624–4627. https://doi.org/10.1021/jf034039+

Park PJ, Je J-Y, Byun HG, Moon S-H, Kim S-K (2004) Antimicrobial activity of hetero-chitosans and their oligosaccharides with different molecular weights. J Microbiol Biotechnol 14(2):317–323

Park Y, Kim M, Park S, Cheong H, Jang M, Nah J, Hahm K (2008) Investigation of the antifungal activity and mechanism of action of LMWS-chitosan. J Microbiol Biotechnol 18(10):1729–1734

Park JK, Chung MJ, Choi HN, Park YI (2011) Effects of the molecular weight and the degree of deacetylation of chitosan oligosaccharides on antitumor activity. Int J Mol Sci 12(1):266–277. https://doi.org/10.3390/ijms12010266

Powell EE, Jonsson JR, Clouston AD (2005) Steatosis: co-factor in other liver diseases. Hepatology 42(1):5–13. https://doi.org/10.1002/hep.20750

Powell E, Jonsson JR, Clouston AD (2010) Metabolic factors and non-alcoholic fatty liver disease as co-factors in other liver diseases. Dig Dis 28:186–191. https://doi.org/10.1159/000282084

Qiao Y, Bai XF, Du YG (2011) Chitosan oligosaccharides protect mice from LPS challenge by attenuation of inflammation and oxidative stress. Int Immunopharmacol 11(1):121–127. https://doi.org/10.1016/j.intimp.2010.10.016

Qin C, Du Y, Xiao L, Li Z, Gao X (2002) Enzymic preparation of water-soluble chitosan and their antitumor activity. Int J Biol Macromol 31(1–3):111–117. https://doi.org/10.1016/S0141-8130(02)00064-8

Qin C, Zhang Y, Liu W, Xu L, Yang Y, Zhou Z (2014) Effects of chito-oligosaccharides supplementation on growth performance, intestinal cytokine expression, autochthonous gut bacteria and disease resistance in hybrid tilapia Oreochromis niloticus ♀×Oreochromis aureus ♂. Fish Shellfish Immunol 40(1):267–274. https://doi.org/10.1016/j.fsi.2014.07.010

Qin P, Zhang B, Guan C, Xu X, Wang L, Gao Q, Yue F (2018) CN108552320A. China

Qu W (2012) Preparation of chitooligosaccharid-chromium(III) complex and hypoglycemic effects on diabetic mice. Liaoning Normal University

Qu D, Han J (2016) Investigation of the antioxidant activity of chitooligosaccharides on mice with high-fat diet. Rev Bras Zootec 45(11):661–666. https://doi.org/10.1590/S1806-92902016001100004

Quintero-Villegas MI, Aam BB, Rupnow J, Sørlie M, Eijsink VGH, Hutkins RW (2013) Adherence inhibition of enteropathogenic escherichia coli by chitooligosaccharides with specific degrees of acetylation and polymerization. J Agric Food Chem 61(11):2748–2754. https://doi.org/10.1021/jf400103g

Rahman MA, Kumar SG, Sang WK, Hye JH, Yu MB, Sung HL et al (2008) Proteomic analysis for inhibitory effect of chitosan oligosaccharides on 3T3-L1 adipocyte differentiation. Proteomics 8(3):569–581. https://doi.org/10.1002/pmic.200700888

Rhoades J, Gibson G, Formentin K, Beer M, Rastall R (2006) Inhibition of the adhesion of enteropathogenic Escherichia coli strains to HT-29 cells in culture by chito-oligosaccharides. Carbohydr Polym 64(1):57–59. https://doi.org/10.1016/j.carbpol.2005.10.025

Riccardi G, Capaldo B, Vaccaro O (2005) Functional foods in the management of obesity and type 2 diabetes. Curr Opin Clin Nutr Metab Care 8:630–635

Roberfroid M, Gibson G, Hoyles L, McCartney A, Rastall R, Rowland I et al (2010) Prebiotic effects: metabolic and health benefits. Br J Nutr 104:S1–S63. https://doi.org/10.1016/j.ajhg.2009.11.016

Robles-Martínez L, Guerra-Sánchez MG, Hernández-Lauzardo AN, Pardo JP, Velázquez-del Valle MG (2014) Effects of chitosan and oligochitosan on development and mitochondrial function of Rhizopus stolonifer. J Basic Microbiol 54(SUPPL.1):42–49. https://doi.org/10.1002/jobm.201300790

Rothhammer V, Borucki DM, Tjon EC, Takenaka MC, Chao C-C, Ardura-Fabregat A et al (2018) Microglial control of astrocytes in response to microbial metabolites. Nature 557:724–728. https://doi.org/10.1038/s41586-018-0119-x

Sa R, Hu W, Xiu Z, Feng K, Jiang A, Ji Y (2017) Effect of active chitosan oligosaccharide coating treatment on quality of fresh-cut apples. Packag Eng 38(17):25–30. https://doi.org/10.19554/j.cnki.1001-3563.2017.17.006

Salah R, Michaud P, Mati F, Harrat Z, Lounici H, Abdi N et al (2013) Anticancer activity of chemically prepared shrimp low molecular weight chitin evaluation with the human monocyte leukaemia cell line, THP-1. Int J Biol Macromol 52(1):333–339. https://doi.org/10.1016/j.ijbiomac.2012.10.009

Salgaonkar N, Prakash D, Nawani NN, Kapadnis BP (2015) Comparative studies on ability of N-acetylated chitooligosaccharides to scavenge reactive oxygen species and protect DNA from oxidative damage. Indian J Biotechnol 14(2):186–192

Sánchez MM, Rivera-Rodríguez G, Moerchbacher B, Acosta N, Heras A (2017) The effect of preparation processes on the physicochemical characteristics and antibacterial activity of chitooligosaccharides. Carbohydr Polym 157:251–257. https://doi.org/10.1016/j.carbpol.2016.09.055

Schwartz M, Hirschberg DL (1995) Central nervous system regeneration and the immune system. Mol Med Today 1(2):60. https://doi.org/10.1016/S1357-4310(95)92274-1

Sechler M, Cizmic A, Avasarala S, Van Scoyk M, Brzezinski C, Kelley N et al (2013) Non-small-cell lung cancer: molecular targeted therapy and personalized medicine – drug resistance, mechanisms, and strategies. Pharmacogenomics Pers Med 6:25–36

Senevirathne M, Ahn CB, Je JY (2011) Hepatoprotective effect of chitooligosaccharides against tert-butylhydroperoxide-induced damage in Chang liver cells. Carbohydr Polym 83(2):995–1000. https://doi.org/10.1016/j.carbpol.2010.09.016

Seyfarth F, Schliemann S, Elsner P, Hipler UC (2008) Antifungal effect of high- and low-molecular-weight chitosan hydrochloride, carboxymethyl chitosan, chitosan oligosaccharide and N-acetyl-d-glucosamine against Candida albicans, Candida krusei and Candida glabrata. Int J Pharm 353(1–2):139–148. https://doi.org/10.1016/j.ijpharm.2007.11.029

Šimůnek J, Koppová I, Filip L, Tishchenko G, BeŁzecki G (2010) The antimicrobial action of low-molar-mass chitosan, chitosan derivatives and chitooligosaccharides on bifidobacteria. Folia Microbiol 55(4):379–382. https://doi.org/10.1007/s12223-010-0063-0

Šimůnek J, Brandysová V, Koppová I, Šimůnek J (2012) The antimicrobial action of chitosan, low molar mass chitosan, and chitooligosaccharides on human colonic bacteria. Folia Microbiol 57(4):341–345. https://doi.org/10.1007/s12223-012-0138-1

Singh P (2016) Effect of chitosans and chitooligosaccharides on the processing and storage quality of foods of animal and aquatic origin: a review. Nutr Food Sci 46(1):51–81. https://doi.org/10.1108/NFS-08-2015-0092

Siow NL, Choi RCY, Cheng AWM, Jiang JXS, Wan DCC, Zhu SQ, Tsim KWK (2002) A cyclic AMP-dependent pathway regulates the expression of acetylcholinesterase during myogenic differentiation of C2C12 cells. J Biol Chem 277(39):36129–36136. https://doi.org/10.1074/jbc.M206498200

Spaans J, Goss G (2014) Drug resistance to molecular targeted therapy and its consequences for treatment decisions in non-small-cell lung cancer. Front Oncol 4:190

Streit WJ, Graeber MB, Kreutzberg GW (1998) Functional plasticity of microglia: a review. Glia 1:301–307

Su P, Han Y, Jiang C, Ma Y, Pan J, Liu S, Zhang T (2017) Effects of chitosan-oligosaccharides on growth performance, digestive enzyme and intestinal bacterial flora of tiger puffer (Takifugu rubripes Temminck et Schlegel, 1850). J Appl Ichthyol 33(3):458–467. https://doi.org/10.1111/jai.13282

Sumiyoshi M, Kimura Y (2006) Low molecular weight chitosan inhibits obesity induced by feeding a high-fat diet long-term in mice. J Pharm Pharmacol 58(2):201–207. https://doi.org/10.1211/jpp.58.2.0007

Suzuki KO, Mikami T, Okawa Y, Tokoro A, Suzuki S, Suzuki M (1986) Antitumor effect of hexa-N-acetylchitohexaose and chitohexaose. Carbohydr Res 151:403–408

Takahashi Y, Soejima Y, Fukusato T (2012) Animal models of nonalcoholic fatty liver disease/nonalcoholic steatohepatitis. World J Gastroenterol 18(19):2300–2308. https://doi.org/10.3748/wjg.v18.i19.2300

Tao Y, Zhang HL, Hu YM, Wan S, Su ZQ (2013) Preparation of chitosan and water-soluble chitosan microspheres via spray-drying method to lower blood lipids in rats fed with high-fat diets. Int J Mol Sci 14(2):4174–4184. https://doi.org/10.3390/ijms14024174

Shen KT, Chen MH, Chan HY, Jeng JH, Wang YJ (2009) Inhibitory effects of chitooligosaccharides on tumor growth and metastasis. Food Chem Toxicol 47(8):1864–1871. https://doi.org/10.1016/j.fct.2009.04.044

Tikhonov VE, Stepnova EA, Babak VG, Yamskov IA, Palma-Guerrero J, Jansson HB et al (2006) Bactericidal and antifungal activities of a low molecular weight chitosan and its N–/2(3)-(dodec-2-enyl)succinoyl/–derivatives. Carbohydr Polym 64(1):66–72. https://doi.org/10.1016/j.carbpol.2005.10.021

Tokoro A, Tatewaki N, Suzuki K, Mikami T, Suzuki S, Suzuki M (1988) Growth- inhibitory effect of hexa-N-acetylchitohexanse and chitohexaose against Meth-A solid tumor. Chem Pharm Bull 36(2):784–790

Tong Q, Yang Y, Xiong Z, Li Z, Yuan W, Wang T (2016) Chitooligosaccharides downregulate TLR4/NF-KB/cox-2 signaling cascade in dextran sulfate sodium-treated mice: a potential mechanism for the anti-colitis effect. Int J Pharmacol 12(7):720–728. https://doi.org/10.3923/ijp.2016.720.728

Tsai GJ, Su WH, Chen HC, Pan CL (2002) Antimicrobial activity of shrimp chitin and chitosan from different treatments and applications of fish preservation. Fish Sci 68(1):170–177. https://doi.org/10.1046/j.1444-2906.2002.00404.x

Tucci P, Bagetta G (2008) How to study neuroprotection? Cell Death Differ 15(6):1084–1085. https://doi.org/10.1038/cdd.2008.32

Tufan T, Arslan C, Sari M, Önk K, Deprem T, Celik E (2015) Effects of chitosan oligosaccharides addition to Japanese Quail's diets on growth, carcass traits, liver and intestinal histology, and intestinal microflora. Kafkas Univ Vet Fak Derg 21(5):665–671. https://doi.org/10.9775/kvfd.2015.13010

Van Ta Q, Kim MM, Kim SK (2006) Inhibitory effect of chitooligosaccharides on matrix metalloproteinase-9 in human fibrosarcoma cells (HT1080). Mar Biotechnol 8(6):593–599. https://doi.org/10.1007/s10126-006-6031-7

Vassar R, Bennett BD, Babu-khan S, Kahn S, Mendiaz EA, Denis P et al (2014) β-secretase cleavage of Alzheimer's amyloid precursor protein by the transmembrane aspartic protease BACE. Science 735(1999):735–742. https://doi.org/10.1126/science.286.5440.735

Vela Gurovic MS, Dello Staffolo M, Montero M, Debbaudt A, Albertengo L, Rodríguez MS (2015) Chitooligosaccharides as novel ingredients of fermented foods. Food Funct 6(11):3437–3443. https://doi.org/10.1039/c5fo00546a

Verbeke KA, Boobis AR, Chiodini A, Edwards CA, Franck A, Kleerebezem M et al (2015) Towards microbial fermentation metabolites as markers for health benefits of prebiotics. Nutr Res Rev 28(1):42–66. https://doi.org/10.1017/S0954422415000037

Vernazza CL, Gibson GR, Rastall RA (2005) In vitro fermentation of chitosan derivatives by mixed cultures of human faecal bacteria. Carbohydr Polym 60(4):539–545. https://doi.org/10.1016/j.carbpol.2005.03.008

Vishu Kumar AB, Varadaraj MC, Gowda LR, Tharanathan RN (2005) Characterization of chito-oligosaccharides prepared by chitosanolysis with the aid of papain and Pronase, and their bactericidal action against *Bacillus cereus* and *Escherichia coli*. Biochem J 391(2):167–175. https://doi.org/10.1042/BJ20050093

Vishu Kumar AB, Varadaraj MC, Gowda LR, Tharanathan RN (2007) Low molecular weight chitosans-preparation with the aid of pronase, characterization and their bactericidal activity towards Bacillus cereus and Escherichia coli. Biochim Biophys Acta Gen Subj 1770(4):495–505. https://doi.org/10.1016/j.bbagen.2006.12.003

Vo TS, Kong CS, Kim SK (2011) Inhibitory effects of chitooligosaccharides on degranulation and cytokine generation in rat basophilic leukemia RBL-2H3 cells. Carbohydr Polym 84(1):649–655. https://doi.org/10.1016/j.carbpol.2010.12.046

Wan J, Jiang F, Xu Q, Chen D, Yu B, Huang Z et al (2017) New insights into the role of chitosan oligosaccharide in enhancing growth performance, antioxidant capacity, immunity and intestinal development of weaned pigs. RSC Adv 7(16):9669–9679. https://doi.org/10.1039/c7ra00142h

Wang X (2016) Comparative study on formula food for special medical purposes and health food. Chin Pharm Aff 30:642–645. https://doi.org/10.16153/j.1002-7777.2016.07.002

Wang X (2017) Suggestions for research and development on formula food for special medical purposes. Food Nutr China 23(4):8–10. https://doi.org/10.1111/j.1469-7610.2010.02280.x

Wang J, Gong X (2001) Advances in the studies on the antitumor activity and immunomodulating action of polysaccharides. Chin J Biochem Pharm 22(1):52–54

Wang X, Du Y, Bai X, Li S (2003) The effect of oligochitosan on broiler gut flora, microovilli density, immune function and growth performance. Acta Zoonutrimenta Sin 15:32–35

Wang Y, Zhou P, Yu J, Pan X, Wang P, Lan W, Tao S (2007) Antimicrobial effect of chitooligosaccharides produced by chitosanase from pseudomonas CUY8. Asia Pac J Clin Nutr 16(SUPPL.1):174–177

Wang JP, Yoo JS, Kim HJ, Lee JH, Kim IH (2009a) Nutrient digestibility, blood profiles and fecal microbiota are influenced by chitooligosaccharide supplementation of growing pigs. Livest Sci 125(2–3):298–303. https://doi.org/10.1016/j.livsci.2009.05.011

Wang R, Wei Y, Ma L, Zhao X, Li H (2009b) Weight-reducing effect of beta-chitoligosaccharide on diet-induced obese rats. J Math Med 22(2):216–218

Wang S, Lu Y, Qian W, Zhao Y, Wang Y, Zheng S, Wang A (2010) Cell charge: potential important target of anti-tumor research. Chin Pharm Bull 26(12):1541–1544

Wang B, Li Z, Luo H, Qu Y (2013) CN 103005255 A. China

Wang J, Jiang M, Xin Y, Geng N, Li X, Xuan S (2016) Effect of chitooligosaccharide on hepatic triglyceride metabolism and related mechanisms. Chin J Hepatol 24:220–224

Wei D (2012) Preparation of chitooligosaccharides chelating vanadium and comparative study of its hypoglycemic function. Liaoning Normal University

Wei P, Ma P, Xu QS, Bai QH, Gu JG, Xi H et al (2012) Chitosan oligosaccharides suppress production of nitric oxide in lipopolysaccharide-induced N9 murine microglial cells in vitro. Glycoconj J 29(5–6):285–295. https://doi.org/10.1007/s10719-012-9392-3

World Health Organization (2014) Global Status Report on Noncommunicable Diseases

Wu H, Yao Z, Bai X, Du Y, Lin B (2008) Anti-angiogenic activities of chitooligosaccharides. Carbohydr Polym 73(1):105–110. https://doi.org/10.1016/j.carbpol.2007.11.011

Wu H, Yao Z, Bai X, Du Y, Ma X (2010) Chitooligosaccharides inhibit nitric oxide mediated migration of endothelial cells in vitro and tumor angiogenesis in vivo. Carbohydr Polym 82(3):927–932. https://doi.org/10.1016/j.carbpol.2010.06.015

Wu H, Aam BB, Wang W, Norberg AL, Sørlie M, Eijsink VGH, Du Y (2012) Inhibition of angiogenesis by chitooligosaccharides with specific degrees of acetylation and polymerization. Carbohydr Polym 89(2):511–518. https://doi.org/10.1016/j.carbpol.2012.03.037

Wu W, Wei W, Lu M, Zhu X, Liu N, Niu Y et al (2017a) Neuroprotective effect of chitosan oligosaccharide on hypoxic-ischemic brain damage in neonatal rats. Neurochem Res 42(11):3186–3198. https://doi.org/10.1007/s11064-017-2356-z

Wu X, Wang J, Shi Y, Chen S, Yan Q, Jiang Z, Jing H (2017b) N-acetyl-chitobiose ameliorates metabolism dysfunction through Erk/p38 MAPK and histone H3 phosphorylation in type 2 diabetes mice. J Funct Foods 28:96–105. https://doi.org/10.1016/j.jff.2016.11.012

Xia W, Liu P, Zhang J, Chen J (2011) Biological activities of chitosan and chitooligosaccharides. Food Hydrocoll 25(2):170–179. https://doi.org/10.1016/j.foodhyd.2010.03.003

Xia Z, Chen J, Wu S (2013) Hypolipidemic activity of the chitooligosaccharides from Clanis bilineata (Lepidoptera), an edible insect. Int J Biol Macromol 59:96–98. https://doi.org/10.1016/j.ijbiomac.2013.04.017

Xiao D (2011) Effects and mechanism of chitosan on growth, intestinal barrier and immunity in piglets. Hunan Agricultural University

Xiong C, Wu H, Wei P, Pan M, Tuo Y, Kusakabe I, Du Y (2009) Potent angiogenic inhibition effects of deacetylated chitohexaose separated from chitooligosaccharides and its mechanism of action in vitro. Carbohydr Res 344(15):1975–1983. https://doi.org/10.1016/j.carres.2009.06.036

Xu Q, Dou J, Wei P, Tan C, Yun X, Wu Y et al (2008) Chitooligosaccharides induce apoptosis of human hepatocellular carcinoma cells via up-regulation of Bax. Carbohydr Polym 71(4):509–514. https://doi.org/10.1016/j.carbpol.2007.06.022

Xu Q, Ma P, Yu W, Tan C, Liu H, Xiong C et al (2010) Chitooligosaccharides protect human embryonic hepatocytes against oxidative stress induced by hydrogen peroxide. Mar Biotechnol 12(3):292–298. https://doi.org/10.1007/s10126-009-9222-1

Xu Y, Zhang Q, Yu S, Yang Y, Ding F (2011) The protective effects of chitooligosaccharides against glucose deprivation-induced cell apoptosis in cultured cortical neurons through activation of PI3K/Akt and MEK/ERK1/2 pathways. Brain Res 1375:49–58. https://doi.org/10.1016/j.brainres.2010.12.029

Xu Q, Liu M, Liu Q, Wang W, Du Y, Yin H (2017a) The inhibition of LPS-induced inflammation in RAW264.7 macrophages via the PI3K/Akt pathway by highly N-acetylated chitooligosaccharide. Carbohydr Polym 174:1138–1143. https://doi.org/10.1016/j.carbpol.2017.07.051

Xu Q, Wang W, Qu C, Gu J, Yin H, Jia Z et al (2017b) Chitosan oligosaccharides inhibit epithelial cell migration through blockade of N-acetylglucosaminyltransferase V and branched GlcNAc structure. Carbohydr Polym 170:241–246. https://doi.org/10.1016/j.carbpol.2017.04.075

Yadav M, Verma MK, Chauhan NS (2018) A review of metabolic potential of human gut microbiome in human nutrition. Arch Microbiol 200(2):203–217. https://doi.org/10.1007/s00203-017-1459-x

Yang TC, Chou CC, Li CF (2005) Antibacterial activity of N-alkylated disaccharide chitosan derivatives. Int J Food Microbiol 97(3):237–245. https://doi.org/10.1016/S0168-1605(03)00083-7

Yang Y, Liu W, Han B, Liu B, Fu C (2006) Protective effects of chitosan oligosaccharide and its derivatives against carbon tetrachloride induced liver damage in mice. HepatolRes 35(3):178–184. https://doi.org/10.1016/j.hepres.2006.04.001

Yang CM, Ferket PR, Hong QH, Zhou J, Cao GT, Zhou L, Chen AG (2012) Effect of chito-oligosaccharide on growth performance, intestinal barrier function, intestinal morphology and cecal microflora in weaned pigs. J Anim Sci 90(8):2671–2676. https://doi.org/10.2527/jas.2011-4699

Yang Y, Tong Q, Luo H, Huang R, Li Z (2016) Chitooligosaccharides attenuate lipopolysaccharide-induced inflammation and apoptosis of intestinal epithelial cells: possible involvement of TLR4/NF-κB pathway. Indian J Pharm Educ Res 50(1):109–115. https://doi.org/10.5530/ijper.50.1.14

Yang F, Luan B, Sun Z, Yang C, Yu Z, Li X (2017) Application of chitooligosaccharides as antioxidants in beer to improve the flavour stability by protecting against beer staling during storage. Biotechnol Lett 39(2):305–310. https://doi.org/10.1007/s10529-016-2248-3

Yang JW, Tian G, Chen DW, Yao Y, He J, Zheng P et al (2018a) Involvement of PKA signalling in anti-inflammatory effects of chitosan oligosaccharides in IPEC-J2 porcine epithelial cells. J Anim Physiol Anim Nutr 102(1):252–259. https://doi.org/10.1111/jpn.12686

Yang X, Zhang J, Chen L, Wu Q, Yu C (2018b) Chitosan oligosaccharides enhance lipid droplets via down-regulation of PCSK9 gene expression in HepG2 cells. Exp Cell Res 366(2):152–160. https://doi.org/10.1016/j.yexcr.2018.03.013

Yoon HJ, Moon ME, Park HS, Im SY, Kim YH (2007) Chitosan oligosaccharide (COS) inhibits LPS-induced inflammatory effects in RAW 264.7 macrophage cells. Biochem Biophys Res Commun 358(3):954–959. https://doi.org/10.1016/j.bbrc.2007.05.042

Yoon NY, Ngo DN, Kim SK (2009) Acetylcholinesterase inhibitory activity of novel chitooligosaccharide derivatives. Carbohydr Polym 78(4):869–872. https://doi.org/10.1016/j.carbpol.2009.07.004

You J (2018) CN 108260802A. China

You J, Ji J (2018a) CN 108703364A. China

You J, Ji J (2018b) CN108013452A. China

You J, Ji J (2018c) CN108013453A. China

You J, Ji J (2018d) CN108618129A. China

Yousef M, Pichyangkura R, Soodvilai S, Chatsudthipong V, Muanprasat C (2012) Chitosan oligosaccharide as potential therapy of inflammatory bowel disease: therapeutic efficacy and possible mechanisms of action. Pharmacol Res 66(1):66–79. https://doi.org/10.1016/j.phrs.2012.03.013

Yu SY, Kwon YI, Lee C, Apostolidis E, Kim YC (2017) Antidiabetic effect of chitosan oligosaccharide (GO2KA1) is mediated via inhibition of intestinal alpha-glucosidase and glucose transporters and PPARγ expression. Biofactors 43(1):90–99. https://doi.org/10.1002/biof.1311

Yuan WP, Liu B, Liu CH, Wang XJ, Zhang MS, Meng XM, Xia XK (2009) Antioxidant activity of chito-oligosaccharides on pancreatic islet cells in streptozotocin-induced diabetes in rats. World J Gastroenterol 15(11):1339–1345. https://doi.org/10.3748/wjg.15.1339

Yuan G, Sun H, Chen X (2018) CN 108633987 A. China

Zhang Q, Yang L (2018) CN 107873933 A. China

Zhang L, Cui H, Liu A, Wang P (2009) Association of hypertemsion, diabetes, dyslipidemia, and metabolic syndrome with overweight/obesity. Chin J Prev Contr Chron Dis 17:6–8

Zhang J, Zhang W, Mamadouba B, Xia W (2012) A comparative study on hypolipidemic activities of high and low molecular weight chitosan in rats. Int J Biol Macromol 51(4):504–508. https://doi.org/10.1016/j.ijbiomac.2012.06.018

Zhang Y, Liu J, Yao J, Ji G, Qian L, Wang J et al (2014) Obesity: pathophysiology and intervention. Nutrients 6(11):5153–5183. https://doi.org/10.3390/nu6115153

Zhang C, Liao Q, Ming J-H, Hu G-L, Chen Q, Liu S-Q, Li Y-M (2017) The effects of chitosan oligosaccharides on OPG and RANKL expression in a rat osteoarthritis model. Acta Cir Bras 32(6):418–428. https://doi.org/10.1590/s0102-865020170060000002

Zhao X, Yu Z, Wang T, Guo X, Luan J, Sun Y, Li X (2016) The use of chitooligosaccharide in beer brewing for protection against beer-spoilage bacteria and its influence on beer performance. Biotechnol Lett 38(4):629–635. https://doi.org/10.1007/s10529-015-2013-z

Zheng J, Cheng G, Li Q, Jiao S, Feng C, Zhao X et al (2018a) Chitin oligosaccharide modulates gut microbiota and attenuates high-fat-diet-induced metabolic syndrome in mice. Mar Drugs 16(2). https://doi.org/10.3390/md16020066

Zheng J, Yuan X, Cheng G, Jiao S, Feng C, Zhao X et al (2018b) Chitosan oligosaccharides improve the disturbance in glucose metabolism and reverse the dysbiosis of gut microbiota in diabetic mice. Carbohydr Polym 190(1):77–86. https://doi.org/10.1016/j.carbpol.2018.02.058

Zhou S, Yang Y, Gu X, Ding F (2008) Chitooligosaccharides protect cultured hippocampal neurons against glutamate-induced neurotoxicity. Neurosci Lett 444(3):270–274. https://doi.org/10.1016/j.neulet.2008.08.040

Zhou Z, Wang J, Li Y (2017) The effects of resistant starch, chitosan oligosaccharide and their compound on high fat diet rats' Glycometabolism and lipid metabolism. Food Res Dev 38(6):189–193

Zopf D, Roth S (1996) Oligosaccharide anti-infective agents. Lancet (London, England) 347(9007):1017–1021. Retrieved from http://www.ncbi.nlm.nih.gov/pubmed/8606566

Zou P, Yang X, Wang J, Li Y, Yu H, Zhang Y, Liu G (2016a) Advances in characterisation and biological activities of chitosan and chitosan oligosaccharides. Food Chem 190(12):1174–1181. https://doi.org/10.1016/j.foodchem.2015.06.076

Zou P, Yang X, Zhang Y, Du P, Yuan S, Yang D, Wang J (2016b) Antitumor effects of orally and intraperitoneally administered chitosan oligosaccharides (COS) on S180-bearing/residual mouse. J Food Sci 81(12):H3035–H3042. https://doi.org/10.1111/1750-3841.13538

The Application of Chitooligosaccharides on Biomaterials

9

Qiming Chen and Liming Zhao

Abstract

Biomaterials are natural or synthetic materials that are suitable for introduction into human body. Chitooligosaccharides has many functions such as bone repair, wound healing, anti-inflammatory and nerve repair. It also has advantages of biocompatibility, biodegradability and no-toxicity. Thus, it widely applied on biomaterials preparation. Although chitooligosaccharides's molecular weight is low compared with chitosan that limited its application on biomaterials, it also has unique advantages such as better solubility. All these are helpful in biomaterials preparation. Chitooligosaccharides has been used in preparation nanoparticles and nanofibers, and further applied in drug and gene delivery, nutrition fortification and wound dress. Hence, we will review the development and preparation of biomaterials based on chitooligosaccharides as well as their applications in different fields in this chapter.

Keywords
Chitooligosaccharides · Biomaterials · Nanoparticles · Nanofibers

9.1 Introduction

Biomaterials are natural or synthetic materials that are suitable for introduction into human body. Biomaterials have become one of the most important base of modern medical engineering, pharmaceutical preparations and so on. The most important or the most basic requirements for biomaterials are biocompatibility and safety. Now, a lot of biomaterials were developed and they showed various functions and

Q. Chen · L. Zhao (✉)
State Key Laboratory of Bioreactor Engineering, School of Biotechnology, East China University of Science and Technology, Shanghai, China
e-mail: zhaoliming@ecust.edu.cn

L. Zhao (ed.), *Oligosaccharides of Chitin and Chitosan*,
https://doi.org/10.1007/978-981-13-9402-7_9

application. Chitooligosaccharides is a linear polysaccharide composed of randomly distributed β-(1→4)-linked D-glucosamine (deacetylated unit) and N-acetyl-D-glucosamine (acetylated unit) with a degree of deacetylation of 90%, polymerization degree 2 to 10 and average molecular weight less than 1.61 kDa (Choi et al. 2016). It is usually prepared from the chitin in shells of shrimp and other crustaceans which is the second most abundant polymer after cellulose on the earth by acid hydrolysis method, enzymatic approach or combination of them. Chitooligosaccharides has various functions such as anti-cancer (Chokradjaroen et al. 2018), boosts immunity (Chi et al. 2017), antidiabetic effect (Yu et al. 2017). It also has good biocompatibility (Richardson et al. 1999), biodegradability (Lee et al. 1995) and no-toxicity (Kim et al. 2001). Chitosan has been prepared as gel (Sacco et al. 2018), quantum dot (Tashkhourian et al. 2018) and nanoparticle (Heidarisasan et al. 2018). It has been widely used in detection (Rezaei et al. 2018), drug carrier (Moghadam et al. 2018), neural repair (Li et al. 2018) and wound healing (Oryan et al. 2018) and so on. Compared with chitosan, there are much less reports about application of chitooligosaccharides on biomaterials. This can contribute to the low molecular weight of chitooligosaccharides which consequently cause it more difficult in biomaterials preparation. However, chitooligosaccharides has obviously better solubility, biocompatibility and biodegradability than chitosan because chitosan is insolube at physiological pH values (7.2–7.4) (Rahmouni et al. 2018). Chitooligosaccharides then has better advantages over chitosan in preparing some biomaterials such as nanoparticle. Hence, the preparation of chitooligosaccharides nanoparticle and nanofibers and their application on drug carriers, gene deliverer and so on will be reviewed in this chapter.

9.2 Chitooligosaccharides Nanoparticle

Nanoparticles are particles between 1 and 100 nanometres (nm) in size. Nanoparticles have various new properties because of surface effect (Xu et al. 1996), volume effect (Choi et al. 2013), quantum size effect (Andrievski and Glezer 2001) and macroscopic quantum tunnelling effect (Materials and Council 1999), which bring various functional changes and better design ability at molecular level. Chitooligosaccharides is a good candidate for nanoparticle preparation because of its free amine groups and cationic nature as well as its solubility, biocompatibility and biodegradability. Nanoparticles are also the main applications of chitooligosaccharides as biomaterials. Chitooligosaccharides nanoparticles have been widely used in drug carriers, gene deliverers, bioimaging, and so on.

Cross-linking method, emulsion-droplet coalescence method, precipitation method and ionic gelation method are common methods for chitooligosaccharides nanoparticles preparation (Liu et al. 2007):

1. **Cross-linking method**. Chitooligosaccharides has amidogens and hydroxy groups on its carbohydrate chain. They can react with cross-linking agents and drug carriers appear. This method, namely cross-linking method, could be used in the coating of medicine that is not lipophilic and the nanoparticles usually have particle size greater than 600 nm. The particle size is oversize. More importantly, the most used cross-linking agent is glutaraldehyde, which has high toxicity. The application of nanoparticles prepared by this method is thus limit.
2. **Emulsion-droplet coalescence method**. Emulsion-droplet coalescence based on emulsion chemical-cross linking and precipitation (Tokumitsu et al. 1999). This method could be used in the coating of soluble medicine and the nanoparticles usually has particle size from 200 to 400 nm. The disadvantages of it is the process complexity and uncontrollability of particle size.
3. **Precipitation method**. This method could be used in the coating of various medicines. However, the particle size is usually over 0.9 μm which is oversize for carriers, and therefore there were only a few reports;
4. **Ionic gelation method.** Ionic gelation method is the most used method for chitooligosaccharides nanoparticles prepared. The reaction condition is mild and it usually used sodium tripolyphosphate as a polyanion crosslinker to induce gelation of Chitooligosaccharides. Sodium tripolyphosphate is a common washing assistant. It can be used as a food additive. The United States Food and Drug Administration lists STPP as "generally recognized as safe." The working concentration is safe for human during the preparation.

Most of the grafts crosslinked on the surface of hydrophobic chitooligosaccharides nanoparticles are fatty acids (Table 9.1). Fatty acid is a carboxylic acid with a long aliphatic chain, which is either saturated or unsaturated. Most naturally occurring fatty acids have an unbranched chain of an even number of carbon atoms, from 4 to 28 (Mcnaught and Wilkinson 2006). Fatty acid used in hydrophobic chitooligosaccharides nanoparticles modification can improve the drug loading capacity and stability. Stearic acid is a saturated fatty acid with an 18-carbon chain. It is widely used in the preparation of hydrophobic surfaces (for chitooligosaccharides, it is also the most used graft (Table 9.1)) because of its nontoxicity, easily availability and hydrophobic properties (Sauthier et al. 2014). The chemical conjugate of the coupling reaction of carboxyl group of stearic acid with amine group of chitooligosaccharides use cross-linking agents such as 1-ethyl-3-(3-dimethylaminopropyl) carbodiimide (Schild and Tirrell 1991). Other fatty acids such as hyaluronic acid (Gwak et al. 2012b) and linoleic acid (Du et al. 2009) are similar with stearic acid in preparation.

9.2.1 Chitooligosaccharides Nanoparticle as Drug Carrier

Many anti-cancer drugs such as doxorubicin and paclitaxel have cytotoxicity in normal tissue, and pervasive cardiotoxic effects or poor water solubility (Barry et al. 2007; Huo et al. 2010). These limit the application of them in therapeutic treatment.

Table 9.1 Application of Chitooligosaccharides in drug carriers

Published time	Functional groups for modification	Molecular weight of chitooligosaccharides	Drug	Reference
2008	Stearic acid	15.0 kDa	Doxorubicin	Hu et al. (2008)
2009	Non	No data	Non	Zhang et al. (2009)
2009	Linoleic acid	45 kDa	Doxorubicin	Du et al. (2009)
2009	Stearic acid	18 kDa	Doxorubicin	Hu et al. (2009b)
2009	Stearic acid	No data	Chlorine e6	Hu et al. (2009a)
2010	Non	5, 11 and 18 kDa	ATP	Du et al. (2010b)
2010	Non	8 kDa	Paclitaxel	Du et al. (2010a)
2010	Stearic acid and poly (lactic-co-glycolic acid)	5 kDa	Hydroxycamptothecin	Zhou et al. (2010)
2010	Pluronic	No data	Doxorubicin	(Yang et al. (2010)
2011	Nanostructured lipid	No data	Flurbiprofen	Luo et al. (2011)
2012	Tpp-hyaluronic acid	No data	DNA	Gwak et al. (2012a)
2012	Non	No data	Erythropoietin	Wang et al. (2012)
2012	Non	No data	Fluorouracil	(Li et al. (2013)
2013	Pegylated and fatty acid	≤200 kDa	Non	(Hu et al. (2013)
2013	Cellulose nanocrystal	5 kDa	Non	Akhlaghi et al. (2013)
2013	Non	6.1, 11.5 and 13.7 kDa	DNA	Delgado et al. (2013)
2013	Non	No data	pCMS-EGFP	Puras et al. (2013)
2016	Non	8.6 kDa	Camptothecin	Tahvilian et al. (2016)
2016	Nanostructured lipid	No data	Curcumin	Li et al. (2016a)
2017	Ursolic acid	6.5 kDa	Nanostructured lipid	Das et al. (2017)
2017	Non	2 kDa	Ulva ohnoi	Fernández-Díaz et al. (2017)
2017	Pluronic	No data	Non	(Wang et al. (2017)

(continued)

Table 9.1 (continued)

Published time	Functional groups for modification	Molecular weight of chitooligosaccharides	Drug	Reference
2018	Non	≤10 kDa	Astaxanthin	Liu et al. (2018a)
2018	Fatty acyl derivatives of CGKRK homing peptides	4–5 kDa	siRNA	El-Sayed et al. (2018)
2018	Beta-lactoglobulin/oleic acid	20 kDa	DHA	Ha et al. (2018)
2018	Silver	150 kDa	Non	Pham et al. (2018)

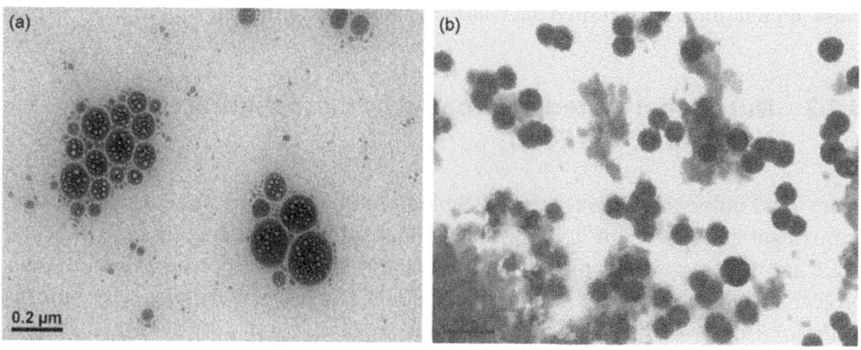

Fig. 9.1 TEM images of chitooligosaccharides nanoparticles loading with ATP before (**a**) and after (**b**) ATP release (Du et al. 2010b)

Drug carriers can control the release of drugs and therefore, can be used to improve electivity, effectiveness, or safety of anti-cancer drugs. They were better candidates compared with existed clinic treatment such as formulated with cremophor eL and ethanol. Chitosan is one of the most commonly studied polymers in drugs delivery because of its biocompatibility, biodegradability and no-toxicity (Mao et al. 2010). However, it still has disadvantage of insolubility at physiological pH values (7.2–7.4). Chitooligosaccharides can form nanoparticles by ionic gelation with polyphosphates (Calvo et al. 1997) and chitooligosaccharides nanoparticles can become a priority of deposition in tumour tissue because of 'enhanced permeability and retention'(EPR) effect (Maeda 2001). Thus, chitooligosaccharides nanoparticles are better candidates for drugs delivery and they can replace chitosan nanoparticles in some cases although the total amount of studies is still low. Here, Table 9.1 summarize all the records of applications of chitooligosaccharides nanoparticles as drug carriers. Figure 9.1 perfectly showed TEM images of chitooligosaccharides nanoparticles loading with ATP. To improve the performance of chitooligosaccharides nanoparticles, most of them were further modified with functional groups such as hydrophobic modification of chitooligosaccharides by esterification and etherification on hydroxyl groups.

9.2.2 Chitooligosaccharides Nanoparticle as Gene Deliverer

DNA and siRNA have been used in demonstrated tremendous therapeutic potential for the treatment of many genetic diseases (Saranya et al. 2011). However, nucleic acid was not stable in vivo (Wang et al. 2010) and the treatment has less effect. Chitooligosaccharides owns positive charges and this is helpful for its interaction with DNA and RNA. Therefore, it protects nucleic acid from nuclease degradation (Delgado et al. 2013) and ultrasonic treatment for in plant cell transformation (Wang et al. 2009). More importantly, Chitooligosaccharides nanoparticle can cross the cell membrane to transmit nucleic acid (Borchard 2001) and has a good transfection efficiency (Chae et al. 2005). All these make Chitooligosaccharides nanoparticles good candidates for gene delivery. Now, Chitooligosaccharides nanoparticles for gene delivery have been applied in animals (Chae et al. 2005) and plant cells (Wang et al. 2009). In future, with the development of related technologies, there should be more applications of chitooligosaccharides nanoparticle in gene delivery.

9.2.3 Nutritional Ingredient Coated by Chitooligosaccharides Nanoparticle in Food

Many food nutrient additives are not stable in food because of their unsaturation or antioxidant and this result in loss of nutritional value and product quality. In Sects. 9.2.1 and 9.2.2, we have introduced that chitooligosaccharides nanoparticles can protect unstable drug or DNA from degradation. The protective effect is also useful in these cases. Docosahexaenoic acid (DHA) is an omega-3 fatty acid that is a primary structural component of the human brain, cerebral cortex, skin, and retina (Guesnet and Alessandri 2011). It is important for infants' brain development as well as reducing the risk of cardiac and immune dysfunction (Ismail et al. 2016). However, it is not stable in oxygen and light during food processing and storage. Ho-Kyung Ha et al. (Ha et al. 2018) tried to design β-lactoglobulin/oleic acid-modified Chitooligosaccharides nanoparticles as an aqueous-based delivery system and coating DHA with it. The result showed that the oxidative stability of DHA increased during 16 days of storage in skim milk. Astaxanthin (Ax) is a type of carotenoid in marine animals (Jia et al. 2017) with various functions such as anti-cancer, antiobesity, and antidiabetic activities (Chuyen and Eun 2015). However, it has a low solubility and bioavailability in aqueous as well as degrading in harsh environmental conditions (Khalid et al. 2017), which is similar with antitumor drug described above. Liu et al. (2018a) used β-lactoglobulin-based nanocomplexes composed of chitooligosaccharidess to coat Astaxanthin. The coated microorganism's stability and slow release ability was improved. Other food ingredients such as goat milk protein (Ha et al. 2017) and plant extract (Fernandez-Diaz et al. 2017) was also tested.

9.3 Chitooligosaccharides Nanofibers

Nanofibers are fibers with the diameters of nanometers scale, and therefore nanofibers has better mechanical properties compared with microfibers (Jiang et al. 2018). Nanofibers are widely useful in areas such as biomedical engineering, filtration, protective clothing (Reneker and Chun 1999). Chitooligosaccharides can be used in nanofiber preparation (Li et al. 2016b; Lu et al. 2016; Rabbani et al. 2016) and the main application of Chitooligosaccharides nanofiber is wound dressing. Skin is the largest organ and functions as a barrier between the environment and internal organs (Wu et al. 2015). The skin is also easily injured by attacks and skin damage is an extremely common disease (Delavary et al. 2011). Wound repair is a complex process and serious problems such as severe infection should be in consideration (Kumar et al. 1999). Dressings made of traditional materials such as wool and polyamide, have been used for wound repair. However, they still have disadvantages such as limited moisture vapor permeability. A good wound dressing should has good biological compatibility and can be degraded in time. If it is helpful for wound repair, it will be better (Kumar et al. 2012). Chitooligosaccharides nanofibers thus become a satisfactory option because of its antimicrobial property and wound healing promotion. There are many methods for making nanofibers, including drawing, electrospinning, self-assembly, template synthesis, and thermal-induced phase separation. Preparing nanofibers by electrospinning have a long history. William Gilbert (1544–1603) firstly reported it (Nascimento et al. 2015; Tucker et al. 2012) and it is the most used method for nanofibers preparation. There have been over 200 kind of polymers used in nanofibers preparation by electrospinning. For preparation of chitooligosaccharides nanofibers, electrospinning is the only methods used in existing studies. Electrospinning is a very simple and effective approach. Figure 9.2 gives description of electrospinning schematic with variations (This file is licensed under the Creative Commons Attribution-Share Alike 3.0 Unported, https://commons.wikimedia.org/wiki/File:Electrospinning_Image_for_Wikipedia.tif). It is important that spinning aid is always necessary for chitooligosaccharides spinning.

9.4 Application of Chitooligosaccharides in Surface Modification

Chitooligosaccharides can be used to form a covalent bond with materials because of their amino/hydroxyl groups. Thus, it has been widely used in surface modification of materials for improvement of performance or functionality. Iron oxide nanoparticles (NPs), Fe_3O_4 and gama-Fe_3O_4, is one of the most prominent features of magnetic nanoparticles (MNPs), has been widely used in biomedical applications such as in vivo magnetic resonance imagining, magnetic-mediated hyperthermia for cancer treatment and tissue-specific delivery of therapeutic agents (Duy Nguyen et al. 2014; Gregorio-Jauregui et al. 2012). However, they tend to aggregate because of magnetic attraction. After surface modification by chitooligosaccharides,

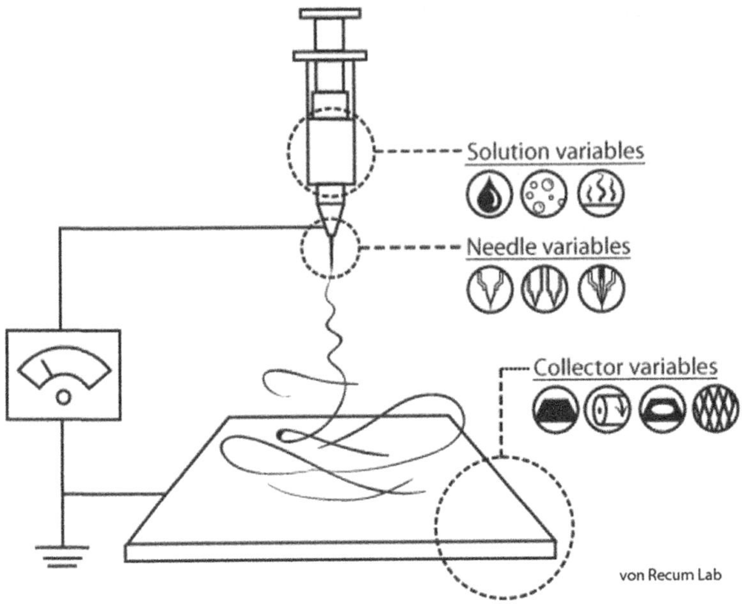

Fig. 9.2 Electrospinning schematic with variations for different processing outcomes. (This file is licensed under the Creative Commons Attribution-Share Alike 3.0 Unported, https://commons.wikimedia.org/wiki/File:Electrospinning_Image_for_Wikipedia.tif)

chemical stability and biocompatibility of MNPs can increase (Safari and Javadian 2014). Fe_3O_4 nanoparticles coated by chitooligosaccharides may serve as a promising platform for the development of new magnetic materials, which could be useful for biomedical applications. Considering the low molecular weight that limited the application in materials, surface modification might be a very good application mode for chitooligosaccharides. Thao et al. (2017) prepared Fe_3O_4 nanoparticles coating with COS (MW = 3705–4071 Da) (Fig. 9.3). These nanoparticles still have superparamagnetic properties. Thus, it can be a new potential magnetic material with high dispersibility for biomedical applications.

9.5 Chitooligosaccharides as Additive in Biomaterial Preparation

Chitooligosaccharides has many interesting functions such as wound healing (Sandri et al. 2017), neural repair (Yang et al. 2007) and bone healing (Wang et al. 2014). Therefore, it can be used as additive in biomaterials biological tissue grafts to improve the therapeutic effect aside from preparing as material itself. Shinn-Jyh Ding et al. (2010) added chitooligosaccharides in calcium silicate cement to improve the handling and enhance osteogenic and immune properties. It slightly prolonged the setting time of CSC and helped in improve cell viability as well as less immune

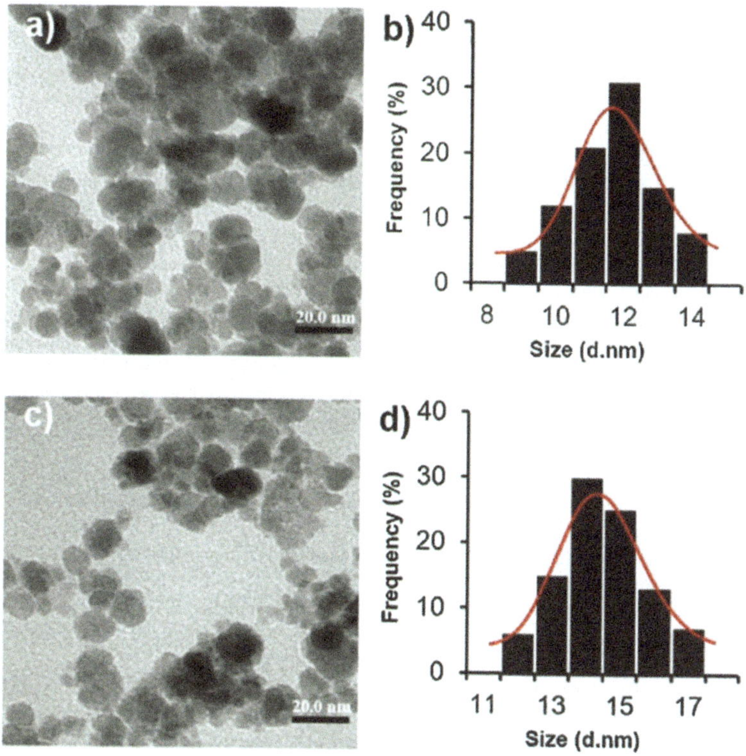

Fig. 9.3 TEM images and particle size histograms of Fe_3O_4 (**a**, **b**) and Fe_3O_4 coating with chitooligosaccharides (**c**, **d**) (Thao Nguyen Le et al. 2017)

response. Liu et al. (2018b) developed an new antibacterial material by synthesizing chitooligosaccharides-N-chlorokojic acid mannich base and it showed good biocompatibility, noncytotoxic and low hemolysis. As described in other chapter, Chitooligosaccharides has various physiological activities. Thus, it can be applied as additive in biomaterials preparation.

9.6 Conclusion

Unlike chitosan, the applications of chitooligosaccharides on biomaterials are rare. The most important reason is that chitooligosaccharides has a low molecular weight. However, chitooligosaccharides still has unique advantages in material preparation such as its better solubility and biocompatibility. Chitooligosaccharides can be used to prepare nanoparticle and further applied in drug and gene delivery as well as protector of nutritional ingredient. It also used in nanofiber preparation for wound healing. In addition, there are also some reports concerning application of chitooligosaccharides as coating materials and additive in biomaterials. Future application of chitooligosaccharides should combine with various functions of chitooligosaccharides.

References

Akhlaghi SP, Berry RC, Tam KC (2013) Surface modification of cellulose nanocrystal with chitosan oligosaccharide for drug delivery applications. Cellulose 20(4):1747–1764

Andrievski RA, Glezer AM (2001) Size effects in properties of nanomaterials. Scr Mater 44(8):1621–1624

Barry E, Alvarez JA, Scully RE, Miller TL, Lipshultz SE (2007) Anthracycline-induced cardiotoxicity: course, pathophysiology, prevention and management. Expert Opin Pharmaco 8(8):1039–1058

Borchard G (2001) Chitosans for gene delivery. Adv Drug Deliv Rev 52(2):145–150

Calvo P, Remuñán-López C, Vila-Jato JL, Alonso MJ (1997) Novel hydrophilic chitosan-polyethylene oxide nanoparticles as protein carriers. J Appl Polym Sci 63(1):125–132

Chae SY, Son S, Lee M, Jang M-K, Nah J-W (2005) Deoxycholic acid-conjugated chitosan oligosaccharide nanoparticles for efficient gene carrier. J Control Release 109(1–3):330–344

Chi XF, Ding XM, Peng X, Li XC, Fang J (2017) Effects of chitosan oligosaccharides supplementation on the cell cycle of immune organs in broilers. Kafkas Univ Veteriner Fakultesi Derg 23(6):1003–1006

Choi H, Lee JP, Ko SJ, Jung JW, Park H, Yoo S, Park O, Jeong JR, Park S, Kim JY (2013) Multipositional silica-coated silver nanoparticles for high-performance polymer solar cells. Nano Lett 13(5):2204

Choi C, Nam J-P, Nah J-W (2016) Application of chitosan and chitosan derivatives as biomaterials. J Ind Eng Chem 33:1–10

Chokradjaroen C, Theeramunkong S, Yui H, Saito N, Rujiravanit R (2018) Cytotoxicity against cancer cells of chitosan oligosaccharides prepared from chitosan powder degraded by electrical discharge plasma. Carbohydr Polym 201:20–30

Chuyen HV, Eun JB (2015) Marine carotenoids: bioactivities and potential benefits to human health. C R C Crit Rev Food Technol 57(12):2600–2610

Das S, Ghosh S, De AK, Bera T (2017) Oral delivery of ursolic acid-loaded nanostructured lipid carrier coated with chitosan oligosaccharides: development, characterization, in vitro and in vivo assessment for the therapy of leishmaniasis. Int J Biol Macromol 102:996–1008

Delavary BM, van der Veer WM, van Egmond M, Niessen FB, Beelen RHJ (2011) Macrophages in skin injury and repair. Immunobiology 216(7):753–762

Delgado D, del Pozo-Rodríguez A, Angeles Solinís M, Bartkowiak A, Rodríguez-Gascón A (2013) New gene delivery system based on oligochitosan and solid lipid nanoparticles: 'in vitro' and 'in vivo' evaluation. Eur J Pharm Sci 50(3):484–491

Ding S-J, Shie M-Y, Hoshiba T, Kawazoe N, Chen G, Chang H-C (2010) Osteogenic differentiation and immune response of human bone-marrow-derived mesenchymal stem cells on injectable calcium-silicate-based bone grafts. Tissue Eng A 16(7):2343–2354

Du Y-Z, Wang L, Yuan H, Wei X-H, Hu F-Q (2009) Preparation and characteristics of linoleic acid-grafted chitosan oligosaccharide micelles as a carrier for doxorubicin. Colloids Surf B: Biointerfaces 69(2):257–263

Du Y-Z, Wang L, Dong Y, Yuan H, Hu F-Q (2010a) Characteristics of paclitaxel-loaded chitosan oligosaccharide nanoparticles and their preparation by interfacial polyaddition in O/W miniemulsion system. Carbohydr Polym 79(4):1034–1039

Du Y-Z, Ying X-Y, Wang L, Zhai Y, Yuan H, Yu R-S, Hu F-Q (2010b) Sustained release of ATP encapsulated in chitosan oligosaccharide nanoparticles. Int J Pharm 392(1-2):164–169

Duy Nguyen H, Dzung Nguyen T, Hai Nguyen D, Nguyen PT (2014) Magnetic properties of Cr doped Fe3O4 porous nanoparticles prepared through a co-precipitation method using surfactant. Adv Nat Sci Nanosci Nanotechnol 5(3):035017

El-Sayed NS, Sharma M, Aliabadi HM, El-Meligy MG, El-Zaity AK, Nageib ZA, Tiwari RK (2018) Synthesis, characterization, and in vitro cytotoxicity of fatty acyl-CGKRK-chitosan oligosaccharides conjugates for siRNA delivery. Int J Biol Macromol 112:694–702

Fernández-Díaz C, Coste O, Malta E-j (2017) Polymer chitosan nanoparticles functionalized with Ulva ohnoi extracts boost in vitro ulvan immunostimulant effect in *Solea senegalensis* macrophages. Algal Res 26:135–142

Gregorio-Jauregui KM, Pineda MG, Rivera-Salinas JE, Hurtado G, Saade H, Ilyina A (2012) One-step method for preparation of magnetic nanoparticles coated with chitosan. J Nanomater.,*2012,(2012-6-27)* 2012(2):4

Guesnet P, Alessandri J-M (2011) Docosahexaenoic acid (DHA) and the developing central nervous system (CNS) – implications for dietary recommendations. Biochimie 93(1):7–12

Gwak SJ, Jung JK, An SS, Kim HJ, Oh JS, Pennant WA, Lee HY, Kong MH, Kim KN, Yoon DH (2012a) Chitosan/TPP-hyaluronic acid nanoparticles: a new vehicle for gene delivery to the spinal cord. J Biomater Sci Polym Ed 23(11):1437–1450

Gwak SJ, Jung JK, An SS, Kim HJ, Oh JS, Pennant WA, Lee HY, Kong MH, Kim KN, Yoon DH, Ha Y (2012b) Chitosan/TPP-hyaluronic acid nanoparticles: a new vehicle for gene delivery to the spinal cord. J Biomater Sci Polym Ed 23(11):1437–1450

Ha HK, Jin WK, Han KS, Yun SS, Lee MR, Lee WJ (2017) Development and characterization of a hydrolyzed goat milk protein/Chitosan Oligosaccharide nano-delivery system. Biotechnol 35(3):208–214

Ha H-K, Lee M-R, Lee W-J (2018) Oxidative stability of DHA in beta-lactoglobulin/oleic acid-modified chitosan oligosaccharide nanoparticles during storage in skim milk. LWT-Food Sci Technol 90:440–447

Heidarisasan S, Ziamajidi N, Karimi J, Abbasalipourkabir R (2018) Effects of insulin-loaded chitosan-alginate nanoparticle on RAGE expression and oxidative stress status in the kidney tissue of rats with type 1 diabetes. Iran J Basic Med Sci 21(10):1035–1042

Hu F-Q, Wu X-L, Du Y-Z, You J, Yuan H (2008) Cellular uptake and cytotoxicity of shell cross-linked stearic acid-grafted chitosan oligosaccharide micelles encapsulating doxorubicin. Eur J Pharm Biopharm 69(1):117–125

Hu FQ, Jiang XH, Huang X, Wu XL, Yuan H, Wei XH, Du YZ (2009a) Enhanced cellular uptake of chlorine e6 mediated by stearic acid-grafted chitosan oligosaccharide micelles. J Drug Target 17(5):384–391

Hu F-Q, Liu L-N, Du Y-Z, Yuan H (2009b) Synthesis and antitumor activity of doxorubicin conjugated stearic acid-g-chitosan oligosaccharide polymeric micelles. Biomaterials 30(36):6955–6963

Hu F, Du Y, Yuan H, Meng P (2013) Pegylated and fatty acid grafted chitosan oligosaccharide, synthesis method and application for drug delivery system

Huo M, Zhang Y, Zhou J, Zou A, Yu D, Wu Y, Li J, Li H (2010) Synthesis and characterization of low-toxic amphiphilic chitosan derivatives and their application as micelle carrier for antitumor drug. Int J Pharm 394(1):162–173

Ismail A, Bannenberg G, Rice HB, Schutt E, Mackay D (2016) Oxidation in EPA- and DHA-rich oils: an overview. Lipid Technol 28(3–4):55–59

Jia J, Gao X, Hao M, Tang L (2017) Comparison of binding interaction between β-lactoglobulin and three common polyphenols using multi-spectroscopy and modeling methods. Food Chem 228:143–151

Jiang S, Chen Y, Duan G, Mei C, Greiner A, Agarwal S (2018) Electrospun nanofiber reinforced composites: a review. Polym Chem 9(20):2685–2720

Khalid N, Shu G, Kobayashi I, Nakajima M, Barrow CJ (2017) Formulation and characterization of monodisperse O/W emulsions encapsulating astaxanthin extracts using microchannel emulsification: insights of formulation and stability evaluation. Colloids Surf B: Biointerfaces 157:355–365

Kim SK, Park PJ, Yang HP, Han SS (2001) Subacute toxicity of chitosan oligosaccharide in Sprague-Dawley rats. Arzneimittelforschung 51(9):769–774

Kumar S, Maiti SK, Kumar N, Samsuzzama MM, Ravindran NA, Balwada AK, Mathew DD (1999) Effect of medical grade chitosan powder in full thickness skin wound healing in rat model. Adv Anim Vet Sci 2(5):270–276

Kumar PTS, Lakshmanan V-K, Biswas R, Nair SV, Jayakumar R (2012) Synthesis and biological evaluation of chitin hydrogel/nano ZnO composite bandage as antibacterial wound dressing. J Biomed Nanotechnol 8(6):891–900

Lee KY, Ha WS, Park WH (1995) Blood compatibility and biodegradability of partially N-acylated chitosan derivatives. Biomaterials 16(16):1211–1216

Li F, Liao H-D, Liu X (2013) Preparation of 5-fluorouracil-chitosan oligosaccharide/selenium nano-microspheres co-carrying anti-tumor drug and sensitizer and their anti-tumor activity. J Control Release 172(1):e24

Li J, Liu D, Tan G, Zhao Z, Yang X, Pan W (2016a) A comparative study on the efficiency of chitosan-N-acetylcysteine, chitosan oligosaccharides or carboxymethyl chitosan surface modified nanostructured lipid carrier for ophthalmic delivery of curcumin. Carbohydr Polym 146:435–444

Li C-W, Wang Q, Li J, Hu M, Shi S-J, Li Z-W, Wu G-L, Cui H-H, Li Y-Y, Zhang Q, Yu X-H, Lu L-C (2016b) Silver nanoparticles/chitosan oligosaccharide/poly(vinyl alcohol) nanofiber promotes wound healing by activating TGF beta 1/Smad signaling pathway. Int J Nanomedicine 11:373–387

Li Y, Yu Z, Men Y, Chen X, Wang B (2018) Laminin-chitosan-PLGA conduit co-transplanted with Schwann and neural stem cells to repair the injured recurrent laryngeal nerve. Exp Ther Med 16(2):1250–1258

Liu C, Tan Y, Liu C, Chen X, Yu L (2007) Preparations, characterizations and applications of chitosan-based nanoparticles. J Ocean Univ China 6(3):237–243

Liu C, Liu Z, Sun X, Zhang S, Wang S, Feng F, Wang D, Xu Y (2018a) Fabrication and characterization of beta-Lactoglobulin-based nanocomplexes composed of chitosan oligosaccharides as vehicles for delivery of Astaxanthin. J Agric Food Chem 66(26):6717–6726

Liu X, Xia W, Jiang Q, Yu P, Yue L (2018b) Chitosan oligosaccharide-N-chlorokojic acid mannich base polymer as a potential antibacterial material. Carbohydr Polym 182:225–234

Lu W, Xu H, Zhang B, Ma M, Guo Y (2016) The preparation of chitosan oligosaccharide/alginate sodium/gelatin nanofibers by spiral-electrospinning. J Nanosci Nanotechnol 16(3):2360–2364

Luo Q, Zhao J, Zhang X, Pan W (2011) Nanostructured lipid carrier (NLC) coated with chitosan oligosaccharides and its potential use in ocular drug delivery system. Int J Pharm 403(1):185–191

Maeda H (2001) SMANCS and polymer-conjugated macromolecular drugs: advantages in cancer chemotherapy. Adv Drug Deliv Rev 46(1):169–185

Mao S, Sun W, Kissel T (2010) Chitosan-based formulations for delivery of DNA and siRNA. Adv Drug Deliv Rev 62(1):12–27

Materials, C. M., Physics,, & Council, N (1999) Condensed-matter and materials physics: basic research for tomorrow's technology. Eur J Phys 21(2):197–202

Mcnaught AD, Wilkinson A (2006) IUPAC compendium of chemical terminology. Encyclopedic dictionary of polymers

Moghadam NH, Salehzadeh S, Rakhtshah J, Tanzadehpanah H, Moghadam AH, Hajibabaei F, Sharifinia S, Asl SS, Saidijam M (2018) Improving antiproliferative effect of the nevirapine on Hela cells by loading onto chitosan coated magnetic nanoparticles as a fully biocompatible nano drug carrier. Int J Biol Macromol 118(Pt A):1220–1228

Nascimento MLF, Araujo ES, Cordeiro ER, de Oliveira AHP, de Oliveira HP (2015) A literature investigation about electrospinning and nanofibers: historical trends, current status and future challenges. Recent Pat Nanotechnol 9(2):76–85

Oryan A, Alemzadeh E, Tashkhourian J, Ana SFN (2018) Topical delivery of chitosan-capped silver nanoparticles speeds up healing in burn wounds: a preclinical study. Carbohydr Polym 200:82–92

Pham DC, Nguyen TH, Ngoc UTP, Le NTT, Tran TV, Nguyen DH (2018) Preparation, characterization and antifungal properties of chitosan-silver nanoparticles synergize fungicide against Pyricularia oryzae. J Nanosci Nanotechnol 18(8):5299–5305

Puras G, Zarate J, Aceves M, Murua A, Díaz AR, Avilés-Triguero M, Fernández E, Pedraz JL (2013) Low molecular weight oligochitosans for non-viral retinal gene therapy. Eur J Pharm Biopharm 83(2):131–140

Rabbani MM, Yang SB, Park S-J, Oh W, Yeum JH (2016) Characterization of pullulan/chitosan oligosaccharide/montmorillonite nanofibers prepared by electrospinning technique. J Nanosci Nanotechnol 16(6):6486–6493

Rahmouni N, Tahri W, Sbihi HM, Nehdi IA, Desbrieres J, Besbes-Hentati S (2018) Improvement of chitosan solubility and bactericidity by synthesis of N-benzimidazole-O-acetyl-chitosan and its electrodeposition. Int J Biol Macromol 113:623–630

Reneker DH, Chun I (1999) Nanometre diameter fibres of polymer, produced by electrospinning. Nanotechnology 7(3):216–223

Rezaei B, Jamei HR, Ensafi AA (2018) An ultrasensitive and selective electrochemical aptasensor based on rGO-MWCNTs/chitosan/carbon quantum dot for the detection of lysozyme. Biosens Bioelectron 115:37–44

Richardson SCW, Kolbe HJV, Duncan R (1999) Potential of low molecular mass chitosan as a DNA delivery system: biocompatibility, body distribution and ability to complex and protect DNA. Int J Pharm 178(2):231–243

Sacco P, Cok M, Asaro F, Paoletti S, Donati I (2018) The role played by the molecular weight and acetylation degree in modulating the stiffness and elasticity of chitosan gels. Carbohydr Polym 196:405–413

Safari J, Javadian L (2014) Chitosan decorated Fe3O4 nanoparticles as a magnetic catalyst in the synthesis of phenytoin derivatives. RSC Adv 4(90):48973–48979

Sandri G, Aguzzi C, Rossi S, Bonferoni MC, Bruni G, Boselli C, Cornaglia AI, Riva F, Viseras C, Caramella C, Ferrari F (2017) Halloysite and chitosan oligosaccharide nanocomposite for wound healing. Acta Biomater 57:216–224

Saranya N, Moorthi A, Saravanan S, Devi MP, Selvamurugan N (2011) Chitosan and its derivatives for gene delivery. Int J Biol Macromol 48(2):234–238

Sauthier G, Segura JJ, Fraxedas J, Verdaguer A (2014) Hydrophobic coating of mica by stearic acid vapor deposition. Colloids Surf A-Physicochem Eng Asp 443:331–337

Schild HG, Tirrell DA (1991) Microheterogeneous solutions of amphiphilic copolymers of N-isopropylacrylamide. An investigation via fluorescence methods. Langmuir 7(7):1319–1324

Tahvilian R, Tajani B, Sadrjavadi K, Fattahi A (2016) Preparation and characterization of pH-sensitive Camptothecin-Cis-Aconityl grafted Chitosan oligosaccharide nanomicelles. Int J Biol Macromol 92:795–802

Tashkhourian J, Nami-Ana SF, Shamsipur M (2018) Designing a modified electrode based on graphene quantum dot-chitosan application to electrochemical detection of epinephrine. J Mol Liq 266:548–556

Thao Nguyen Le T, Thi Hiep N, Dong Quy H, Tuong Vi T, Ngoc Thuy N, Dai Hai N (2017) Development of new magnetic nanoparticles: Oligochitosan obtained by gamma-rays and -coated Fe3O4 nanoparticles. Appl Surf Sci 422:863–868

Tokumitsu H, Ichikawa H, Fukumori Y (1999) Chitosan-gadopentetic acid complex nanoparticles for gadolinium neutron-capture therapy of cancer: preparation by novel emulsion-droplet coalescence technique and characterization. Pharm Res 16(12):1830–1835

Tucker N, Stanger JJ, Staiger MP, Razzaq H, Hofman K (2012) The history of the science and Technology of Electrospinning from 1600 to 1995. J Eng Fibers Fabr 7:63–73

Wang FH, Liu J, Tang DY, Xue CG, Tong CY, Wu XY, Liu XM (2009) Chitosan nanoparticle as gene vehicle mediated by particle bombardment in onion cell transformation. J Hunan Univ 36(5):67–70

Wang J, Lu Z, Wientjes MG, Au JL (2010) Delivery of siRNA therapeutics: barriers and carriers. AAPS J 12(4):492–503

Wang T, Hu Y, Leach MK, Zhang L, Yang W, Jiang L, Feng Z-Q, He N (2012) Erythropoietin-loaded oligochitosan nanoparticles for treatment of periventricular leukomalacia. Int J Pharm 422(1):462–471

Wang D, Zhang Y, Hong Z (2014) Novel fast-setting chitosan/β-dicalcium silicate bone cements with high compressive strength and bioactivity. Ceram Int 40(7, Part A):9799–9808

Wang N, Chen C, Hu X (2017) Chitosan oligosaccharide/Pluronic polymers nano-aggregates as a potential oral drug delivery system. J Control Release 259:e147–e148

Wu Z, Tang Y, Fang H, Su Z, Xu B, Lin Y, Zhang P, Wei X (2015) Decellularized scaffolds containing hyaluronic acid and EGF for promoting the recovery of skin wounds. J Mater Sci-Mater Med 26(1)

Xu T, Zhang Z, Zhao J, Xue Q (1996) Study on the structure of surface-modified MoS2 nanoparticles. Mater Res Bull 31(4):345–349

Yang Y, Liu M, Lin S, Ding F, Gu X (2007) Effect of chitooligosaccharide on neuronal differentiation of neurospheres. Med J Commun 55(7):429–436

Yang L, Guo C, Jia L, Liang X, Liu C, Liu H (2010) Dual responsive copolymer micelles for drug controlled release. J Colloid Interface Sci 350(1):22–29

Yu S-Y, Kwon Y-I, Lee C, Apostolidis E, Kim Y-C (2017) Antidiabetic effect of chitosan oligosaccharide (GO2KA1) is mediated via inhibition of intestinal alpha-glucosidase and glucose transporters and PPAR expression. Biofactors 43(1):90–99

Zhang X, Jiang T, Li Y, Wang T, Wang S (2009) Evaluation of Chitosan Oligosaccharide nanoparticles prepared by A novel microemulsion-coacervation method. In: International symposium on crystal engineering and drug delivery system 2009)

Zhou Y-Y, Du Y-Z, Wang L, Yuan H, Zhou J-P, Hu F-Q (2010) Preparation and pharmacodynamics of stearic acid and poly (lactic-co-glycolic acid) grafted chitosan oligosaccharide micelles for 10-hydroxycamptothecin. Int J Pharm 393(1):144–152

The Application of Chito/Chitin Oligosaccharides as Plant Vaccines

10

Xiaoming Zhao, Mengyu Wang, Wenxia Wang, Qishun Liu, Jianguo Li, and Heng Yin

Abstract

Chito/chitin oligosaccharides has positive effects on triggering the resistance mechanism of plants. In 1980, it was first reported that chito/chitin oligosaccharide could induce plant immunity. Since then, chito/chitin oligosaccharide has been considered as a potent elicitor of plant immunity that is used in many crops, such as rice, wheat, cucumber, pepper, oilseed rape, tomato, bean, tobacco, flower crops, fruit crops and so on. This part summarized the study about the effect of the chito/chitin oligosaccharide application in lab/field experiments, In the studies, researchers found that chito/chitin oligosaccharide stimulates a variety of plant responses, which are mainly about the effects of chito/chitin oligosaccharide on inducing plant disease resistance, promoting plant growth and preventing plant from abiotic stress, such as cold stress. These effects demonstrate the ability of chito/chitin oligosaccharide to control plant fungi and virus diseases, enhance plant growth and yield, as well as prolong the shelf life of flowers and fruits. Meanwhile, the applications of chito/chitin oligosaccharide derivatives chemical constitution and the effect of used as plant disease vaccine are also reviewed, and the complexity mechanism of chito/chitin oligosaccharide activated plant immunity is discussed.

Plant induced disease resistance and the mechanism has been studied for a long time, but its application in agricultural production has not fully developed. Considering the excellent physical properties – high solubility, nontoxicity, biocompatibility and the cost advantage – rich in production, low cost, the excellent

X. Zhao (✉) · W. Wang · Q. Liu · H. Yin (✉)
Dalian Institute of Chemical Physics, Chinese Academy of Sciences, Dalian, China
e-mail: zhaoxm@dicp.ac.cn; yinheng@dicp.ac.cn

M. Wang
Dalian Maritime University, Dalian, China

J. Li
Agricultural Technology Promotion Center of Ankang City, Ankang, Shaanxi, China

L. Zhao (ed.), *Oligosaccharides of Chitin and Chitosan*,
https://doi.org/10.1007/978-981-13-9402-7_10

effects on plants, chito/chitin oligosaccharide showed huge economic benefits and will have a good application prospect.

Keywords
Plant resistance · Plant immune elicitor · Plant diseases · Plant vaccines · Growth regulation

10.1 Introduction

The study of chitin oligosaccharides began in the 1960s. Ayers et al. (1976) found that phytoalexins can be induced by low molecular weight chitin oligosaccharides of the fungal cell wall. Albersheim et al. (Albersheim and Darvill 1985) proposed chitin oligosaccharides not only regulate the growth, development and reproduction of plants, but also stimulate the immune system of plants. Since then, the physiological and biochemical reactions of chitosan have been studied and it act as a stimulator of plant defense against injury (Doares et al. 1995) and pathogen infections (Yu et al. 2012).

Yin et al. (2010) reported that chitooligosaccharides (COS) is one of the plant vaccines, which effectively control crop diseases at 50 PPM. In recent years, studies have found that COS can not only induce plant disease resistance, but also stimulate plant resistance to abiotic stress, including drought and cold environment. These results indicate that there will be a bright prospect of the application and research of COS, and it is a multifunctional plant immune elicitor that has excellent positive effects on agricultural production.

10.2 The Application of Chito/Chitin Oligosaccharides to Protect and Control Diseases on Crop

10.2.1 Control of Crop Fungi Diseases by Chito/Chitin Oligosaccharides Elicitors

10.2.1.1 Control of Fungi Diseases in Rice

Rice blast and rice sheath blight pose a huge threat to rice production, causing billions of kilograms yield losing per year. Ning et al. (2004) sprayed 5 μg/mL COS to rice H7R (resistance, positive control) and H7S (susceptibility), respectively. After inoculation with rice blast (*Magnaporthe grisea* 01-19B small species) spores for 8 h, it was showed that COS obviously increased the rice resistance to rice blast by 50% and promoted lesions reduction and decreased infection rate. Meanwhile, COS induced the hypersensitive responses (HR) together with H_2O_2 production in rice.

Bai et al. (2010) shows the effect of COS on rice sheath blight. The results showed that there was no obvious inhibitory effect on the mycelium and sclerotia of *Rhizoctonia solani*. However, after spraying COS on adult rice leaves, the incidence of rice disease index was significantly lower than that of the control. Strongest resistance effect was 65.56% from 50 μg/mL COS treatment, meanwhile, the activities of enzymes such as rice plant peroxidase (POD), polyphenol oxidase (PPO) and β-1, 3-glucanase were increased significantly.

10.2.1.2 Control of Fungi Diseases in Wheat

According to the results of field experiments, Zhang et al. (2008) reported that COS has a good protection and control effect on wheat sheath blight, rising from 88.40% to 90.60%.

Liu et al. (2001) showed that the COS pretreating wheat seeds can eliminate wheat yellowing seedlings and embryonic cells of deoxynivalenol (DON) induced by *Fusarium graminearum*, indicating that COS can increase the resistance of plants to pathogenic toxins. .

10.2.1.3 Control of Fungi Diseases in Cucumber

Ben-Shalom et al. (2003) used chitosan and chitin oligomers, respectively, to control the cucumber diseases. As shown in Fig. 10.1, after inoculation of chitosan for 24 h, the disease index of chitosan was 0.45, the protective effect was 87.14%, and the control disease index was 3.5, which was not significantly different from chitin oligomer.

The inhibitory effect of chitosan on spore germination and germ tube length of *Botrytis cinerea* was studied. The results indicate that chitin has no effect on the fungus. However, the inhibition rate of chitosan at low concentrations (20~30 μg/mL) was as high as 50%, while the inhibition rate of chitosan (50 μg/mL) significantly controlled the growth of *Botrytis cinerea*. After 24 h, the spore length of chitosan treatment at 10 μg/mL was 2 μm, while the average length of water-pretreated spores was 15 μm.

The inhibitory effect of COS on powdery mildew of cucumber (*Sphaerotheca fuliginea*) was studied with transmission electron microscopy (Ma et al. 2004). Ultrastructural observations showed that mycelial growth was significantly inhibited after COS pretreating. It was found that pretreating COS before 5–7 days on the cucumber seedlings leaves reduced lesions, delayed sporulation and prolonged latency for 10 days, indicating that COS has the ability to induce systemic acquired resistance of cucumber seedlings to powdery mildew (Ma et al. 2005).

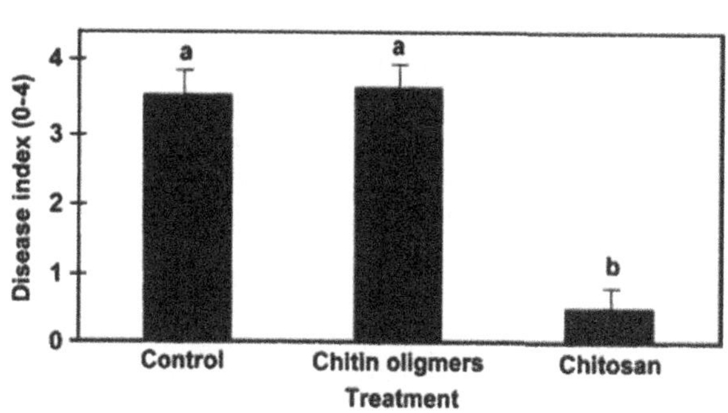

Fig. 10.1 The effect of chitosan and chitin oligomers on *B. cinerea*

10.2.1.4 Control of Fungi Diseases in Pepper

Xiao et al. (2009) reported that COS can induce the resistance of pepper to powdery mildew at a concentration of 25–125 μg/mL, and the best effect is 80% after 1 day of pretreatment at 50 μg/mL. At the same time, they found that COS induced systemic acquired resistance of pepper to powdery mildew. After pretreating COS, the activities of polyphenol oxidase (PPO), peroxidase (POD) and phenylalanine ammonia lyase (PAL) significantly increased compared with control. The results showed that the enzyme activity was increased with the systemic acquired resistance after COS pretreatment.

Xu et al. (2006) found that spraying COS in the field had a great effect on *Phytophthora capsici*, and its detraction effect was as high as 73.2% at 40 μg/mL. The growth rate method showed that COS inhibits mycelial growth at an EC50 of 100 μg/mL.

10.2.1.5 Control of Fungi Diseases in Oilseed Rape (Brassica napus H)

It was found that the action of COS had time-dependent inhibitory effects on *Sclerotinia sclerotiorum* (Yin et al. 2008). Pretreating of 50 μg/mL COS for 3 days before the inoculation of the fungus in oilseed rape resulted in an optimal control effect of up to 72.1%. *BnPDF 1.2*, one of the most important resistance genes can be up-regulated, which was detected by semi-quantitative RT-PCR. And one of the most important enzyme lipoxygenases (LOX) in the jasmonic acid pathway showed a significant increasing. All of the above indicated that COS may induce rapeseed against *Sclerotinia sclerotiorum* via a JA/ET-mediated pathway.

10.2.1.6 Control of Fungi Diseases in Tomato

In 2001, scientists at the Shanxi Institute of Plant Protection demonstrated that COS played a vital role in tomato early blight. The control effects of COS on fungus were 67.25% at 20 μg/mL, 71.29% at 25 μg/mL, and 79.40% at 40 μg/mL, respectively; The effects of COS from the second treatment (7 days after the first treatment), were 67.99% at 20 μg/mL, 77.50% at 25 μg/mL, and 80.15% at 40 μg/mL, compared to 70.63% of the control; 7 days after the second treatment, the effects of COS were 70.82% at 20 μg/mL, 77.37% at 25 μg/mL, 84.47% at 40 μg/mL, and 73.27% in the control group (unpublished).

10.2.1.7 Control of Fungi Diseases in Watermelon

Xu et al. (2003) showed that COS had a great effect on watermelon seedlings disease-resistant and promoting the watermelon seedlings growth. The control effects of COS on *Fusarium oxysporum* were 63.98% in laboratory and 71.82% in the field testing. In addition, plant height, root length and fresh weight from COS treatment increased significantly compare to control.

10.2.2 Control of Crop Virus Diseases by Chito/Chitin Oligosaccharides Elicitors

10.2.2.1 Control of Virus Diseases in Bean

COS-induced plant resistance is not only resistant to plant fungal diseases, but also significantly resistant to plant viral diseases. Pospieszny et al. (Pospieszny and Atabekov 1989) first reported chitosan can inhibit infection of alfalfa mosaic virus (AMV) on bean. And then different plant mosaic virus, such as tobacco necrosis virus (TNV), tobacco mosaic virus (TMV), peanut stunt virus (PSV), cucumber mosaic virus (CMV) and potato X virus were reported (Table 10.1) (Pospieszny et al. 1991). It is worth noting that treatment of bean leaves with chitosan reduced the amount of local necrosis caused by AMV and TMV infection. The inhibition of TMV infection depends on the chitosan concentration, which is a negative collection. Furthermore, TMV proliferation was completely inhibited after inoculation of 0.01% chitosan by tobacco for 6–8 h prior to protoplast treatment by ELISA. It showed chitosan not only inhibits viral infection, but also inhibits virus proliferation in cells. These findings indicate that chitosan inhibits viral virus infection in plants, and chitosan concentrations are associated with chitosan treatment effects.

10.2.2.2 Control of Virus Diseases in Tobacco

Shang et al. (2005) reported COS induced tobacco resistance to TMV. 50 μg/mL COS inhibited TMV infection to 84.73% before TMV inoculation in 24 h. COS pretreatment reduced starch staining in semi-smoke leaves by KI-I staining. In addition, the chlorophyll content from treatment with 50 μg/mL COS was 8.67 μg/g, which was higher than that of TMV treatment and virus A treatment.

Table 10.1 Induced resistance of chitosan to different viruses in plants

Plant	Virus	Effects	Inhibition
Phaseolus vulgaris	AMV-L	Decrease the number of necroses	+ + + +
Phaseolus vulgaris	TNV		+ + + +
Chenopodium quinoa	TNV		+ +
Chenopodium quinoa	CMV		+ + +
Nicotiana tabacum Var. Samsun	TMV		+ + +
Nicotiana tabacum Xanthi nc	TMV		+ +
Nicotiana glutinosa	TMV		+ +
Nicotiana paniculata	PSV		+ +
Phaseolus vulgaris	ALMV-S		+ + + +
Phaseolus vulgaris	PSV	Decrease the number of infected plants	+ + + +
Pisum sativum	ALMV		+ + +
Pisum sativum	PSV		+ + +
Lycopersicum esculentum	PVX		+ + +

All plants were inoculated after 0.1% chitosan treated for 1 day

Plants systemically infected were examined by ELISA after inoculated 10–12d

Shang et al. (2006b) reported COS inhibit TMV proliferation in tobacco. The results showed that 50 μg/mL COS pretreatment induced tobacco system resistance. The viral disease was delayed by 4–7 days and the average severity was reduced to 82.9%, and the TMV content of tobacco plants after COS treatment was reduced. The OD of COS group was 0.400 after 10 days of inoculation, which was 23% of the water treatment group, meanwhile, the OD of the new leaves in COS group was 0.190, which was 38.7% of the water treatment group. All these results indicate that COS inhibits TMV proliferation in tobacco. The induction effect of COS to long-distance movement of TMV in tobacco was reported by Shang et al. (2006a). The results show that the long-distance downward movement and upward movement of the TMV are delayed or suppressed.

Shang et al. (2007) reported COS induction decreased TMV-CP gene expression significantly in tobacco. It was reported by Shang et al. (2008) that COS can inactivate TMV particles in vitro, indicating that TMV pathogenicity is reduced by COS. Using JEM-1230 transmission electron microscopy, it was found that 300 μg/mL COS treatment directly broke 80% TMV particles into 50–150 nm micro-fragments. Shang et al. (2010) found that the activity of tobacco defense-related enzymes superoxide dismutase (SOD), POD, PAL and pathogenesis-related protein 1a(PR-1a) induced by 50 μg/mL COS. Zhao et al. (2004) showed that COS has effective antiviral activity on tobacco virus disease and the control effect is 77.9%.

10.2.2.3 Control of Virus Diseases in Tomato and Pepper

Liaoning Institute of Pesticide Testing found that tomato virus disease can be inhibited by COS treatment at 40 μg/mL, 50 μg/mL and 60 μg/mL, and the control effects were 51.6%, 64.7% and 68.2%, respectively (unpublished). Zhao et al. (2004) report the control effect of COS on tomato virus disease. The control effects of 40 μg/mL, 50 μg/mL and 60 μg/mL COS were 74.45%, 70.8%, 63.05%, and the growth rate was 16.7%, 15.0% and 10%, respectively.

In 2001, Hainan Institute of Plant Protection showed that COS can inhibit capsicum virus disease, and the effect of 50 μg/mL COS treatment was 88.69%. Zhao et al. (2004) reported the control effects of 40 μg/mL, 50 μg/mL and 60 μg/mL COS were 56.9%, 69.8%, 77.0%, and the growth rate were 3.9%, 10.1% and 18.2%, respectively.

10.2.3 Control of Crop Bacterial Diseases by Chito/Chitin Oligosaccharides Elicitors

Some experiments on the inhibition of bacterial growth by chitosan have been completed, which was reported by Rabea et al. (2003) (Table 10.2). It should be noted that the inhibition of chito/chitin oligosaccharide depending on the type of bacteria. For example, chitosan significantly inhibits *Corinebacterium michiganence* at a concentration of 10 ppm lower, while the effective concentration for inhibiting the growth of *Bacillus cereus* is 1000 ppm lower.

Table 10.2 The minimum inhibiting concentrations (MIC) of chitosan

Bacteria	MIC(ppm)
Agrobacterium tumefaciens	100
Bacillus cereus	1000
Corinebacterium michiganence	10
Erwinia sp.	500
Erwinia carotovora subsp.	200
Escherichia coli	20
Klebsiella pneumoniae	700
Micrococcus luteus	20
Pseudomonas fluorescens	500
Staphylococcus aureus	20
Xanthomonas campestris	500

In 2001, the Institute of Vegetables and Flowers of the Chinese Academy of Agricultural Sciences showed that Chinese cabbage soft rot was inhibited by field COS. The control effect was 66.90% at 60 μg/mL, 62.26% at 50 μg/mL and 47.60% at 40 μg/mL COS. Liaoning Institute of Pesticide Inspection reported that after 60 μg/mL COS treating, the control effect from Chinese cabbage soft rot was 78.62, and the growth rate was 16.67% in 2001.

10.2.4 Mechanisms of Chito/Chitin Oligosaccharides Induced Plant Immunity

The mechanism of COS activated plant immunity is mainly divided into three parts: COS signal recognition, signal transduction and defense response.

The phenomenon of COS binding to plant cell wall and membrane has been observed, but the binding protein has still not identified. Recently, Liu et al. further confirmed the combining of COS to wheat plasma membrane through quartz crystal microbalance, and identified three COS binding proteins from the plasma membrane by affinity chromatography, including one cell wall receptor kinase protein and two G-type lectin serine/threonine protein kinases (Liu et al. 2018). These results put forward the COS recognition study, contributing a better understanding of the plant immune mechanism of COS.

After plant sensing COS signal, the early signaling molecules, such as reactive oxygen species (ROS), nitric oxide (NO) and calcium ion will be triggered burst. These molecules accumulation are very important for defense signal transduction (Yin et al. 2010). Srivastava et al. reported that nitric oxide production occurs in 10mins after chitosan treatment, while ROS increased already by 5 min. They proved that nitric oxide production occurs downstream of reactive oxygen species in guard cells during stomatal closure induced by chitosan (Srivastava et al. 2009). Lu et al. found that COS could increase free calcium ions and regulated the expression of calreticulin, which is necessary for inhibition effect of COS on TMV (Hang et al.; Bai et al. 2010). These early signaling molecules, ROS in especially, is the important marker used for analysis COS signal pathway. Prodhan et al. observed that in the SA-deficient *nahG* mutant, chitosan treatment did not induce either reactive oxygen species or stomatal closure, implying that SA is a crucial element in chitosan induced stomatal closure (Prodhan et al. 2017).

Initial signal will be further amplified by hormone signal. COS activated hormone signal transduction typically involve multiple pathways, which are usually parallel and crossover. Salicylic acid (SA) and jasmonic acid/ethylene (JA/ET) mediated signaling pathways play important roles in plant immunity regulation. When plants are induced against viral diseases such as tobacco mosaic virus (TMV) or rice black streaked dwarf virus (SRBSDV), COS pretreatment could induce the SA pathway of the challenged plants to resist the pathogen (Jia et al. 2016; Yang et al. 2017). The latest study found that when plants are invaded by hemibiotrophic bacteria such as *Pseudomonas syringae* (*Pst* DC3000), COS simultaneously up-regulate the SA and JA pathways to stimulate plant against disease invasion (Jia et al. 2018). Similarly, COS triggered innate immunity of oilseed rape resistance against *Sclerotinia Sclerotiorum* though JA pathway (Yin et al. 2013). In summary, COS induced different signal pathways to achieve disease resistance against different types of plant pathogen, indicating the complexity of its mechanism of action, and its role as a broad-spectrum plant immune regulator in a variety of crops.

The final activated defense response of plant reflected in reinforcement of cell wall, expression of pathogen related protein and accumulation of antimicrobial secondary metabolites. All these response will help plant to counter the pathogen. Zhao et al. applied chitosan in the citrus fruit, the treatment can promote PAL pathway to accumulate lignin and maintain cell wall structure integrity by reducing cell wall degradation related enzymes activities (Zhao et al. 2018). Lucini et al. explored the mechanisms of chitosan in grape. They found that chitosan increased the level of phenylpropanoid and triterpenoids phytoalexins (Luigi et al. 2018). Badiali et al. reported that COS is effective in enhancing xanthone biosynthesis of *Hypericum perforatum* root cultures, which is positive related to the antifungal activity (Badiali et al. 2018).

10.3 The Application of Chito/Chitin Oligosaccharides to Regulate Growth on Crop

10.3.1 The Application of Chito/Chitin Oligosaccharides in Vegetable Crops Growth Regulation

Chitosan has been widely used in vegetable production, and chitosan reactions have been studied in many plant species (Table 10.3). Adding 1.0% (w/v) high molecular weight (600–900 kDa) chitosan to the soil can increase the plant height, canopy diameter and leaf area of the pepper (Chookhongkha et al. 2012). Foliar application of chitosan at an optimum concentration of 75 mg/L on Indian spinach (*Basella alba* L.) resulted in increasing of plant height, leaf number, shoot number, leaf area

Table 10.3 Agronomic responses of vegetable crops to chitosan application

Crop	Chitosan type and concentration	Method of application	Effects
Artichoke	Chitosan A (MW = 400 kDa) and Chitosan B (MW = 149 kDa) from Primex (Reykjavik, Iceland) at 3% or 4% (w/v) in 1% or 5% (v/v) acetic acid at 1% (v/v)	Seed coating after harvest	4% (w/v) chitosan B was better for coating
			Enhanced seed germination and quality
			Increased plant growth
Bell pepper	Chitosan at 0.5% or 1.0% (w/v) Crab-shell chitosan (10 mg/mL) in 0.04 N HCl	Fruit coating in post-harvest application on stem scar of fruit after harvest	Addition of lemon grass oil enhanced the antimicrobial activity
			Chitosan alone was more effective at extending the fruit shelf life
			Protection against *Botrytis cinerea*
Broccoli	Medium MW chitosan (98% DD) at 2% (w/v) in 1% (v/v) acetic acid and a glycerol/chitosan weight ratio of 0.28	Floret coating in post-harvest	Inhibited bacterial growth
			Inhibited the yellowing and opening florets
Celery	Chitosan at 3 mg/mL	Root dipping before transplanting	Reduced severity of Fusarium yellows
Chili	High MW (600–900 kDa) chitosan at 1.0% (w/v)	Soil supplement before transplanting	Increased height, canopy diameter, leaf number, leaf width and length
Cucumber	Chitosan (85% DD;	Foliar spray prior to inoculation	Controlled grey mold (*B. cinerea*)
	MW = 1000,000 kDa) at 0.1% (w/v)		Increased POD activity
Habanero pepper	Chitosan at 100 and 500 mg/L	Spraying on 3-month-old plant	Reduced *Phytophthora capsici* growth in vitro
			Prevented spread of the oomycetes
Indian spinach	Chitosan at 75 mg/L	Plant spraying during growing period	Increased height and leaf number, branch number, leaf area, leaf and stem fresh weight
Lemon balm	Chitosan [2.26% (w/v) of polymeric chitosan] at 0.025% (w/v)	Chitosan [2.26% (w/v) of polymeric chitosan] at 0.025% (w/v)	Controlled infection of pathogenic fungi; *Alternaria alternata*, *Fusarium avenaceum*, *Fusarium culmorum*, *Epicoccum purpurascens*, *S. sclerotiorum*

(continued)

Table 10.3 (continued)

Crop	Chitosan type and concentration	Method of application	Effects
Lotus root	Chitosan at 1.2% (w/v) in 1% (v/v) acetic acid, 2% (v/v) ascorbic acid and 1% (v/v) citric acid	Root slice dipping with a modified atmosphere application	Prevented browning on the surface of lotus root slices
Mentha piperita	Chitosan (200 mg/L) in 0.1 M acetic acid	Adding to cell culture	Enhanced production of secondary metabolites (mentol)
Ocimum basilicum L.	Chitosan (200 mg/L) in 5% (v/v) 1 N HCl	Adding to cell culture	Enhanced accumulated cell biomass
Ocimum gratissimum L.	Chitosan (50 mg/L) in 5% (v/v) 1 N HCl	Adding to cell culture	Enhanced accumulated cell biomass
Ocimum sanctum L.		Adding to cell culture	Enhanced accumulated cell biomass
Okra	Chitosan (100 or 125 mg/L)	Foliar application at 25, 40 and 55 d after sowing	Increased plant height, leaf number/plant, and relative and absolute growth rate and yield
			Increased NR activity
			Increased net photosynthesis rate
Oregano	COS (50, 200 and 500 mg/L)	Plant spraying 2 weeks before anticipated flowering time	Increased plant height and growth
			Up-regulated polyphenol contents
Sweet basil	Chitosan (30 cP) at 0.1% (w/v)	Seed soaking and root dipping before transplanting	Increased phenolic and terpenic compounds, especially rosmarinic acid and eugenol
			Increased antioxidant activity
			Enhanced growth

DD deacetylation, *MW* molecular weight, *NO* nitric oxide, *NR* nitrate reductase, *PA* phosphatidic acid, *PLC/DGK* phospholipase C (PLC) and diacylglycerol kinase (DGK), *PLD* phospholipase D, *POD* peroxidase, *PPO* polyphenol oxidase, *PVX* potato virus X

and fresh and dry weight. However, the application of chitosan in excess of 75 mg/L leads to a decrease in plant growth (Mondal et al. 2012). In contrast, optimal okra (*Hibiscus esculentus* L.) growth enhancement requires a higher chitosan concentration. Foliar application of 100–125 mg/L chitosan increased okra fruit yield, as well as plant height, leaf number, relative growth rate, photosynthetic rate and nitrate reductase (NR) activity (Mondal et al. 2012). Chitosan oligosaccharide increased the polyphenol content in the Greek oregano (*Origanum vulgare* L. ssp. hirtum) and stimulated plant growth at certain chitosan concentrations (200 and 500 mg/L). The content of phenolic compounds in several Basil plants was also increased by chitosan seed soaking and immersion. 1% (w/v) chitosan resulted in increased growth and secondary metabolite content, and rosmarinic acid and eugenol levels increased by 2.5% and 2%, respectively. (Hyun-Jin et al. 2005). The addition of an appropriate chitosan concentration to the cell cultures of *O. basilicum* L., *Ocimum tenuiflorum* L. and *Ocimum gratissimum* L. increased the cell biomass (Mathew 2012). The antibacterial activity of chitosan can inhibit various vegetable pathogens including *Sclerotinia, Pythium, Botrytis cinerea, Phytophthora capsici, Alternaria, Cladosporium, Epicoccum purpurascens* and *Fusarium avenaceum*. In addition, chitosan coating reduced plant weight loss (Ali et al. 2015). Using chitosan as a coating on broccoli (*Brassica oleracea* L. var. Italica) reduced the number of mesophilic and psychrophilic bacteria and extended the florescence of broccoli (Moreira et al. 2011). More results of chitosan applied on vegetables crops are summarized in Table 10.3 (Pichyangkura and Chadchawan 2015).

Zhang et al. (2002) proved that chitosan can promote the growth of cucumber in agriculture to a large extent. The results showed that plants treated with COS were disease-resistant than untreated plants. Guo et al. used different concentrations of COS to tobacco seedlings. The results showed that 0.01 mg/L COS promoted the growth of tobacco seedlings, increased seedling height, chlorophyll content, net photosynthetic rate (Pn), stomatal conductance (Gs), and concentration of intercellular CO_2 concentration (Ci), transpiration rate (Tr) in tobacco seedlings, while stomatal limitation (Ls) decreased (Table 10.4) (Guo et al. 2009).

10.3.2 The Application of Chito/Chitin Oligosaccharides in Flower Crops Growth Regulation

The seed germination percentage and the growth enhancement were found in Dendrobium treated with chitosan (Kananont et al. 2010; Nge et al. 2006; Pornpienpakdee et al. 2010). Spraying Dendrobium 'Eiskul' once a week with chitosan causes early flowering and more total inflorescences compared to untreated plants (Limpanavech et al. 2008). The application of 1.0% (w/v) chitosan to the soil led plants early flowering, such as Wishbone flower (*Torenia fournieri Linden ex E, Fourn.*), Persian Violet (*Exacum affine Balf.*), Elatior begonia (*Begonia hiemalis*

Table 10.4 Effects of times and concentrations of Chitooligosaccharide on the growth of tabaco

Chitosan concentration mg·L^{-1}	Stem length cm		Stem diameter cm		Areas of function leaves cm^2		Length of root cm	
	Treat once	Treat twice	Treat once	Treat twice	Treat once	Treat twice	Treat once	Treat twice
Ck	10.6b	10.2b	0.338a	0.341a	36.12a	38.75a	12.6a	11.9a
0.001	10.9b	11.1b	0.329a	0.322a	38.19a	37.63a	12.1a	12.5a
0.01	15.3d	18.5e	0.341a	0.338a	47.51b	50.03b	14.6b	15.9b
1	13.2c	16.1d	0.3218a	0.326a	41.93a	47.77b	14.1b	15.3b
100	8.9a	8.3a	0.332a	0.334a	37.16a	35.28a	11.9a	11.5a

$p < 0.05$

Fotsch.), Gloxinia [*Sin-ningia speciosa (Lodd.) Hiern*], garden lobelia (*Lobelia erinus* L.)and Monkey-Flower (Ohta et al. 2004). Gladiolus *communis* L. corn treated with chitosan were increased the number of flowers. It also extends the florescence and increases the number of corollas (Gianfranco et al. 2009). Similarly, the application of high MW chitosan resulted in an increasing in chlorophyll content, the premature flowering, the number of inflorescence branches. In addition, the number and weight of bulbs has also increased significantly (Salachna and Zawadzińska 2014). Chitosan can also increase the secondary metabolite levels in many species. When the eustoma flower buds are placed in a chitosan solution, the anthocyanins accumulated in the petals (Uddin et al. 2001).The effect of chitosan on flowers and ornamental plants are summarized in Table 10.5 (Pichyangkura and Chadchawan 2015).

10.3.3 The Application of Chito/Chitin Oligosaccharides in Fruit Crops Growth Regulation

The effects of chito/chitin oligosaccharide on fruit crops have been studied (Table 10.6). Foliar spraying of nanoparticle chitosan can enhance coffee plant growth (Sang et al. 2013) and strawberries yield without affecting fruit quality (El-Miniawy et al. 2013). In dragon fruit, chitosan spraying reduced the frequency and severity of anthrax infection by 22% and 66%, respectively. This is consistent with increased lignin accumulation and β-1,3-glucanase and chitinase activities (Ali et al. 2014). In papaya, fruits soaked with 1.5–2.0% (w/v) chitosan reduced the 70% of the disease caused by infection with *Colletotrichum gloeosporioides*. Peroxidase (POD), chitinase, β- 1,3-glucanase activity and the level of total phenol compound were enhanced (Ali et al. 2012). Similarly, coating the chitosan on peaches induced resistance to brown rot fungi, which is associated with defense-related enzymes such as catalase (CAT), POD, β-1,3-glucanase and chitinase increased (Ma et al. 2013). Treatment of pears with different MW chitosan resulted in different responses, COS of 6 kDa increasing chitinase and β-1,3-glucanase activity, while high MW chitosan (350 kDa) increased POD activity, respectively (Meng et al. 2010). These results implied there is a chitosan size-dependent response in various plant.

Coating chitosan inhibits browning in apples (Qi et al. 2011), litchi (Zhang and Quantick 1997), loquat(Ghasemnezhad et al. 2011) and rambutan (Gustavo et al. 2009), fresh litchi fruits (Dong et al. 2004; Reuck et al. 2009), longan (Jiang and Li 2001) and tomato (Badawy and Rabea, 2009), which extends the shelf life of the fruit. Chitosan also prevents weight loss in apricot (Ghasemnezhad et al. 2010), pitaya fruit (Chutichudet and Chutichudet 2011), litchi (Dong et al. 2004; Lin et al. 2011; Zhang et al. 2002), loquat (Ghasemnezhad et al. 2011), papaya (Ali et al. 2011), strawberry (Park et al. 2010; Pilar et al. 2008) and prickly pears (Ochoa-Velasco and Guerrero-Beltrán 2014). This is another measure that can extend the shelf life of the fruit. Some results were shown in Table 10.6 (Pichyangkura and Chadchawan 2015).

Table 10.5 Agronomic responses of flower and ornamental crops to chitosan application

Crop	Chitosan type and concentration	Method of application	Effects
Begonia hiemalis Fotsch	Chitosan at 1% (w/v)	Seedling pretreatment and soil application	Promoted seedling growth induced earlier flowering
Chrysanthemum	Chitosan at 0.01% to 0.05% (w/v)	Twice spraying on symptomatic plants	Protected against *Oidium chrysanthemi* and *Puccinia horiana*
Dendrobium 'Eiskul'	P (polymeric chitosan)-70 (10 mg/L) or P-90 (20 mg/L)	Adding to medium adding to medium	Enhanced PLB multiplication
		Plant dipping before exflasking	Enhanced plantlet regeneration from PLB
	O (oligomeric chitosan)-80 or P-80 chitosan at 10 mg/L		Enhanced growth after exflasking
	P-70 chitosan at 20 mg/L		
	O-80 at 50 mg/L	Plant spraying once a week	Induced early flowering
			Increased accumulative inflorescence number
			Enlarged chloroplasts, affected *ycf2* gene expression and increased number of silica bodies
Dendrobium bigibbum var. compactum	P-70, P-80, P-90, O-70 and O-80 at 10 mg/L	Adding to medium	Improved protocorm growth
Dendrobium formosum	P-70 at 10 or20 mg/L	Adding to medium	Improved protocorm growth
Dendrobium phalaenopsis	Chitosan (10 kDa) from fungal cell wall at 20–25 mg/L	Adding to medium	Enhanced growth in vitro
Freesia	Medium- and high-MW chitosan (50 kDa and 970 kDa, respectively) (85% DD) at 0.5% (w/v)	Corm soaking for 20 min before planting	Increased plant height, relative chlorophyll content
			Increased corm weight and number
			Induced early flowering
			Increased number of inflorescence shoots
Garden lobelia	Chitosan at 1% (w/v)	Seedling pretreatment and soil application	Promoted seedling growth
			Induced earlier flowering

(continued)

Table 10.5 (continued)

Crop	Chitosan type and concentration	Method of application	Effects
Gladiolus	Commercial chitosan (biorend) at 1.5% (w/v) together with hot water treatment	Corm dipping before planting	Accelerated corm emergence
			Increased number of flowers
			Extended vase life
			Increased number of cormlets
Gloxinia	Chitosan at 1% (w/v)	Seedling pretreatment and soil application	Promoted seedling growth.
			Induced earlier flowering
Lisianthus	Chitosan flake at 1% (w/v)	Adding to soil at sowing time	Enhanced growth
			Shortened flowering time
			Increased number and weight of flowers
	Chitosan (0.25 M)	Putting floral bud into chitosan solution supplemented with sugar	Accumulated anthocyanin in petals in vitro
Monkey flower	Chitosan at 1% (w/v)	Seedling pretreatment and soil application	Promoted seedling growth
			Induced earlier flowering
Persian violet	Chitosan at 1% (w/v)	Seedling pretreatment and soil application	Promoted seedling growth
			Induced earlier flowering
Rose	Chitosan at 0.01–0.02% (w/v)	Plant spraying twice weekly after rose powdery mildew symptoms	Protected against *Sphaerotheca pannosa var. rosae*, *Peronospora sparsa* and *Diplocarpon rosae*
Wishbone flower	Chitosan at 1% (w/v)	Seedling pretreatment and soil application	Promoted seedling growth
			Induced earlier flowering

DD deacetylation, *MW* molecular weight, *PLB* protocorm-like body

Chitosan also decreased the respiration rates of fruits, such as apple (Qi et al. 2011), Indian jujube (Zhong and Xia, 2007), litchi (Lin et al. 2011; Reuck et al. 2009), longan (Jiang and Li 2001), mango (Zhu et al. 2010), papaya (Hewajulige et al. 2009), peach (Li and Yu 2001) and strawberry (Pilar et al. 2008). Meanwhile, chitosan induces accumulation of phenolic compounds in many fruits (Ali et al. 2012; Badawy and Rabea 2009; Ghasemnezhad et al. 2011; Ghasemnezhad et al.

Table 10.6 Agronomic responses of fruit crops to chitosan application

Crop	Chitosan type and concentration	Method of application	Effects
Apples	Shrimp chitosan (97% DD) at 1% (w/v) in 2% (w/v) ascorbic acid and 0.5% (w/v) CaCl2	Slice coating	Decreased respiration rate
			Retarded enzymatic browning
	Medium-MW chitosan (>60% DD) at 2.0 g/L	Slice coating	Decreased infected area
			Decreased bacterial contamination
Apricot	0.5% low-MW chitosan in 0.5% (v/v) glacial acetic acid	Fruit coating in post-harvest	Increased total phenolics
			Content and antioxidant activity
Banana	Chitosan at 1% (w/v) in 0.5% (v/v) acetic acid	Hand dipping in post-harvest	Delayed ripening (changes in
			Peel color, firmness and soluble solids)
			Decreased disease severity
Cherries	Chitosan at 1% (w/v) in 1% (v/v) glacial acetic acid	Fruit dipping in post-harvest	Promoted vitamin C synthesis
			Increased anthocyanin and total phenolic content
Citrus	Low-MW chitosan (15 kDa) at 0.1 and 0.2% (w/v) in acetic acid	Fruit dipping in post-harvest	Slowed Murcott tangor decay at 15 °C
			Inhibited growth of *Penicillium*
			digitatum and *Penicillium italicum*, [when used at 0.2% (w/v)]
			Improved firmness, titratable acidity, ascorbic acidity and water content
Coconut	Chitosan at 10 mg/mL	Adding to cell culture	Accumulation of H_2O_2, increased β-1,3-glucanase activity
			Activated a 46-kDa MAPK-like protein
			Stimulated receptor-like kinases, Verticillium-like protein, and mitochondrial alternate oxidase 1b
	Chitosan at 200 mg/L	Adding to endosperm cell culture	Enhanced accumulation of p-hydroxybenzoic acid, p-coumaric acid and ferulic acid
			Increased PAL, p-coumaroyl-CoA ligase and p-hydroxybenzaldehyde dehydrogenase activity

(continued)

Table 10.6 (continued)

Crop	Chitosan type and concentration	Method of application	Effects
Coffee (Robusta)	High-MW (600 kDa) chitosan nanoparticles (prepared by nanospray drier) at 10 mg/L.	Foliar spray on seedling	Enhanced leaf chlorophyll content
			Increased net photosynthesis rate
			Increased nutrient uptake
Dragon fruit	Low-MW chitosan from crab shell (50 kDa, 75–85% DD) encapsulated as droplets of 200, 600 and 1000 nm diameter	Fruit dipping in post-harvest	Reduced anthracnose symptoms
			Maintained postharvest quality (decreased weight loss, increased fruit firmness and titratable acidity)
			Increased total phenolics, flavonoids, lycopene and antioxidants
			Delayed increase of respiration rate
	Submicron-sized chitosan at 1% (w/v) in 0.5% (v/v) glacial acetic acid, prepared by nano emulsifying in an alcohol-free system	Spray on 6-month-old plant until run off	Reduced anthracnose incidence and severity by 22% and 66%, respectively, when treated with 600-nm droplet size
			Increased lignin and β −1,3-glucanase and chitinase activity
	Chitosan solution at 1–3% (w/v)	Fruit dipping in post-harvest	Extended postharvest life
			Decreased weight loss and stomatal conductance
	Chitosan (MW = 12.36 ± 0.17 kDa, 95–98% DD and 630 mPas viscosity) at 0.2%, 0.5% or 1% (w/v) in acetic acid	Fruit coating in post-harvest	Retarded water loss and reduced sensory quality
			Inhibited microorganism growth.
Grape	Chitosan application at 500 L/ha	Spraying on plants during pre-bunch closure and veraison stage	Protected detached chardonnay leaves against *Botrytis cinerea* and reduced lesion diameter
			Increased POD and PAL activities
	Crab shell chitosan at 75–150 mg/L	Excised leaf incubation	Induced LOX, PAL and chitinase activities
			Highly protected against grey mould caused by *B. cinerea*

(continued)

Table 10.6 (continued)

Crop	Chitosan type and concentration	Method of application	Effects
	Chitosan at 50 and 100 mg/L	Adding to liquid cultures	Improved production of mono-glucosylated stilbenes
			Accumulation of other polyphenols, such as catechins
	Crab shell chitosan at 1% (w/v) in 1% (v/v) acetic or lactic acid	Fruit dipping in post-harvest	Reduced grey mold at cold and ambient storage temperatures
			Decreased CO_2 and O_2 exchange
	Chitosan at 1.0 and 2.5 mg/mL	Fruit dipping in post-harvest	Reduced *B. cinerea* growth
			Inhibited lesions by *B. cinerea*
	Crab shell chitosan at 1% (w/v) in 0.5% (v/v) glacial acetic acid	Cluster spray before harvest 1–2 days	Eeduced incidence of grey mold and increased PAL activity
	Chitosan at 1 g/L in 0.5% (v/v) acetic acid	Preharvest: Cluster spraying until run off 10 days before harvest postharvest: fruit dipping in post-harvest	Preharvest and postharvest
			Decreased weight loss, SOD activity and decay index
	Chitosan (85% DD) at 1.0–2.5% (w/v)	Adding to wound	Protect against *Colletotrichum sp.*
Grapevine	Chitosan at 1% (w/v)	Dipping of stem cutting before planting	Improved rooting of cuttings
			Increased number of new canes formed, longer canes, more internodes
			Increased leaf chlorophyll content in normal and drought conditions
Kiwifruit	Chitosan (~85% DD, MW = 20–30 kDa) applied at 1.0 kg/ha	Spray on filed grown plants every 2–4 weeks	Reduced disease symptoms
			Increased fruit fresh weight after 3 years of application
Litchi	600 g/L medium-MW chitosan (400 kDa) at 600 g/L in citric acid	Fruit dipping in post-harvest	Maintained anthocyanin and oxidative enzyme compositions

(continued)

Table 10.6 (continued)

Crop	Chitosan type and concentration	Method of application	Effects
	Crab shell chitosan at 1% (w/v) in 2% (v/v) acetic acid and 0.05–0.5% (w/v) ascorbic acid	Fruit dipping in post-harvest	Lowered respiration rate and sarcocarp temperature
			Restrained PPO and weight loss
			Reduced moisture loss and lower heat of respiration during storage
	Crab-shell chitosan at 1% or 2% (w/v) in 2% (w/v) l-glutamic acid	Fruit coating in post-harvest	Delayed changes in anthocyanin, flavonoid and total phenolic contents
			Delayed increase in PPO activity
			Reduced weight loss, and partially inhibited increase in POD activity
			Inhibited fruit decay during storage
	Crab-shell chitosan at 1 g/L with MAP	Fruit dipping in post-harvest	Reduced PPO and POD activity
			Retained membrane integrity, anthocyanin content and prevented decline of pericarp color during storage
	Chitosan (90–95% DD, 690–100 mPas viscosity) at 1%, 2% or 3% (w/v) in 12.5% (v/v) acetic acid	Fruit coating (peeled) in post-harvest	Retarded weight loss and sensory quality decline, with higher total soluble solids, titratable acid and ascorbic acid levels
			Suppressed the increased PPO and POD activities

2010; Kerch et al. 2011; Liu et al. 2007; Meng and Tian 2010; Zhang and Quantick 1997) and increased or maintained their antioxidant activities (Ghasemnezhad et al. 2011; Ghasemnezhad et al. 2010; Ma et al. 2013).

10.4 The Application of Chito/Chitin Oligosaccharides to Activate Plant Cold Resistance

As a kind of abiotic stress, cold stress can cause serious damage to plant growth. Low temperature stress can seriously damage the integrity of plant cell membrane, causing different degrees of dehydration of plants, directly or indirectly affecting the change of osmotic potential in plant cells, and finally forming osmotic stress; at the same time, it will change the structure of chloroplast, limit the synthesis of chlorophyll and affect the plant photosynthesis. Furthermore, cold stress will reduce the metabolism of roots, stems, leaves, and reduce nitrogen absorption and nutrient transport capacity, leading to slow growth and development of crops, leading to plant death in severe cases. Therefore, enhancing rice tolerance to cold stress is crucial for higher crop yield.

In the study of plant cold resistance, COS can enhance the ability of plants such as rice and wheat to resist low temperature stress, and stimulate the cold resistance of plants. Zhou et al. (2018) studied the effects of soaking seeds with 0 mg/L, 50 mg/L, 100 mg/L, 200 mg/L COS on the cold resistance of rice seedlings. The results showed that 100 mg/L soaking treatment had the best effect. COS increased the content of soluble sugar, soluble protein and proline as well as SOD activity, while decreased H_2O_2 content and MDA content, respectively. The photosynthesis of rice seedlings was increased and the chlorophyll b content of seedlings increased by 164.3%. For the potential mechanism, COS can significantly increase the expression levels of *psbA* genes in rice seedling leaves under low temperature conditions, suggesting that COS may accelerate the repair process of photosynthetic system II (promotes degradation of damaged D1 protein and synthesis of new D1 protein) and accelerate the accumulation of chlorophyll b to induce the cold resistance of the rice seedlings.

Zhang Yang et al.(2019) studied the effect of 150 mg/L COS treatment on rice. COS increased the proline and soluble sugar content in leaves as well as the root of seedlings. At the same time, COS can effectively enhance the activity of glutamine synthetase and glutamate synthase, reduce the activity of glutamate dehydrogenase, make glutamic acid accumulation, increase the activity of the Δ1-pyrroline-5-carboxylic acid synthase and the δ-ornithine aminotransferase, inhibit the expression of the proline dehydrogenase gene. In a word, COS may induce rice seedlings cold tolerance by acting on regulating glutamate metabolism and proline metabolism. Kuang et al. (2009) treated the seed of rice (*Oryza sativa* L.) with 25 mg/L, 50 mg/L, 100 mg/L, 200 mg/L COS. It was found that each treatment improved the cold resistance of rice seedlings and the effect of 100 mg/L COS treatment was more significant. Compared to control, the survival rate of seedlings increased by 15.54% from 100 mg/L COS treatment under cold stress; and electrolyte leakage and MDA content decreased by 12.88% and 6.98%, respectively; The activities of SOD, CAT and POD increased by 37.94%, 22.75% and 21.35%, respectively; soluble sugar, soluble protein and proline content increased by 15.82%, 12.92%, and 24.47%, respectively.

In the study of COS effects on cold resistance of wheat, Zou et al. (Zou et al. 2017) studied the effects of COS with different degrees of polymerization on the wheat under cold stress. The dry weight and fresh weight of seedlings increased significantly from COS treatments; the content of MDA decreased by 65.8% and 67.9% respectively; the soluble sugar content increased by 17.8% and 11.9% respectively. After 24 h of low temperature stress, chlorophyll content increased by 35.4% and 28.3% respectively; proline content increased by 37.7% and 43.8%, respectively; SOD activity increased by 54.8% and 61.2% respectively; APX activity increased by 60.5% and 73.0% respectively; there was no significant increase in

Table 10.7 The effect of chitosan on crisp pear in fruit setting (April.25th, 2009)

Treatment	Number of flower buds	Number of fruit setting	fruit setting ratio(%)	The increase of fruit setting(times)
Chitosan Concentration (75 μg/ml)	426	124	29.11	9.90
CK	450	12	2.67	–

CAT activity. After 48 h of low temperature stress, the chlorophyll content increased by 33.9% and 29.2% respectively; the proline content increased by 59.0% and 66.0% respectively; the SOD activity increased by 45.9% and 48.0% respectively; the CAT activity increased by 50.7%, and the APX activity increased by 81.4%, respectively. The study also showed that the degree of polymerization plays an important role in COS activity.

In order to study the effect of COS on cold-tolerance activity, COS was applied to eggplant (*Solanurn melogena L.*) seedlings at 4 leaf stage with 1/1500, 1/1000, 1/500 and 1/100(w/v) for 2 days and then exposed to 5 °C for 3 days (Kuang et al. 2009). The results showed that cold tolerance of eggplant seedlings can be improved by COS treatment. Compared to control, the activities of superprotective enzymes such as SOD, POD and CAT increased, while the content of MDA decreased in treated plants. On the other hand, COS can effectively promote the increase of proline and soluble sugar content in eggplant seedling leaves under low temperature stress. COS showed the most effective results at a concentration of 1/1000 (w/v).

DICP (Dalian Institute of Chemical Physics, Chinese Academy of Sciences) studied the role of COS in the cold resistance of plants in Pucheng County, Shaanxi Province. The results showed that the crisp pear tree treated with 75 mg/L COS showed cold tolerance and high seed setting rate before the cold in the spring. The results for 2009 show that trees treated with COS have a seed setting rate 9.9 times higher than that of untreated trees.

In addition to the higher fruit setting rate, young fruits of trees treated with COS grew faster (Table 10.7). In the study report of April 25, 2009, the young fruit treated with COS was 1.42 cm in diameter. If the young fruit encounters cold weather, treating COS can protect the fruit from cold damage.

From April 2 to 6, 2010, the fruit period suffered from cold weather. Young fruit treated with COS showed strong cold tolerance. And the treated fruit has a much lower chilling rate than untreated trees. The survey showed that the chilling injury rate of trees treated with COS was 20.5%, while that of untreated trees was 80.25%. The area of cold injury in the fruit is 1–5%, and that of untreated trees is 10–30%.

In summary, COS have the potential to be used as an environmentally friendly cold-resistant agent in the field of crop under cold stress.

10.5 Chito/Chitin Oligosaccharides Derivatives as Plant Disease Vaccine

Chito/chitin oligosaccharide have smaller molecular weights, better water solubility and easier absorption than glycan. In recent years, more studies have turned to oligosaccharide plant elicitor. However, chito/chitin oligosaccharide with elicitor activity would be hydrolyzed by endo- or exonuclease when they act on higher plants, producing smaller fragments without elicitor activity (Raetz 1990). Therefore, it is important to increase the stability of the elicitor-active oligosaccharide. By chemically modifying the structure of the oligosaccharide, it can improve its stability, and even increase the biological activity or expand the application range. At present, oligosaccharide-modified disease vaccines mainly focus on the derivatives of chito/chitin oligosaccharide.

Lu et al. (Hang 2009) synthesized COS-ruthenium complex and studied its activity of against tobacco mosaic virus resistance. The highest inhibition rate was 55% in the concentration range of 1–100 μg/mL, which was slightly lower than the inhibition rate of the positive control COS with inhibition of 63.9%. The optimal concentration of chitosan oligosaccharide-ruthenium complex was 10 μg/mL, which was lower than that of COS 50 μg/mL (Table 10.8).

Copper ion is a bactericidal metal ion, and it is also a trace element necessary for plant growth. The complexation of COS with copper ions will allow the complexes to mutually promote these properties of both. Liu et al. (Xiaoyu 2005) synthesized COS-copper complexes with low molecular chitosan and copper sulfate, and studied their inhibitory activities against 32 plant pathogenic fungi. The results showed that low molecular chitosan and their copper complexes can inhibit the spore germination, mycelial growth and cell membrane permeability changes, while copper and COS have synergistic effects. The authors also studied the effects of complexes on *Botrytis cinerea* and *Cercospora fuligena* (Tables 10.9 and 10.10). The low molecular chitosan and COS-copper treatment of the uninoculated tomato leaves increased the activities of PAL, PPO, and the COS-copper improved the three enzyme activities better than low molecular chitosan. For susceptible plants, low molecular chitosan and COS-copper can inhibit the growth of pathogenic bacteria in vitro, and can be used as an elicitor in plants to induce the enzyme activity of related defense enzymes in plants to rapidly increase, thereby improving plant disease resistance.

Table 10.8 The effects of different concentrations of chitosan oligosaccharide-ruthenium complex on tobacco mosaic virus ($P < 0.05$)

Treatment	TMV spots	Inhibition rate(%)
Chitosan oligosaccharide-ruthenium complex 1 μg/ml	43 ± 18^{c}	37.0
25 μg/ml	31 ± 13^{b}	55.0
50 μg/ml	38 ± 18^{cd}	44.0
100 μg/ml	42 ± 19^{c}	37.7
COS 50 μg/ml	24 ± 14^{b}	63.9
CK	68 ± 26^{a}	–

Table 10.9 Effect of oligosaccharide and complex on defensive enzymatic of tomato leaves infected by *Botrytis cinerea*

Treatment	Defense enzyme activity		
	PAL(U/g)	PO(U/g)	PPO(U/g)
CK water	297.98	0.17	1.67
CK oligosaccharide	321.81	0.22	1.76
CK oligosaccharide-copper complex	392.30	0.26	1.79
Inoculation with *Botrytis cinerea*	509.27	0.91	2.37
Inoculation with *Botrytis cinerea* + oligosaccharide	522.10	0.98	2.87
Inoculation with *Botrytis cinerea* + oligosaccharide-copper complex	863.62	1.02	1.96

Table 10.10 Effect of oligosaccharide and complex on defensive enzymatic of tomato leaves infected by *Cercospora fuligena*

Treatment	Defense enzyme activity		
	PAL(U/g)	PO(U/g)	PPO(U/g)
CK water	297.98	0.17	1.67
CK oligosaccharide	321.81	0.22	1.76
CK oligosaccharide-copper complex	392.30	0.26	1.79
Inoculation with *Cercospora fuligena*	326.15	0.30	2.07
Inoculation with *Cercospora fuligena* + oligosaccharide	571.02	0.43	2.20
Inoculation with *Cercospora fuligena* + oligosaccharide-copper complex	1341.81	0.90	5.87

Both PAL and PPO are copper-containing enzymes. The lack of copper ions can significantly reduce their activity, while the proper amount of copper ions can induce the increase of their enzyme activity. The COS-copper complex can release copper ions which enhance the activity of the defense enzyme.

Schiff base and Schiff base metal complexes have a variety of biological activities, including bactericidal activity, antiviral activity, etc. As a phosphorus-containing analog of natural amino acids, α-aminophosphonic acid is the third type of amino acid found in living organisms after aminocarboxylic acid and sulfamic acid, and has plant growth regulating activity, anti-plant virus activity and other biological activities. It is one of the important directions for the research and development of new pesticides in the world today. Yang (Nan 2010) synthesized 30 kinds of COS Schiff base-metal complexes and 15 kinds of COS Schiff base phosphonates by microwave synthesis method (Fig. 10.2). In order to study the anti-TMV activity induced by this kind of compounds, Xanthi-NN or N. glutinosa, a non-systemic infective tobacco containing N gene, was selected as a receptor for anti-TMV optimal agent, combined with the activity changes of defense enzymes before and after infection with TMV in common tobacco.

It was found that COS Schiff base-metal complex (S-COS-M) agent has a significant inhibitory effect on TMV, and the preventive effect is better than the therapeutic effect. However, a negative number appears in the therapeutic effect of 5-bromo salicylaldehyde COS Schiff base zinc (II) agent. The author speculates that

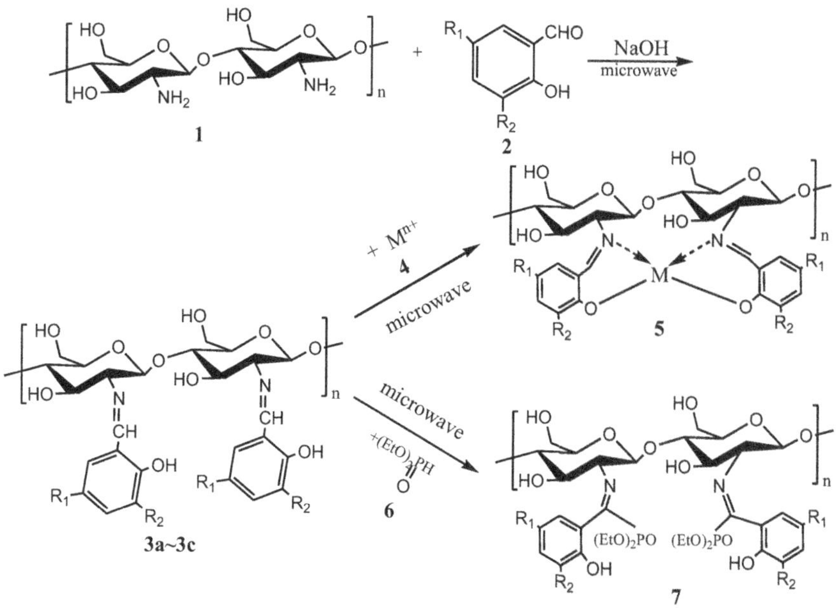

3a:R^1=H, R^2=H; **3b**: R^1=Cl, R^2=H; **3c**:R^1=Cl, R^2=Cl; **3d**:R^1=Br, R^2=H; **3e**: R^1=NO_2, R^2=H;

M^{n+}= Cu^{2+}, Co^{2+}, Ag^{+}, Ni^{2+}, Mn^{2+}, Zn^{2+}

Fig. 10.2 Synthesis route of COS Schiff base-metal complex and COS Schiff base phosphonate

it may be that when the TMV virus is inoculated, the damage of the quartz sand to the tobacco leaves is too great, which leads to serious damage in the late stage, and the physiological system in the tobacco does not produce defensive functions in a short period of time. It is also possible that the leaves of the tobacco absorb the 5-bromo salicylaldehyde chitosan oligosaccharide Schiff base zinc (II) too slowly, resulting in the reduction of 5-bromo salicylaldehyde COS when the quartz sand is washed in the late inoculation. It can be seen (Table 10.11) that the inhibitory effect of the agent on the TMV virus in the prophylactic effect generally shows a tendency of copper salt > silver salt > cobalt salt > nickel salt > zinc salt > manganese salt, and the therapeutic effect generally shows copper salt > Silver salt > cobalt salt > nickel salt > manganese salt > zinc salt trend. The effect of copper complex agent is always the best in both prevention and therapeutic effects. For the complex agents of different Schiff base substituents, the preventive effect and therapeutic effect are roughly as follows: 3, 5-dichlorosalicylaldehyde COS Schiff base>5-chlorosalicylaldehyde COS Schiff base >5-bromo salicylaldehyde COS Schiff base > salicylaldehyde COS Schiff base >5-nitrosalicylaldehyde COS Schiff base. In general, 3,5-dichlorosalicylaldehyde COS Schiff base-copper (II) complex has the most significant inhibitory effect on TMV. The inhibition rate in its preventive effect is as high as 78.86%, and the therapeutic effect reaches 44.09%.

Table 10.11 Effects of S-COS-M complexes on lesion numbers of inoculated Xanthi-NN leaves

Test reagent(100 μg/ml)	Preventive effects		Therapeutic effects	
	Lesions numbers in one cm^2 leave	Inhibition rate (%)	Lesions numbers in one cm^2 leave	Inhibition rate (%)
Salicylaldehyde COS Schiff base copper (II)	0.1249	66.20abcd	0.239 0	38.68 cd
Salicylaldehyde COS Schiff base cobalt (II)	0.1457	60.58abcd	0.252 6	35.19bcd
Salicylaldehyde COS Schiff base silver (I)	0.1155	68.75bcd	0.242 5	37.78 cd
Salicylaldehyde COS Schiff base nickel (II)	0.1694	54.17abc	0.288 1	26.10abcd
Salicylaldehyde COS Schiff base manganese (II)	0.2101	43.16a	0.323 5	17.02ab
Salicylaldehyde COS Schiff base zinc (II)	0.2003	45.81ab	0.338 0	13.28a
5-chlorosalicylaldehyde COS Schiff base copper (II)	0.1016	72.52 cd	0.230 0	40.99d
5-chlorosalicylaldehyde COS Schiff base cobalt (II)	0.1473	60.16abcd	0.252 4	35.26bcd
5-chlorosalicylaldehyde COS Schiff base silver (I)	0.1265	65.77abcd	0.231 1	40.70d
5-chlorosalicylaldehyde COS Schiff base nickel (II)	0.154 5	58.19abcd	0.257 2	34.02bcd
5-chlorosalicylaldehyde COS Schiff base manganese (II)	0.1635	55.77abcd	0.268 4	31.14abcd
5-chlorosalicylaldehyde COS Schiff base zinc (II)	0.1604	56.61abcd	0.326 5	16.25ab
3,5-Dichlorosalicylaldehyde COS Schiff base copper (II)	0.0781	78.86d	0.217 9	44.09d
3,5-Dichlorosalicylaldehyde COS Schiff base cobalt (II)	0.0979	73.52 cd	0.244 3	37.33 cd
3,5-Dichlorosalicylaldehyde COS Schiff base silver (I)	0.0944	74.46 cd	0.239 7	38.50 cd
3,5-Dichlorosalicylaldehyde COS Schiff base nickel (II)	0.1287	65.17abcd	0.281 8	27.70abcd
3,5-Dichlorosalicylaldehyde COS Schiff base manganese (II)	0.1717	53.55abc	0.312 1	19.93abc
3,5-Dichlorosalicylaldehyde COS Schiff base zinc (II)	0.1500	59.41abcd	0.308 9	20.76abc
5-nitrosalicylaldehyde COS Schiff base copper (II)	0.1429	61.35abcd	0.2564	34.22bcd
5-nitrosalicylaldehyde COS Schiff base cobalt (II)	0.1694	54.16abc	0.2500	35.87bcd
-nitrosalicylaldehyde COS Schiff base silver (I)	0.1525	58.74abcd	0.2447	37.23 cd

(continued)

Table 10.11 (continued)

Test reagent(100 μg/ml)	Preventive effects		Therapeutic effects	
	Lesions numbers in one cm² leave	Inhibition rate (%)	Lesions numbers in one cm² leave	Inhibition rate (%)
5-nitrosalicylaldehyde COS Schiff base nickel (II)	0.2115	42.77a	0.2922	25.03abcd
5-nitrosalicylaldehyde COS Schiff base manganese (II)	0.2005	45.76a	0.2710	30.48abcd
5-nitrosalicylaldehyde COS Schiff base zinc (II)	0.2821	23.69a	0.3629	6.90a
5-bromo salicylaldehyde COS Schiff base copper (II)	0.1029	72.17 cd	0.2667	31.59bcd
5-bromo salicylaldehyde COS Schiff base cobalt (II)	0.1053	71.52 cd	0.2635	32.40bcd
5-bromo salicylaldehyde COS Schiff base silver (I)	0.1416	61.70abcd	0.2900	25.60abcd
5-bromo salicylaldehyde COS Schiff base nickel (II)	0.1570	57.51abcd	0.3333	14.49a
5-bromo salicylaldehyde COS Schiff base manganese (II)	0.1632	55.86abcd	0.3391	13.01a
5-bromo salicylaldehyde COS Schiff base zinc (II)	0.1747	52.74ab	0.4780	–
Water CK	0.3696	–	0.3898	–

Note: Date in a column followed by the same small letters are not significantly different at $P_{0.05}$ by Dunca's multiple range test, respectively

The activity study of 15 kinds of COS Schiff base phosphonate (S-COS-P) showed that the drug had good inhibitory effect on TMV compared with the water. Salicylaldehyde chitosan oligosaccharide Schiff base phosphonate, 5-nitrosalicylaldehyde COS Schiff base phosphonate and 5-bromosalicylaldehyde COS Schiff base phosphonate agent have better effects than 5% di (octylamine ethyl) glycinate AS diluted at 800-folds. Wherein the 5-bromo salicylaldehyde COS Schiff base phosphonate is more effective than the 300-fold diluted Ningnanmycin aqueous solution. COS Schiff base phosphonate (S-COS-P) has a significant anti-TMV effect on tobacco, and its efficacy is basically 200 μg/mL > 100 μg/mL > 50 μg/mL, but 3,5-dichloro-salicylaldehyde COS Schiff base phosphonate does not exhibit regularity. On the whole, the 5-bromo salicylaldehyde COS Schiff base phosphonate had the best effect at the concentration of 200 μg/mL, and the inhibition rate of TMV to the preventive effect reached 90.79% (Table 10.12).

Hu et al. (2018) carboxylate the hydroxyl group at the C6 position of the GlcNAc residue of COS to obtain an anionic oligouronic acid with more than 70% carboxyl, and form a stable complex with iodine. In the process, oligomeric acid iodine (oligomeric N-acetyl-D-glucuronoic acid iodine) was obtained. The inhibitory effect of oligo-acid iodine on TMV showed good fast-acting and sustained effect. After 17 days of inoculation, the inhibitory effect increased rapidly from 45.08% to 74.48%. When inoculated for 21–35 days, the inhibitory effect on TMV exceeded

Table 10.12 Effects of S-COS-P on lesion numbers of inoculated *N.glutinosa* leaves

Test reagent	Preventive effects	
	Lesions numbers in one cm^2 leave	Inhibition rate (%)
Salicylaldehyde COS Schiff base phosphate 50 μg/ml	7.67	43.95abcdef
Salicylaldehyde COS Schiff base phosphate 100 μg/ml	8.03	41.31abcde
Salicylaldehyde COS Schiff base phosphate 200 μg/ml	7.85	42.66abcde
5-nitrosalicylaldehyde COS Schiff base phosphate 50 μg/ml	6.44	52.90bcdef
5-nitrosalicylaldehyde COS Schiff base phosphate 100 μg/ml	5.34	60.96cdef
5-nitrosalicylaldehyde COS Schiff base phosphate 200 μg/ml	5.89	56.97cdef
5-bromo salicylaldehyde COS Schiff base phosphate 50 μg/ml	3.54	74.15def
5-bromo salicylaldehyde COS Schiff base phosphate 100 μg/ml	2.99	78.12ef
5-bromo salicylaldehyde COS Schiff base phosphate 200 μg/ml	1.26	90.79f
5-chlorosalicylaldehyde COS Schiff base phosphate 50 μg/ml	9.85	28.01abcd
5-chlorosalicylaldehyde COS Schiff base phosphate 100 μg/ml	11.08	19.02abc
5-chlorosalicylaldehyde COS Schiff base phosphate 200 μg/ml	7.80	42.99abcde
3,5-dichlorosalicylaldehyde COS Schiff base phosphate 50 μg/ml	4.56	66.65def
3,5-dichlorosalicylaldehyde COS Schiff base phosphate 100 μg/ml	8.89	35.02abcde
3,5-dichlorosalicylaldehyde COS Schiff base phosphate 200 μg/ml	12.66	7.49a
300 times diluted Ningnanmycin water	4.54	66.79def
800-fold diluted 5% bis(octylamine ethyl) glycinate AS	8.35	38.98abcde
Water(CK)	13.68	–

Note: Date in a column followed by the same small letters are not significantly different at $P_{0.05}$ by Dunca's multiple range test, respectively

80%. The EC50 value of oligo-acid iodine to TMV was 15.62 mg/L, and the inhibitory activity against TMV was higher than that of 3% COS solution. Three hundred milligram per liter of oligo-acid iodine can be used as the optimal reagent for inhibiting TMV activity.

Zou et al. (Xing et al. 2016) applied Sulfated COS to wheat seedlings to study their effects on plant defense responses under salt stress. Exogenous COS treatment was found to reduce the malondialdehyde content, increase the chlorophyll content

Table 10.13 Effects of SCOS on growth parameters of wheat seedlings

	Shoot length (cm)	Rroot length (cm)	Wet wt (g)	Dry wt (g)
CK	23.6 ± 0.6bc	16.1 ± 0.4b	0.337 ± 0.031c	0.040 ± 0.005b
NaCl	19.1 ± 0.5d	13.3 ± 0.7c	0.260 ± 0.019d	0.027 ± 0.002c
SCOS + NaCl	24.7 ± 1.1a	18.4 ± 1.1a	0.358 ± 0.032a	0.056 ± 0.005a
COS + NaCl	24.1 ± 1.2ab	17.9 ± 0.7a	0.344 ± 0.026ab	0.053 ± 0.003a
SA + NaCl	23.1 ± 0.8c	16.6 ± 1.2b	0.280 ± 0.017d	0.039 ± 0.002b

Values are the mean ± SD of three replicates. Different letters indicate significant differences at $P < 0.05$

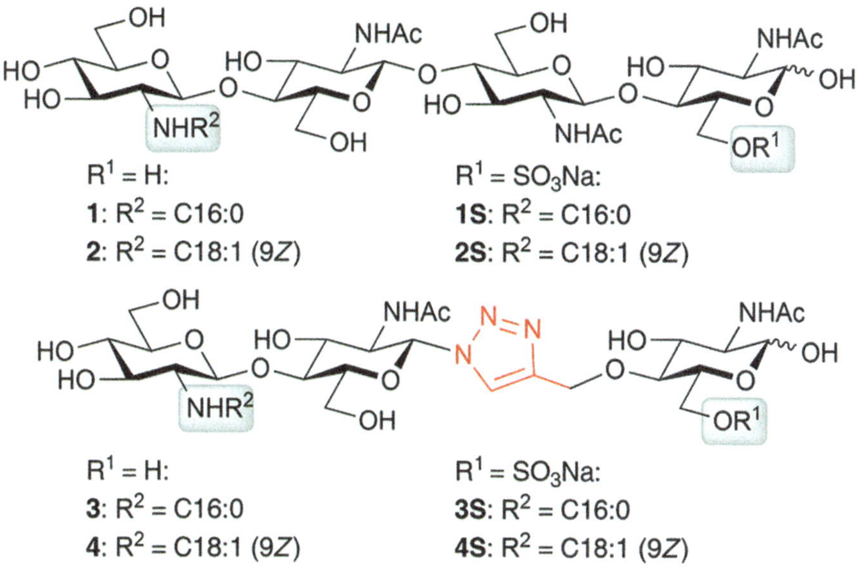

Fig. 10.3 Natural LCOs (1, 1S, 2, 2S) and the targeted LCO analogues(3, 3S, 4, 4S)

and regulate the fluorescence characteristics of wheat seedlings under salt stress. In addition, COS is capable of modulating the activity of antioxidant enzymes including superoxide dismutase, catalase, peroxidase, ascorbate peroxidase, glutathione reductase and dehydroascorbate reductase. Further studies find that COS can alleviate the damage of salt stress by regulating the activity of plant antioxidant enzymes. In addition, the effect of COS on the photochemical efficiency of wheat seedlings was related to the enhancement of antioxidant enzymes, which prevented the structural degradation of photosynthetic tissues under NaCl stress (Table 10.13).

Tanaka et al. (2015) found that Lipo-chitooligosaccharides (LCOs) (Fig. 10.3) promote root growth in C4 crops. Treatment of corn and foxtail seedlings with LCO significantly enhanced root growth. Further studies have found that LCO is involved in root growth promotion by up-regulating related genes. In addition, some stress-related genes are down-regulated after LCO treatment, which may indicate that some resources are shifting from plant defense responses to growth promotion. At the same time, LCO can also activate the promoter activity of the upregulated gene. These data indicate that LCO can directly affect maize root growth and gene expression. Berthelot et al. (2017) synthesized a lipid-tetrasaccharide analog in which the central GlcNAc residue was replaced by a triazole unit, by a copper-catalyzed azide-alkyne cycloaddition process using a functioned N-acetyl-glucosamine as feedstock. However, it was found that the plasto-tetraose tetrasaccharide analog did not bind at all to the lipid-chitooligosaccharide receptor LYR3. Therefore, it is important that the third GlcNAc unit from the non-reducing end of the LCO for binding to LysM-RLK LYR3.

Nanoclusters are plant bioactive substances that have been developed in recent years. Ho et al. (2015) developed a new method for the preparation of silver core-shell nanoparticles. The amino group of COS was first modified with 3,4-dihydroxyphenylacetic acid, and then the core-shell typed silver nanoparticles were prepared by protecting the silver ions with a mussel adhesion mechanism, as shown in Fig. 10.4. The core-shell silver nanoparticles exhibit strong growth inhibitory effects on phytopathogenic fungi such as *Phytophthora capsici* and *Phytophthora nicotianae* (Fig. 10.5).

Sun et al. (2015) synthesized a COS quaternary ammonium salt silver nanoparticles solution using a novel COS quaternary ammonium salt as a reducing agent and stabilizer, and the particle size is mainly at 7–12 nm (Fig. 10.6). It was found that the 25 μg/mL COS quaternary ammonium salt silver nanoparticles solution had the best inhibitory effect on the TMV, and the inhibition rate was 74.0%. This activity is 41.5% and 24.4% higher than the 50 μg/mL COS solution and the 2% ningnammycin aqueous solution, respectively. For ordinary tobacco K326, the COS quaternary ammonium silver nanoparticles solution can alleviate the decrease of chlorophyll content of tobacco infected by virus and increase SOD, POD and activity of PPO. In addition, it can also reduce the content of MDA and increase the content of soluble protein. These results indicated that the COS quaternary ammonium silver nanoparticles solution can improve the resistance of tobacco to TMV (Table 10.14).

Xing et al. (2016) synthesized a series of oleoyl-chitosan and prepared a nano-dispersion system. The nanoparticles were spherical and the particle size was about 296.9 nm. Mycelial growth experiments showed that *Nigrospora sphaerica*, *Botryosphaeria dothidea*, *Nigrospora oryzae* and *Alternaria tenuissima* were sensitive to this nano-dispersion system, while *Gibberella zeae* and *Fusarium culmorum*

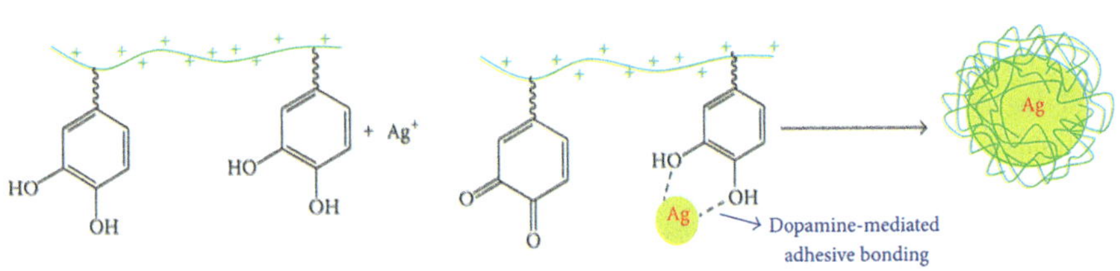

Fig. 10.4 Preparation of Ag core-DHPAC shell via mussel adhesion mechanism

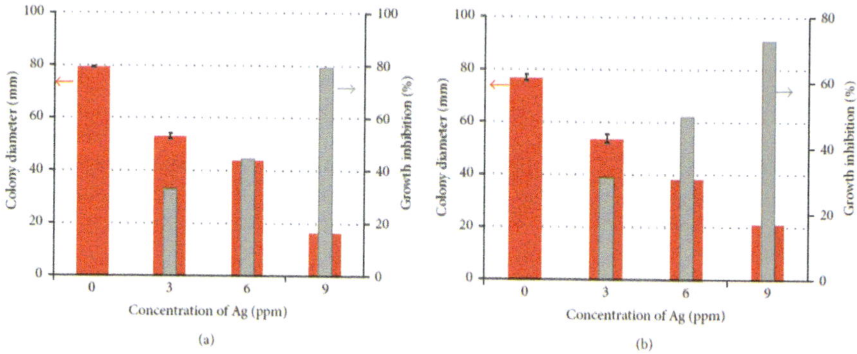

Fig. 10.5 Growth inhibition of Ag core-DHPAC shell NCs on the plant-pathogenic fungi: *P. capsici* (**a**) and *P. nicotianae* (**b**)

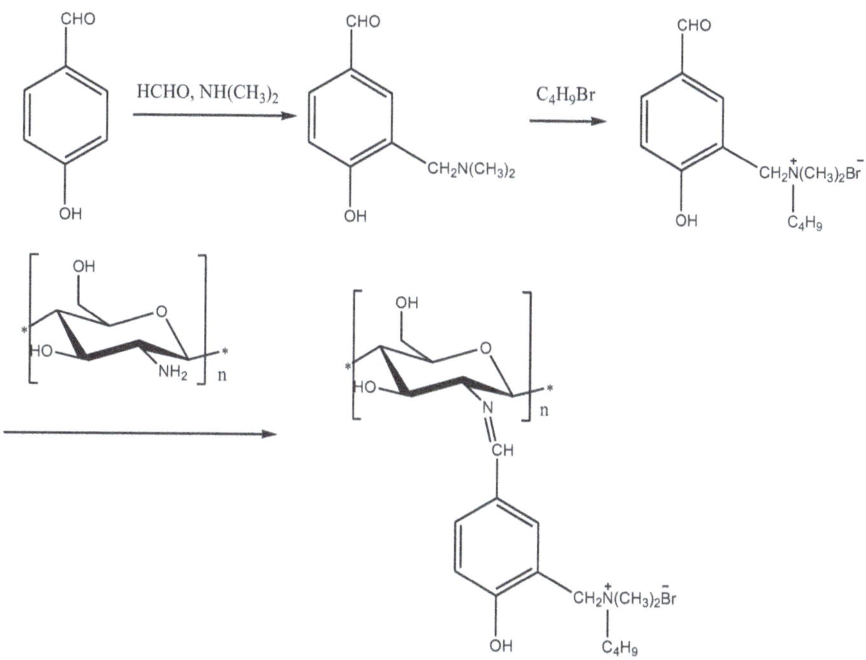

Fig. 10.6 Synthesis route of chitosan quaternary ammonium salt Schiff base derivatives

are resistant to this system, as shown in Fig. 10.7. As the concentration of nanoparticles in the chitosan-sensitive fungal system increases, the antifungal index increases. Fatty acid analysis showed that the plasma membrane of chitosan-sensitive fungi showed lower level of unsaturated fatty acids than chitosan-resistant fungi.

Table 10.14 Inhibition of the test reagent against TMV diseases on *Nicotiana tabacum* cv. Xanthi-nc

Test reagent	Concentration (μg/mL)	Lesion numbers in one cm^2 leave		Inhibition rate %
		Reagent treatment	Water treatment	
COS quaternary ammonium salt derivative silver nanoparticles	50	0.109	0.309	64.7 bc
	25	0.0732	0.282	74.0 a
	10	0.119	0.436	72.7 ab
COS quaternary ammonium salt derivative	50	0.133	0.292	54.5 d
	25	0.146	0.329	55.6 d
COS	50	0.174	0.365	52.3 d
2% ningnanmycin AS	Dilute 260 times	0.122	0.301	59.5 cd

Data in a column followed by different small letters are significant difference at 5%
Note: Different letters indicate significant differences at $P < 0.05$

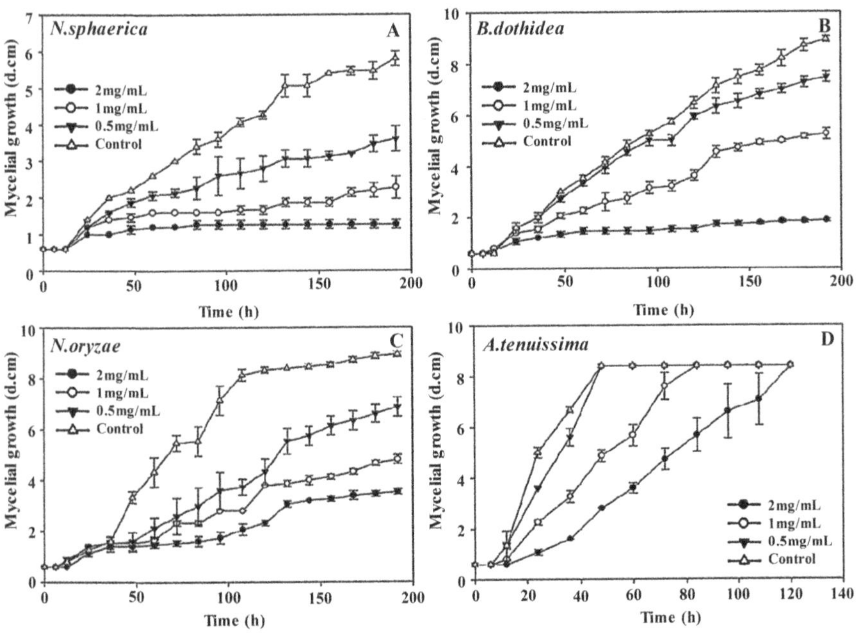

Fig. 10.7 Effect of O-chitosan nanoparticles on mycelium growth of chitosan-sensitive fungi

10.6 Conclusion and Perspective

Chito/chitin oligosaccharide has positive effects on triggering the resistance mechanism of plants. In 1980, it was first reported that chito/chitin oligosaccharide could induce plant immunity. Since then, chito/chitin oligosaccharide has been considered as a potent elicitor of plant immunity that is used in many crops, such as rice, wheat, cucumber, pepper, oilseed rape, tomato, bean, tobacco, flower crops, fruit crops and so on. During the study, researchers found that chito/chitin oligosaccharide stimulates a variety of plant responses, including induction of disease resistance, enhanced plant growth and yield, as well as the shelf life of flowers and fruits. The complexity of how plants perceive and respond to chito/chitin oligosaccharide has been investigated and cellular response models have been proposed. However, more information about these processes requires further study.

Now days, people are committed to the research of high-efficiency chemical inducers while studying the signal transmission mechanism, so as to reduce the environmental and food pollution caused by a large number of pesticides. Chito/chitin oligosaccharide is a naturally occurring compound that has been used as seed, leaf and fruit chemicals in vegetable production. Because of the excellent physical properties – high solubility, nontoxicity, biocompatibility, rich in production, low cost – COS is considered an effective plant regulation substance by researchers. It is believed that with the deepening of research, the application of COS will soon will have a good application prospect.

References

Albersheim P, Darvill AG (1985) Oligosaccharins. Sci Am 253(3):58–64

Ali A, Mohamed MTM, Siddiqui Y (2012) Control of anthracnose by chitosan through stimulation of defence-related enzymes in Eksotika II papaya (Carica papaya L.). J Biol Life Sci 3(1):114–126

Ali A, Muhammad MTM, Sijam K, Siddiqui Y (2011) Effect of chitosan coatings on the physicochemical characteristics of Eksotika II papaya (*Carica papaya* L.) fruit during cold storage. Food Chem 124(2):620–626

Ali A, Noh NM, Mustafa MA (2015) Antimicrobial activity of chitosan enriched with lemongrass oil against anthracnose of bell pepper. Food Packag Shelf Life 3(3):56–61

Ali A, Zahid N, Manickam S, Siddiqui Y, Alderson PG, Maqbool M (2014) Induction of lignin and pathogenesis related proteins in dragon fruit plants in response to submicron chitosan dispersions. Crop Prot 63(5):83–88

Ayers A, Ebel J, Valent B, Albersheim P (1976) Host-pathogen interactions: X. fractionation and biological activity of an elicitor isolated from the mycelial walls of Phytophthora megasperma var. sojae. Plant Physiol 57(5):760–765

Badawy MEI, Rabea EI (2009) Potential of the biopolymer chitosan with different molecular weights to control postharvest gray mold of tomato fruit. Postharvest Biol Technol 51(1):110–117

Badiali C, Angelis GD, Simonetti G, Brasili E, Tobaruela EDC, Purgatto E, Yin H, Valletta A, Pasqua G (2018) Chitosan oligosaccharides affect xanthone and VOC biosynthesis in Hypericum perforatum root cultures and enhance the antifungal activity of root extracts. Plant Cell Rep:1–14

Bai CW, Jiang XL, Ding HX, Zuo X (2010) Resistance induced by chitosan oligosaccharide to rice sheath blight. Guizhou Agric Sci 38(8):103–106

Benshalom N, Ardi R, Pinto R, Aki C, Fallik E (2003) Controlling gray mould caused by Botrytis cinerea in cucumber plants by means of chitosan. Crop Prot 22(2):285–290

Berthelot N, Brossay A, Gasciolli V, Bono JJ, Baron A, Beau JM, Urban D, Boyer FD, Vauzeilles B (2017) Synthesis of lipo-chitooligosaccharide analogues and their interaction with LYR3, a high affinity binding protein for nod factors and Myc-LCOs. Org Biomol Chem 15:7802

Chookhongkha, N., Miyagawa, S., Jirakiattikul, Y., Photchanachai, S. (2012) Chilli growth and seed productivity as affected by chitosan. In: International conference on advances in computer and electrical engineering, pp 146–147–148–149

Chutichudet B, Chutichudet P (2011) Effects of chitosan coating to some postharvest characteristics of Hylocercus undatus (Haw) Brit. and rose fruit. Int J Agric Res 6(1):82–92

Doares SH, Syrovets T, Weiler EW, Ryan CA (1995) Oligogalacturonides and chitosan activate plant defensive genes through the octadecanoid pathway. Proc Natl Acad Sci 92(10):4095–4098

Dong H, Cheng L, Tan J, Zheng K, Jiang Y (2004) Effects of chitosan coating on quality and shelf life of peeled litchi fruit. J Food Eng 64(3):355–358

El-Miniawy SM, Ragab ME, Youssef SM, Metwally AA (2013) Response of strawberry plants to foliar spraying of chitosan. Res J Agric Biol Sci 9(6):366–372

Ghasemnezhad M, Nezhad MA, Gerailoo S (2011) Changes in postharvest quality of loquat (Eriobotrya japonica) fruits influenced by chitosan. Hortic Environ Biotechnol 52(1):40–45

Ghasemnezhad M, Shiri MA, Sanavi M (2010) Effect of chitosan coatings on some quality indices of apricot (Prunus armeniaca L.) during cold storage. Br Heart J 74(6):584–591

Gianfranco R, Franka Mlikota G, Dennis M, Mackey BE, Smilanick JL (2009) Effect of chitosan dissolved in different acids on its ability to control postharvest gray mold of table grape. Phytopathology 99(9):1028–1036

Guo W, Zhao X, Du YG (2009) Effects of oligochitosan on the seed germination seedling growth and photosynthetic characters of cucumber. Chin Agric Sci Bull

Gustavo MC, Keiko S, Clara PZ, Pérez-Flores LJ, Sepúlveda-Sánchez JD (2009) Effect of lactobacillus plantarum and chitosan in the reduction of browning of pericarp Rambutan (Nephelium lappaceum). Food Microbiol 26(4):444–449

Hang LU (2009) Study of two oligosaccharides inducing resistance to tobacco mosaic virus. Acta Agric Boreali-Occidentalis Sin

Hewajulige IGN, Sultanbawa Y, Wijeratnam RSW, Wijesundara RLC (2009) Mode of action of chitosan coating on anthracnose disease control in papaya. Phytoparasitica 37(5):437–444

Ho VA, Le PT, Nguyen TP, Nguyen CK, Nguyen VT, Tran NQ (2015) Silver core-shell nanoclusters exhibiting strong growth inhibition of plant-pathogenic Fungi. J Nanomater 2015(1):13

Hu X, Zhinqing T, Liang L, Shikui W, Xiangmin Z, Zhimin Z (2018) Inhibition activity of oligoglucuronic acids-iodine (OA-I) on TMV and its effect on tomato growth. J Chin Agric Univ 23(8):41–49

Hyun-Jin K, Feng C, Xi W, Rajapakse NC (2005) Effect of chitosan on the biological properties of sweet basil (Ocimum basilicum L.). J Agric Food Chem 53(9):3696–3701

Jia X, Meng Q, Zeng H, Wang W, Yin H (2016) Chitosan oligosaccharide induces resistance to tobacco mosaic virus in Arabidopsis via the salicylic acid-mediated signalling pathway. Sci Rep 6:26144

Jia X, Zeng H, Wang W, Zhang F, Yin H (2018) Chitosan oligosaccharide induces resistance to Pseudomonas syringae pv. tomato DC3000 in *Arabidopsis thaliana* by activating both salicylic acid- and jasmonic acid-mediated pathways. Mol Plant Microbe Interact 31:1271. MPMI-03-18-0071-R-

Jiang Y, Li Y (2001) Effects of chitosan coating on postharvest life and quality of longan fruit. Food Chem 73(2):139–143

Kananont N, Pichyangkura R, Chanprame S, Chadchawan S, Limpanavech P (2010) Chitosan specificity for the in vitro seed germination of two Dendrobium orchids (Asparagales: Orchidaceae). Sci Hortic 124(2):239–247

Kerch G, Sabovics M, Kruma Z, Kampuse S, Straumite E (2011) Effect of chitosan and chitooligosaccharide on vitamin C and polyphenols contents in cherries and strawberries during refrigerated storage. Eur Food Res Technol 233(2):351–358

Kuang YJ, Peng HE, Gui-Ping YE, Qin CQ (2009) Effects of Oligochitosan on Cold Resistance of Eggplant Seedlings. North Hortic 48:617

Li H, Yu T (2001) Effect of chitosan on incidence of brown rot, quality and physiological attributes of postharvest peach fruit. J Sci Food Agric 81(2):269–274
Limpanavech P, Chaiyasuta S, Vongpromek R, Pichyangkura R, Khunwasi C, Chadchawan S, Lotrakul P, Bunjongrat R, Chaidee A, Bangyeekhun T (2008) Chitosan effects on floral production, gene expression, and anatomical changes in the Dendrobium orchid. Sci Hortic 116(1):65–72
Lin B, Du Y, Liang X, Wang X, Wang X, Yang J (2011) Effect of chitosan coating on respiratory behavior and quality of stored litchi under ambient temperature. J Food Eng 102(1):94–99
Liu D, Jiao S, Cheng G, Li X, Pei Z, Pei Y, Yin H, Du Y (2018) Identification of chitosan oligosaccharides binding proteins from the plasma membrane of wheat leaf cell. Int J Biol Macromol 111:1083. S0141813017330714
Liu J, Tian S, Meng X, Xu Y (2007) Effects of chitosan on control of postharvest diseases and physiological responses of tomato fruit. Postharvest Biol Technol 44(3):300–306
Liu X, Du Y, Bai X (2001) Relieving effects of oligoglucosamine on the inhibition induced by deoxynivalenol in wheat embryo cells. Acta Bot Sin 43(4):370–374
Luigi L, Greta B, Youssef R, Giuseppe C, Luigi B, Marco T (2018) Chitosan treatment elicited defence mechanisms, pentacyclic triterpenoids and stilbene accumulation in grape (Vitis vinifera L.) bunches. Phytochemistry:1–8
Ma Q et al (2004) Induction of oligosaccharide to ultrastructure of cucumber resistance to powdery mildew fungus. Acta Phytopathologica Sinica 34(6):525–530. https://doi.org/10.1300/J064v24n01_09
Ma Q, Sun H, Du YG, Zhao XM, Shang HS (2005) Effect of oligosaccharide on the resistance induction of cucumber against Spharotheca fuliginea. J North Sci-Technol Univ Agric For 33:79
Ma Z, Yang L, Yan H, Kennedy JF, Meng X (2013) Chitosan and oligochitosan enhance the resistance of peach fruit to brown rot. Carbohydr Polym 94(1):272–277
Mathew R (2012) Effect of methyl jasmonate and chitosan on growth characteristics of Ocimum basilicum L., Ocimum sanctum L. and Ocimum gratissimum L. cell suspension cultures. Afr J Biotechnol 11(21):4759
Meng X, Yang L, Kennedy JF, Tian S (2010) Effects of chitosan and oligochitosan on growth of two fungal pathogens and physiological properties in pear fruit. Carbohydr Polym 81(1):70–75
Meng XH, Tian SP (2010) Effects of preharvest application of antagonistic yeast combined with chitosan on decay and quality of harvested table grape fruit. J Sci Food Agric 89(11):1838–1842
Mondal MMA, Malek MA, Puteh AB, Ismail MR, Ashrafuzzaman M (2012) Effect of foliar application of chitosan on growth and yield in okra. Aust J Crop Sci 6(5):918–921
Moreira MDR, Roura SI, Ponce A (2011) Effectiveness of chitosan edible coatings to improve microbiological and sensory quality of fresh cut broccoli. LWT Food Sci Technol 44(10):2335–2341
Nan Y (2010) Microwave synthesis of chitooligosaccharide schiff base derivatives and their induced resistance to tobacco mosaic virus. Henan Agricultural University
Nge K, Nwe N, Chandrkrachang S, Stevens W (2006) Chitosan as a growth stimulator in orchid tissue culture. Plant Sci 170(6):1185–1190
Ning W, Chen F, Mao B, Li Q, Liu Z, Guo Z, He Z (2004) N-acetylchitooligosaccharides elicit rice defence responses including hypersensitive response-like cell death, oxidative burst and defence gene expression. Physiol Mol Plant Pathol 64(5):263–271
Ochoa-Velasco CE, Guerrero-Beltrán JÁ (2014) Postharvest quality of peeled prickly pear fruit treated with acetic acid and chitosan. Postharvest Biol Technol 92(3):139–145
Ohta K, Morishita S, Suda K, Kobayashi N, Hosoki T (2004) Effects of chitosan soil mixture treatment in the seedling stage on the growth and flowering of several ornamental plants. J Jpn Soc Hortic Sci 73(1):66–68
Park SI, Stan SD, Daeschel MA, Zhao Y (2010) Antifungal coatings on fresh strawberries (Fragaria × ananassa) to control Mold growth during cold storage. J Food Sci 70(4):M202–M207
Pichyangkura R, Chadchawan S (2015) Biostimulant activity of chitosan in horticulture. Sci Hortic 196:49–65

Pilar H-M, Eva A, Valeria Del V, Dinoraz V, Rafael G (2008) Effect of chitosan coating combined with postharvest calcium treatment on strawberry (Fragaria×ananassa) quality during refrigerated storage. Food Chem 110(2):428–435

Pornpienpakdee P, Singhasurasak R, Chaiyasap P, Pichyangkura R, Bunjongrat R (2010) Improving the micropropagation efficiency of hybrid Dendrobium orchids with chitosan. Sci Hortic 124(4):490–499

Pospieszny H, Atabekov JG (1989) Effect of chitosan on the hypersensitive reaction of bean to alfalfa mosaic virus. Plant Sci 62(1):29–31

Pospieszny H, Chirkov S, Atabekov J (1991) Induction of antiviral resistance in plants by chitosan. Plant Sci 79(1):63–68

Prodhan MY, Issak M, Nakamura T, Munemasa S, Nakamura Y, Murata Y (2017) Chitosan signaling in guard cells requires endogenous salicylic acid. Biosci Biotechnol Biochem 81(8):1536–1541

Qi H, Hu W, Jiang A, Tian M, Li Y (2011) Extending shelf-life of fresh-cut 'Fuji' apples with chitosan-coatings. Innovative Food Sci Emerg Technol 12(1):62–66

Rabea EI, Badawy ET, Stevens CV, Smagghe G, Steurbaut W (2003) Chitosan as antimicrobial agent: applications and mode of action. Biomacromolecules 4(6):1457

Raetz CR (1990) Biochemistry of endotoxins. Annu Rev Biochem 59:129–170

Reuck KD, Sivakumar D, Korsten L (2009) Integrated application of 1-methylcyclopropene and modified atmosphere packaging to improve quality retention of litchi cultivars during storage. Postharvest Biol Technol 52(1):71–77

Salachna P, Zawadzińska A (2014) Effect of chitosan on plant growth, flowering and corms yield of potted freesia. J Ecol Eng 15:97

Sang NV, Minh HD, Anh DN (2013) Study on chitosan nanoparticles on biophysical characteristics and growth of Robusta coffee in green house. Biocatalysis Agric Biotechnol 2(4):289–294

Shang W, Wu Y, Zhao X, Du Y, Shang H (2006a) Effect of chito--oligosaccharide induction on long – distance movement of TMV in tobacco. Acta Botan Boreali-Occiden Sin 26

Shang W, Zhao X, Du Y, Shang H (2005) First report on oligosaccharide induced resistance to plant virus. J North Sci-Technol Univ Agric For 33:73

Shang WJ, Wu YF, Shang HS, Zhao XM, Du YG (2008) The inactivating effect of chito-oligosaccharides on TMV particles in vitro. Chin J Virol 24(1):76–78

Shang WJ, Wu YF, Zhao XM, Du YG, Shang HS (2010) Changes of defensive enzymes and PR-1a gene expression of tobacco induced by chito-oligosaccharides. Acta Phytopathol Sin 40(1):99–102

Shang WJ, Yun-Feng WU, Shang HS, Zhao XM, Yu-Guang DU (2007) Inhibitory effect to TMV-CP gene expression in tobacco induced by chito-oligosaccharides. Acta Phytopathol Sin 37:637

Shang WJ, Yun-Feng WU, Zhao XM, Yu-Guang DU, Shang HS (2006b) Induced resistance to TMV multiplication in tobacco with Chito-oligosaccharides. J North Sci-Technol Univ Agric For (Natural Science Edition) 34(5):88–92

Srivastava N, Gonugunta VK, Puli MR, Raghavendra AS (2009) Nitric oxide production occurs downstream of reactive oxygen species in guard cells during stomatal closure induced by chitosan in abaxial epidermis of Pisum sativum. Planta 229(4):757–765

Sun C, Xu C, Zhao M, Huang Y, Lu Z, You F, Zhou F (2015) Synthesis of silver nanoparticles using quaternized oligochitosan derivative and its induced resistance against TMV in tobacco plant. Chin J Pestic Sci

Tanaka K, Cho SH, Lee H, An QP, Batek JM, Cui S, Jing Q, Khan SM, Joshi T, Zhang ZJ (2015) Effect of lipo-chitooligosaccharide on early growth of C4 grass seedlings. J Exp Bot 66:5727

Uddin AFMJ, Hashimoto F, Kaketani M, Shimizu K, Sakata Y (2001) Analysis of light and sucrose potencies on petal coloration and pigmentation of lisianthus cultivars (in vitro). Sci Hortic 89(1):75–84

Xiao ZJ, Jiang XL, Xiao-Xia LI, Xi-Li MO (2009) Preliminary study on the inducement of resistance against powdery mildew of pepper by Oligochitosan. Hubei Agric Sci 48:617

Xiaoyu L (2005) Preparation of water-soluble oligosaccharides and oligosaccharides-copper and their biological activities. Shenyang Agricultural University

Xing K, Shen X, Zhu X, Ju X, Miao X, Tian J, Feng Z, Peng X, Jiang J, Qin S (2016) Synthesis and in vitro antifungal efficacy of oleoyl-chitosan nanoparticles against plant pathogenic fungi. Int J Biol Macromol 82:830

Xu J, Zhao X, Bai X, Du Y (2006) Effect of Oligochitosan on controlling pepper Phytophthora blight in field and on Phytophthora Capsici in vitro. Chin Agric Sci Bull

Xu Z, Li L, Li C (2003) Study on Zhongshengmycin and Amino-Oligosaccharins to Fusarium oxysporum on Watermelon. Chin Veg 3:10

Yang A, Yu L, Chen Z, Zhang S, Shi J, Zhao X, Yang Y, Hu D, Song B (2017) Label-free quantitative proteomic analysis of chitosan oligosaccharide-treated rice infected with southern rice black-streaked dwarf virus. Viruses 9(5):115

Yin H, Bai X, Du Y (2008) The primary study of oligochitosan inducing resistance to Sclerotinia sclerotiorum on *Brassica napus*. J Biotechnol 136(4):S600–S601

Yin H, Yan L, Zhang HY, Wang WX, Hang L, Kai G, Zhao X, Du Y (2013) Chitosan oligosaccharides–triggered innate immunity contributes to oilseed rape resistance againstSclerotinia Sclerotiorum. Int J Plant Sci 174(4):722–732

Yin H, Zhao X, Du Y (2010) Oligochitosan: a plant diseases vaccine – a review. Carbohydr Polym 82(1):1–8

Yu T, Yu C, Chen F, Sheng K, Zhou T, Zunun M, Abudu O, Yang S, Zheng X (2012) Integrated control of blue mold in pear fruit by combined application of chitosan, a biocontrol yeast and calcium chloride. Postharvest Biol Technol 69(3):49–53

Zhang D, Quantick PC (1997) Effects of chitosan coating on enzymatic browning and decay during postharvest storage of litchi (Litchi chinensis Sonn.) fruit. Postharvest Biol Technol 12(2):195–202

Zhang W, Sui X, Xia W, Zhang Y, Lin C (2002) Preparation of chitoligmer and its application to cucumber growth. J Funct Polym 15:199

Zhang WG, Wang Y, Wei B, Lu HE, Zhang L, Liang J, Ming LI, Zhang L (2008) Efficacy of a chitooligosacchaide aqua in controlling wheat Rhizoctonia root rot. Hubei Agric Sci 47(4):433–434

Zhang Y, Fan LQ, Zhao MY, Chen QM., Qin Z, Feng ZH, Fujiwara T, Zhao LM (2019) Chitooligosaccharide plays essential roles in regulating proline metabolism and cold stress tolerance in rice seedlings. Acta Physiol Plant proof correction

Zhao X, Du Y, Bai X (2004) The field experiment of oligochitosan controlling plant virus disease. Chin Agric Sci Bull 20:245

Zhao Y, Deng L, Zhou Y, Yao S, Zeng K (2018) Chitosan and Pichia membranaefaciens control anthracnose by maintaining cell structural integrity of citrus fruit. Biol Control 124:92

Zhong Q, Xia W (2007) Effect of 1-methylcyclopropene and/or chitosan coating treatments on storage life and quality maintenance of Indian jujube fruit. LWT Food Sci Technol 40(3):404–411

Zhou J, Chen Q, Zhang Y, Fan L, Zhao L (2018) Chitooligosaccharides enhance cold tolerance by repairing photodamaged PS II in rice. J Agric Sci 156(7):1–12

Zhu X, Wang Q, Cao J, Jiang W (2010) Effects of chitosan coating on postharvest quality of mango (Mangifera indica L. CV. Tainong) fruits. J Food Process Preserv 32(5):770–784

Zou P, Tian X, Dong B, Zhang C (2017) Size effects of chitooligomers with certain degrees of polymerization on the chilling tolerance of wheat seedlings. Carbohydr Polym 160:194–202

The Application of Oligosaccharides in Breeding Industry

11

Qingsong Xu, Lunxue Wang, and Yuguang Du

Abstract

As the increasing demand for antibiotic substitution at home and abroad, animal husbandry will face serious challenges. How to ensure the healthy growth of breeding animals in the post-antibiotic era has become the most concerned issue in animal husbandry. Oligosaccharides, as one of the alternatives to antibiotics, have attracted more and more attentions. Chitooligosaccharides (COS) is one of the important functional oligosaccharides, which has become a research hotspot in recent years. Because COS contains reactive chemical structure, such as hydroxyl and amino groups, it is characterized as meaningful ingredient for animal health. Thus, the aim of this chapter is to summarize and discuss the results of investigations in which the application of COS in livestock, poultry and ruminants. Numerous studies have shown that COS feed additives have good effects in sow, piglet, broiler, laying hen and calf breeding. It not only can regulate the physiological function of the body and promote metabolism, but also has the functions of antibacterial and enhancing immunity. More importantly, it can accelerate the growth speed, improve the animal performance, promote the quality of products and reduce the mortality rate. Collectively, the introduction and summarization of this chapter reveal the application scope, effect and mechanism of COS in breeding animals, which will provide theoretical support for the development of COS feed additives, and accelerate its industrialization process.

Q. Xu (✉)
Key Laboratory of Biotechnology and Bioresources Utilization of Ministry of Education, Dalian Minzu University, Dalian, China

L. Wang
Dalian Institute of Chemical Physics, Chinese Academy of Sciences, Dalian, China

Y. Du
Institute of Process Engineering, Chinese Academy of Sciences, Beijing, China

L. Zhao (ed.), *Oligosaccharides of Chitin and Chitosan*,
https://doi.org/10.1007/978-981-13-9402-7_11

Keywords
Chitosan oligosaccharides · Feed additive · Livestock · Poultry · Ruminant

11.1 Introduction

Animal husbandry is an important part of agriculture, and the growth rate of animal products increased year on year. However, the development of animal husbandry will face several challenges in the future. (1) The situation of animal epidemic prevention and control is still grim. It not only affects consumption, but also farming confidence. (2) The extensive application of antibiotics in animal husbandry has led to drug resistance in modern husbandry. Antibiotic prohibition is the trend of future development, and the research and development of antibiotic substitutes are urgent. (3) The hidden danger of animal product quality and safety exists for a long time. The quality and safety of livestock products involve people's life and health, and safety is no trivial matter, and responsibility is as great as heaven. (4) The construction of modern livestock and poultry breeding system is lagging behind. Most of the major improved breeds of livestock and poultry in China, such as pigs, dairy cows and poultry, depend on imports from abroad, and the ability of independent breeding needs to be improved urgently.

In recent years, people have developed antibiotic substitutes such as feed enzymes, feed microbial agents, acidifiers, antimicrobial peptides, oligosaccharides, Chinese herbal medicines and organic metal trace elements (Fig. 11.1). These additives bring new hope to the animal industry. Oligosaccharides, as one of the most important feed additives, usually are not digested by digestive enzymes in animal intestines, and cannot be used by harmful microorganisms such as *Salmonella* and *Escherichia coli*. When oligosaccharides are used by beneficial bacteria, the intestinal beneficial bacteria such as *Lactobacillus* and *Bifidobacterium* multiply in large quantities, and produce more lactic acid, acetic acid, butyric acid and propionic acid, which reduce the intestinal pH value and inhibit the reproduction of acid-resistant harmful bacteria. Functional oligosaccharides have been widely applied in animal husbandry and gradually accepted by people. At present, the commonly used species include fructo-oligosaccharides, xylo-oligosaccharides, mannan-oligosaccharides, chitosan oligosaccharides, maltose, arabinose, galactose, and so on.

Chitosan oligosaccharides (COS), known as oligomeric β-(1-4) -2-amino-2-deoxy-D-glucose, is linked by glucosamine through β-1,4 glycoside bonds. COS is derived from crab and shrimp shells by enzymatic degradation. It is naturally non-toxic, low molecular weight, soluble in water and easily absorbed and utilized by animals (Liaqat and Eltem 2018). Compared with antibiotics, it has the advantages of no drug resistance, no drug residues, no destruction and killing of beneficial bacteria (Swiatkiewicz et al. 2015). Currently, the efficacy of COS on animal health is presented in this chapter.

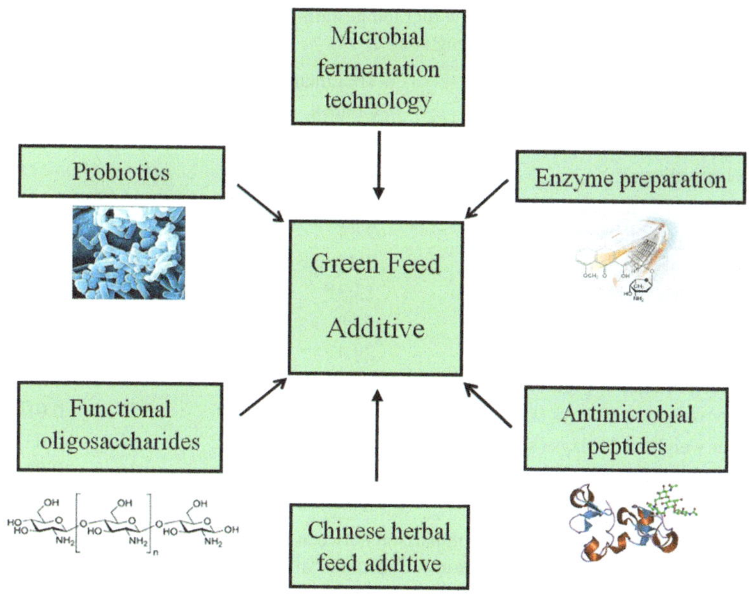

Fig. 11.1 Green feed additive used as antibiotic substitutes

11.2 The Application of Oligosaccharides in Swine Industry

11.2.1 The Application of COS in Sow

In the swine breeding industry, litter size is one of the most significant economic traits and varies among different individual sows. Nevertheless, reproductive traits in sow are usually complex, from ovulation, fertilization, and implantation until to the birth of piglets, every process may influence the litter size of sow, and various efforts have been made to investigate the factors influencing litter size, such as genetic factors, optimizing nutrition, management of sows and husbandry. It is well known that sow diets and health during the whole gestation are important for fetal survival and sow reproductive performance. Nutrition optimization for increased litter size has in turn promoted the prolificacy of animals over the past decades. For example, some amino acid and functional oligosaccharides have been applied to improve the pregnant sow reproductive performance (Wan et al. 2016).

11.2.1.1 Effect of COS on the Reproductive Performance of Sows

To improve the litter size of piglets, Cheng et al. (2015) evaluated the effect of COS supplementation on the reproductive performance of sows. The results showed that COS supplementation could increase the total number of piglets born by 18.5%, the number of piglets born alive by 19.2%, and the live born litter weight by 31.3% (Table 11.1). To our knowledge, these findings innovatively revealed that litter size and litter birth weight were improved in sows through the COS intervention (Cheng

Table 11.1 Effects of supplementing the diet fed gestating sows with COS (40 mg/kg) on reproduction and lactation performance[a] (Cheng et al. 2015)

Parameters	Treatment		SEM	*P* value
	Control	COS		
Reproductive performance				
No. of observations (n)	18	20	–	–
Average parity of sows	3.06	3.00	–	–
Total number of piglets born (n)	9.83	11.65	0.62	0.045
Total number of piglets born alive (n)	8.44	10.45	0.49	0.007
Live born litter weight (kg)	13.31	17.47	0.84	0.001
Born individual weight (kg)	1.58	1.68	0.04	0.112
Pregnant sow weight change (kg)	29.4	32.6	2.9	0.440
Growth performance of piglets				
Total litter weight at 12 days (kg)	34.35	41.26	1.79	0.010
Total litter weight at 21 days (kg)	51.52	60.52	2.91	0.035
Daily litter weight gain 0–21 days (kg/day)	1.91	2.15	0.26	0.064
Weaning individual weight (kg)	6.54	6.87	0.24	0.357

[a]Values are means with pooled SEM, comparison between the two groups, *P* values less than 0.05 indicate significant difference; comparison between the two group, *P* values less than 0.01 indicate extremely significant difference

et al. 2015). Another research obtained similar results that COS supplementation could lead to a higher fetal survival rate ($P < 0.05$) on day 35 of pregnancy compared with the litters from control group. COS supplementation not only increased the number of piglets born viable ($P < 0.05$) but also decreased the number of mummified and stillborn piglets born per litter ($P < 0.05$). The birth weights, including average individual weight of pigs born alive within litters and total weight of pigs born alive per litter, were greater ($P < 0.05$) in multiparous sows fed the COS-supplemented diet than in those fed the control diet (1.32 vs. 1.20 kg; and 17.28 vs. 14.50 kg, respectively) (Wan et al. 2016).

11.2.1.2 Effect of COS on the Antioxidant Capacity and Immunity of Sows

Dietary COS supplementation not only increased serum immunoglobulin (IgA, IgG and IgM) concentrations (Table 11.2) but also increased the total antioxidant capacity (T-AOC) at 35 days of gestation (Wan et al. 2016). Moreover, the serum IgG concentrations significantly increased in COS than in the control group at 85 days of gestation. The results demonstrated that dietary COS supplementation had the obvious effect on the antioxidant capacity and immunity of sows.

11.2.1.3 Effect of COS on the Reproductive Hormones of Sows

Dietary supplementing with COS up-regulated the expression levels of leptin at days 35 and 85 of gestation (Table 11.3) (Wan et al. 2016). Meanwhile, Dietary COS supplementation has the tendency to increase the concentration of progesterone, estradiol, luteinizing hormone of sows at days 35 and 85 of gestation, of which the difference was not significant ($P > 0.05$).

Table 11.2 Effects of dietary COS supplementation on serum immune responses of sows at days 35 and 85 of gestation (Wan et al. 2016)

Items	Treatments		*P*-value
	CON	COS	
Day 35 of gestation			
IL-1 (pg mL^{-1})	284.48 ± 14.63	359.95 ± 7.79[a]	0.002
TNF-α (pg mL^{-1})	386.69 ± 27.70	522.63 ± 41.04[a]	0.025
IgG (μg mL^{-1})	233.18 ± 28.18	343.40 ± 35.06[a]	0.040
IgA (μg mL^{-1})	48.52 ± 7.04	77.88 ± 3.34[a]	0.006
IgM (μg mL^{-1})	87.81 ± 5.31	117.05 ± 3.42[a]	0.002
Day 85 of gestation			
IL-1 (pg mL^{-1})	299.80 ± 14.79	278.53 ± 24.50	0.425
TNF-α (pg mL^{-1})	439.00 ± 37.41	457.61 ± 55.86	0.789
IgG (μg mL^{-1})	240.77 ± 24.72	424.77 ± 33.20[a]	0.002
IgA (μg mL^{-1})	82.35 ± 6.99	102.13 ± 7.02[b]	0.081
IgM (μg mL^{-1})	111.86 ± 4.60	114.32 ± 10.22	0.832

[a]$P < 0.05$ versus the CON group. CON, a corn-soybean basal diet; COS, the basal diet supplemented with 100 mg kg^{-1} COS

[b]Values are means of six replicates per treatment

Table 11.3 Effects of dietary COS supplementation on serum hormone concentrations of sows at days 35 and 85 of gestation[b] (Wan et al. 2016)

Items	Treatments		*P*-value
	CON	COS	
Day 35 of gestation			
Leptin (ng mL^{-1})	8.96 ± 1.07	15.36 ± 1.24[a]	0.003
Progesterone (ng mL^{-1})	2.95 ± 0.13	3.62 ± 0.37	0.122
Estradiol (ng mL^{-1})	0.94 ± 0.06	1.18 ± 0.15	0.169
Luteinizing hormone (mIU mL^{-1})	5.54 ± 0.46	6.61 ± 0.50	0.154
Day 85 of gestation			
Leptin (ng mL^{-1})	11.24 ± 1.36	16.01 ± 1.32[a]	0.030
Progesterone (ng mL^{-1})	1.39 ± 0.16	1.74 ± 0.15	0.149
Estradiol (ng mL^{-1})	0.87 ± 0.04	0.90 ± 0.12	0.852
Luteinizing hormone (mIU mL^{-1})	4.91 ± 0.26	6.24 ± 0.76	0.137

[a]$P < 0.05$ versus the CON group. CON, a corn-soybean basal diet; COS, the basal diet supplemented with 100 mg kg^{-1} COS

[b]Values are means of six replicates per treatment

11.2.1.4 Effect of COS on the Amniotic Fluid Composition in Sows

Supplementing sows with COS had enhanced the fetal survival rate and size at 35 days (Wan et al. 2017b). Dietary COS supplementation had both elevated ($P < 0.05$) CAT, SOD and T-AOC activities and up-regulated ($P < 0.05$) IgG, IgM and IL-10 concentrations in the amniotic fluid. Furthermore, COS had regulated ($P < 0.05$) the amniotic fluid's citrate, lysine, hypoxanthine and glucose levels (Table 11.4) (Wan et al. 2017a, b). It could be concluded that COS supplementation

Table 11.4 OPLS-DA coefficients obtained from the NMR data of amniotic fluid metabolites from the COS and CON groups (Wan et al. 2017a, b)

Metabolites	OPLS-DA coefficient (r)
	COS versus CON[a]
Lysine	+0.773
Citrate	+0.798
Glucose	+0.868
Hypoxanthine	−0.806

[a]CON: A corn-soybean basal diet; COS: COS (the basal diet supplemented with 100 mg/kg COS)

elevated amniotic fluid antioxidant status and metabolic profile modifications characterizing improvements in fetal survival and growth in a sow model.

11.2.1.5 Effect of COS on the Amino Acids Transport in Sows

Few studies investigated the effects of COS on the placental nutrition transport capacity. Xie et al. (2016a, b, c) investigated the effects of COS supplementation during late gestation on placental amino acids transport and antioxidant defense capacity of sows (Xie et al. 2016c). The results demonstrated that maternal dietary supplementing with COS activated mTOR signaling pathway, and promoted amino acids transporters expression in placenta. These results indicated that COS supplement in the diet during late gestation increased placental amino acids transport, which may contribute to the development of fetus during gestation. Xie et al. (2015) also investigated the effect of COS supplementing sows during gestation and lactation on the levels of amino acids in sow milk. The results indicated that supplementation of the maternal diet with COS improved the concentration of amino acids in sow milk (Table 11.5). In conclusion, dietary supplementation of the diet with COS during late gestation and lactation promoted the growth rate of suckling piglets by providing more amino acids for growth and utilization of piglets (Xie et al. 2015).

11.2.1.6 Effect of COS on the Milk Oligosaccharides in Sows

As a nutritional link between mother and newborn piglets, milk is vital to the health and growth of piglets. Milk oligosaccharides have several important biological activities, such as preventing pathogens from sticking to the intestinal epithelial and serving as nutrients for intestinal beneficial bacteria. Therefore, it is valuable to examine the variation of the contents of porcine milk after COS supplementation. In the milk of lactating sows, 17 different PMOs were identified in the trials (Cheng et al. 2015). Among them, a tetrasccharide $Hex_2GlcNAcNeuAc$ and a trisaccharide Hex_3 were 150% and 60% higher in the COS group than in the control group (Fig. 11.2). These results showed that the COS supplementation could regulate the oligosaccharides in porcine milk (Cheng et al. 2015). The milk oligosaccharides included prevention of pathogen binding to the intestinal epithelium and promotion of growth of beneficial bacteria. Immune competence and health status of piglets could be improved through regulating the composition and content of milk oligosaccharides by COS.

Table 11.5 The concentration of amino acids in the milk from sows consuming control or COS diets (Xie et al. 2015)

Item	Dietary treatment		
μmol/L	Control	COS	*P*-value
Asp	6.535 ± 0.28[b]	7.93 ± 0.55[a]	0.036
Thr	3.29 ± 0.17	3.92 ± 0.31	0.087
Ser	3.93 ± 0.18	4.82 ± 0.40	0.052
Glu	16.89 ± 0.64[b]	21.56 ± 1.36[a]	0.009
Gly	2.625 ± 0.09	3.08 ± 0.24	0.102
Ala	2.86 ± 0.14[b]	3.44 ± 0.23[a]	0.045
Cys	0.32 ± 0.02[b]	2.75 ± 0.19[a]	<0.001
Val	3.82 ± 0.21	4.51 ± 0.32	0.089
Met	1.25 ± 0.09	1.47 ± 0.12	0.128
Ile	2.67 ± 0.16[b]	3.26 ± 0.23[a]	0.047
Leu	8.01 ± 0.41[b]	9.61 ± 0.59[a]	0.041
Tyr	3.00 ± 0.19[b]	3.70 ± 0.285[a]	0.049
Phe	3.20 ± 0.17[b]	4.01 ± 0.30[a]	0.029
Lys	5.82 ± 0.32	6.98 ± 0.49	0.064
NH_3	1.38 ± 0.07[b]	1.90 ± 0.11[a]	0.001
His	1.95 ± 0.11[b]	2.42 ± 0.15[a]	0.020
Arg	3.57 ± 0.145	4.23 ± 0.33	0.078

[a, b]Means without a common superscript in the same row differ ($P < 0.05$)
Dietary treatment: control = basal diet, COS = basal diet + COS (30 mg/kg basal diet)
Data are expressed as mean ± EM, n = 8

11.2.1.7 Effect of COS on the Gene Expression of Fetus and Placenta of Sows

Fetal loss and intrauterine growth restriction are major problems in sow, but there were very few effective ways to cope with this problem. Wan et al. (2017a, b) investigated how COS affected fetal survival and growth in a sow model. COS supplementation had increased BMP2, BMP4, and PPARγ in the fetuses of sows (Wan et al. 2017a, b). It could be concluded that COS inclusion promoted BMP2, BMP4, and PPARγ expression for the improvements in fetal survival and growth. In addition, Wan et al. (2016) also investigated the effects of dietary COS supplementation on some important genes expression in the placenta and fetus of sows (Wan et al. 2016). These results showed that dietary COS supplementation not only increased the expression levels of VEGFA and leptin in the placenta but also regulated the expression of critical fetal development-related genes (TGF-β, STAT3 and FGFR2) in the fetus at 35 days of gestation. Collectively, the above research results furthered our understanding of the molecular mechanisms underlying the significant effects of COS on fetal development and reproductive performance in pregnant sows.

11.2.1.8 Effect of Dietary COS Supplementation on the Pig Ovary Transcriptome

With the rapid development of bioinformatics analysis and sequencing technique, RNA-Seq technology provides a platform for measuring large-scale gene

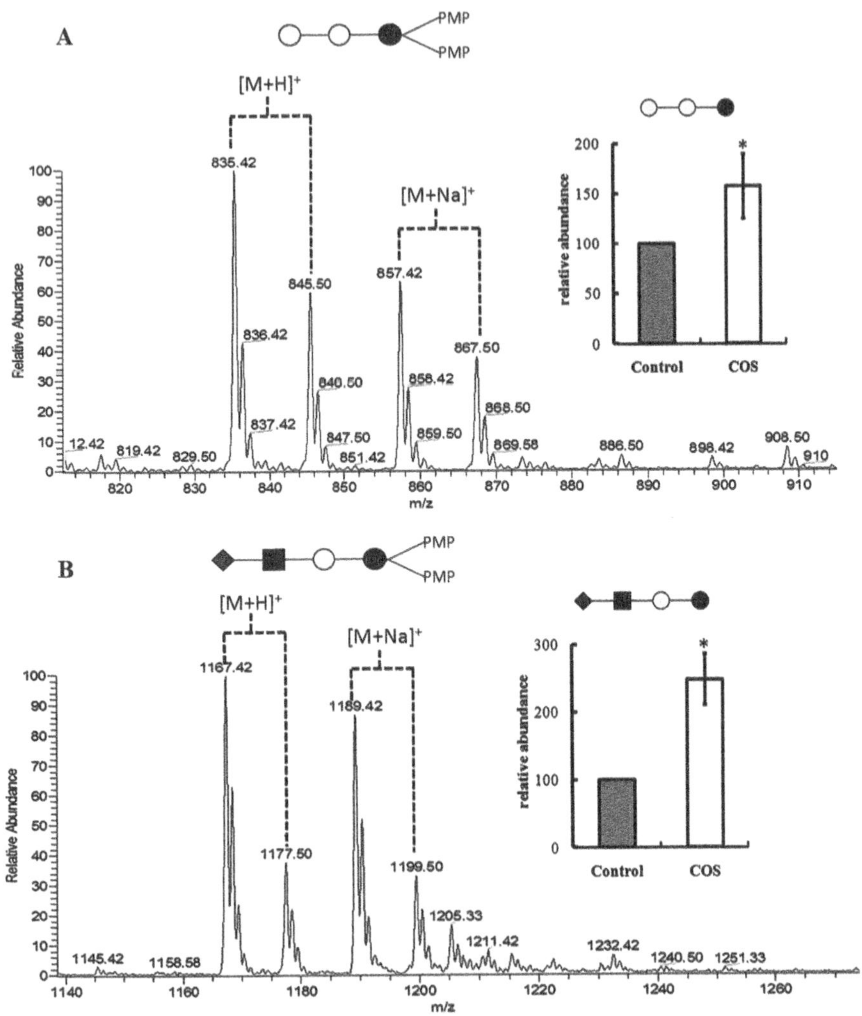

Fig. 11.2 Mass spectrum and statistical results of glycans found in milk of sows fed a control diet and diet supplemented with COS during gestation. (Cheng et al. 2015)

expression pattern. Xu et al. (2018a, b) identified 486 differentially expressed genes (DEGs) associated with litter size from COS supplementation sow ovaries, and a large number of these DEGs were related to progesterone-mediated oocyte maturation, cell cycle, oocyte meiosis, metabolic pathways, or hematopoietic cell lineage and so on (Fig. 11.3) (Xu et al. 2018a). These novel findings not only deepened our understanding of the molecular mechanisms underlying the significant effects of COS on fetal survival and reproductive performance in pregnant sows, but also established a theoretical basis for developing functional carbohydrates feed additives such as COS for the breeding industry.

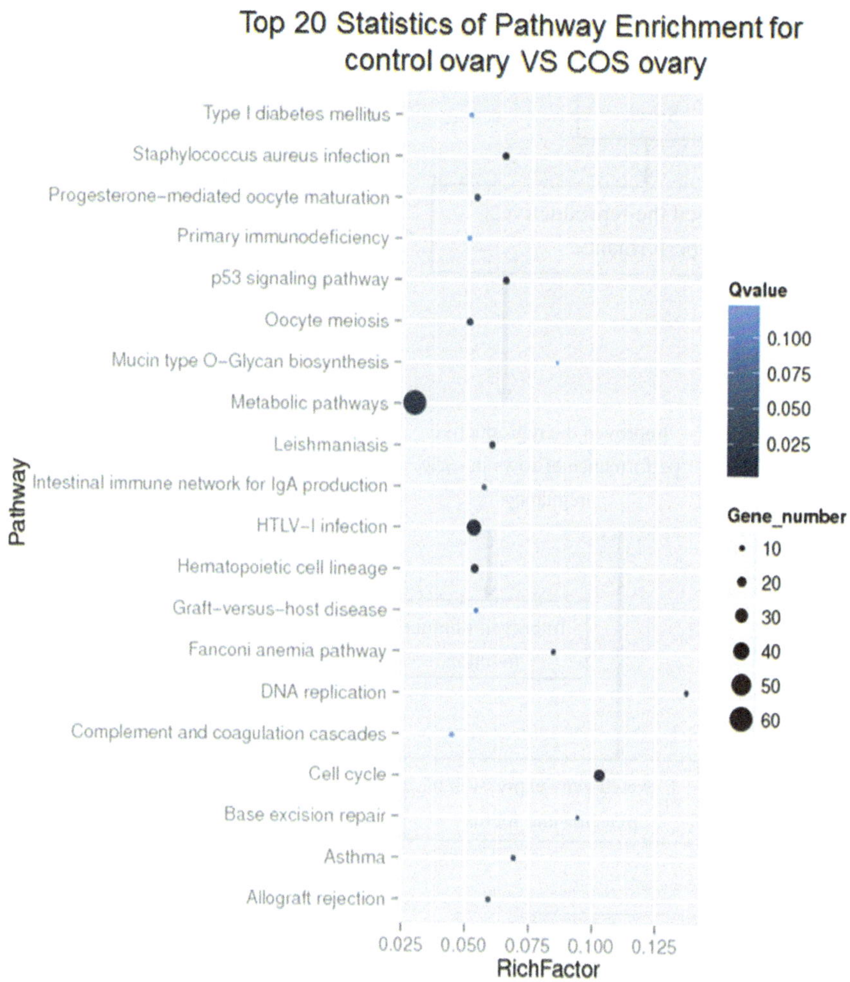

Fig. 11.3 Effect of COS on KEGG enrichment pathway analysis from DEGs. (Xu et al. 2018a, b)

From above investigations, dietary supplementation COS could improve the reproductive performance of sows, including total number of piglets born, total number of piglets born alive, the average weights for piglets born alive and the litter weighs for piglets born alive (Fig. 11.4). The mechanism of COS improving reproductive performance of sows can be explained from the two aspects. (1) During early pregnancy, dietary COS supplementation ameliorated the health condition through improving the antioxidant capacity and immune function, this was beneficial to embryo implantation. At the same time, COS can enhance the expression of some important genes, such as leptin, VEGFA, STAT3, TGF-β, and FGFR2. Therefore, embryonic survival rate, crown-to-rump length, and viable fetuses weight were improved through the COS intervention. Dietary COS could regulate

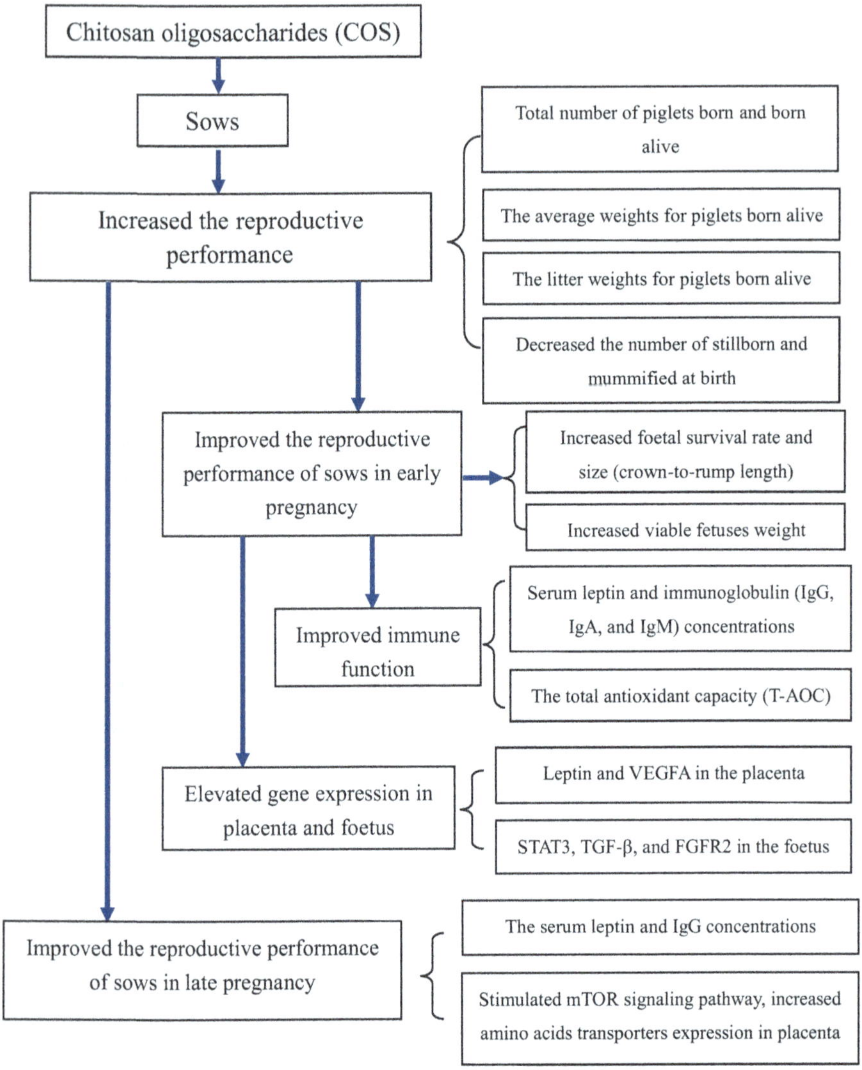

Fig. 11.4 Representative pictures of possible mechanism of chitosan oligosaccharides regulating reproductive performance in sows

endocrine status through increasing the concentrations of leptin, which was contributed to reproductive performance. (2) During late pregnancy, dietary COS supplementation promoted the health condition of sows through improving the amniotic fluid composition, that was beneficial to fetus growth and development. Meanwhile, COS supplementation in the diet during late gestation activated mTOR signaling pathway and elevated amino acids transporters expression in placenta so as to promote placental amino acids transport, which may contribute to development of fetus during gestation.

11.2.2 The Application of COS on Suckling Piglets

The low survival rate of suckling piglets has always been one of the important factors that hinder the healthy development of pig industry, and it is also an important index reflecting the production level of intensive pig farms, which directly affects the economic benefits of farmers. Due to the influence of physiological conditions and environmental factors of suckling piglets, a large number of suckling piglets often die, which brings serious losses to breeding production. It is reported that the mortality of suckling piglets in poor domestic conditions is relatively high, reaching about 30%, and the mortality of piglets in small and medium-sized pig farms is about 20–25%. Therefore, it is necessary to analyze the causes of death of suckling piglets and take measures to reduce risks and improve benefits.

11.2.2.1 Effect of COS on the Growth Performance of Suckling Piglets

The growth rate of suckling piglets is a major factor in determining weaning weight. However, the effect of COS on the offspring of supplemented sows was unknown. Xie et al. (2015) investigated the effect of COS supplementing sows during gestation and lactation on the growth rate of the suckling piglets (Xie et al. 2015). The results indicated that supplementation of the maternal diet with COS elevated daily gain and weaning weight (Table 11.6). Collectively, dietary supplementation of the diet with COS during late gestation and lactation promoted the growth rate of suckling piglets.

11.2.2.2 Effect of COS on the Plasma Glucose and Hepatic Gluconeogenesis of Suckling Piglets

COS can lower blood glucose levels in diabetic rats and was widely used as a dietary supplement, but the effect of COS on the offspring of supplemented sows was unknown. Xie et al. (2015) investigated the effect of COS supplementing sows during gestation and lactation on the levels of plasma glucose on suckling piglets (Xie et al. 2015). The results showed that maternal supplementation with COS elevated the mRNA expression levels and activities of PEPCK-M, PEPCK-C and G6Pase in

Table 11.6 Growth performance of suckling piglets from sows consuming control or COS diets (Xie et al. 2015)

	Dietary treatment		
Item	Control	COS	*P*-value
Total birth weight/pen (kg)	16.75 ± 1.03	16.81 ± 1.05	0.987
Average birth weight of piglet/pen (kg)	1.56 ± 0.08	1.69 ± 0.06	0.145
Total weaning weight/pen (kg)	45.59 ± 2.42	48.93 ± 2.40	0.336
Average weaning weight of piglet/pen (kg)	4.06 ± 0.14[b]	4.49 ± 0.14[a]	0.038
Average daily gain/pen (g)	176.02 ± 6.02[b]	202.39 ± 6.44[a]	0.005

[a,b]Means without a common superscript in the same row differ ($P < 0.05$)
Dietary treatment: control=basal diet, COS=basal diet+ COS (30 mg/kg basal diet)
Data are expressed as mean±SEM, n = 20

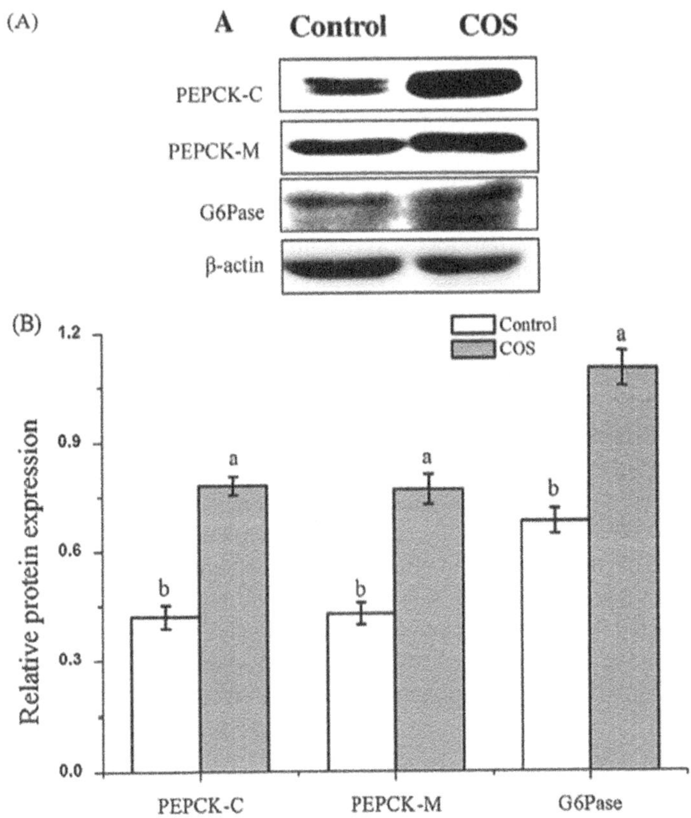

Fig. 11.5 The relative protein expression of PEPCK-C, PEPCK-M and G6Pase in the liver of suckling piglets. Values are means (n = 4) with their standard errors represented by vertical bars. [a, b]Means values with different letters were significantly different ($P < 0.05$). (Xie et al. 2015)

the liver of piglets compared with piglets from control fed sows (Fig. 11.5). In addition, the levels of plasma glucose were higher and hepatic glycogen was lower in piglets from COS group compared with that in the control group. Collectively, dietary supplementation of the diet with COS during late gestation and lactation decreased piglet hypoglycemia by activating hepatic gluconeogenesis and promoted the growth rate of suckling piglets.

11.2.2.3 Effect of COS on the Development of Small Intestinal Villi of Suckling Piglets

Small intestine is the most important organ for piglets' nutrient absorption. Xie et al. (2016a, b, c) investigated how maternal supplementation with COS affected the development of small intestinal villi of suckling piglets. The results showed that piglets from the COS groups had greater villus length and ratio of villus length to crypt depth in the jejunum and ileum (Fig. 11.6) (Xie et al. 2016a). Therefore, supplementing COS promoted the development of the small intestine in suckling

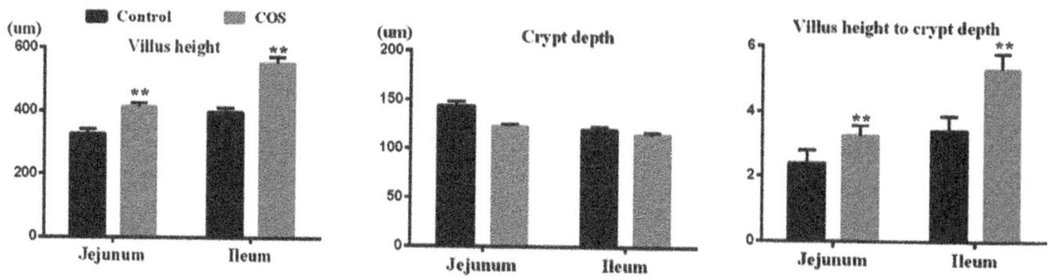

Fig. 11.6 Villus height, crypt depth, and ratio of villus height to crypt depth in the jejunum and ileum of piglets that suckled sows that consumed a basal corn-soybean diet with no COS (Control) and the basal diet supplemented with COS (30 mg/kg basal diet; COS) diets during late gestation and lactation. (Data are presented as means with the SEM; n = 8; $**P < 0.01$. (Xie et al. 2016a, b, c))

piglets. This may reflect that chitosan oligosaccharides can improve the nutrient absorption capacity of suckling piglets.

11.2.2.4 Effect of COS on Antioxidant Ability of Suckling Piglets

COS is a vital antioxidant in food chemistry. Nevertheless, how the antioxidant ability was transmitted from sows to piglets still remains unclear. Xie et al. (2016a, b, c) aimed to explored maternal supplementation with COS affected the antioxidant capacity of suckling piglets (Xie et al. 2016a, b, c). The results showed that piglets from the COS groups had greater plasma glutathione peroxidase activity. qRT-PCR results demonstrated that the relative mRNA level of glutathione peroxidase 1 (GPx1) and Cu/Zn-superoxide dismutase (SOD) significantly elevated in COS groups pigs duodenum and colon, respectively, however, GPx1 and Mn-SOD obviously reduced in the liver compared with those in the control group. In conclusion, supplementing COS in sow diet improved the intestinal total antioxidant capacity to some extent, despite inconsistent results being acquired in the liver.

11.2.2.5 Effect of COS on the Lipid Accumulation and Lipid Metabolism in the Suckling Piglets

Previous studies showed that COS could decrease lipid accumulation in liver in rats and mice. Xie et al. (2016a, b, c) investigated whether maternal COS supplementation influences hepatic lipid metabolism by regulating the expression of circadian clock genes in piglets. The results showed that maternal COS supplementation elevated hepatic and plasma cholesterol accumulation (Table 11.7) and increased the mRNA expression level of negative-regulated element period 1 (Per 1), and decreased the expression of the positive elements, circadian locomotor output cycles kaput (CLOCK), and brain muscle Arnt-like 1 (BMAL1) in the suckling piglets on day 14 (Xie et al. 2016b). These variations may elevate the hepatic cholesterol accumulation and regulate hepatic bile acid metabolism, and reduced the relative lipid metabolism-associated genes expression levels in the liver. These results indicated that maternal COS supplementing during late gestation and lactation did not change hepatic lipid metabolism of newborn piglets, but obviously affected the expression

Table 11.7 Contents of HDL-C, LDL-C, total CHO, and triglyceride in the plasma and liver in piglets (Xie et al. 2016b)

	Day 0			Day 14		
	Dietary treatment			Dietary treatment		
Item	Control	COS	*P*-value	Control	COS	*P*-value
In the plasma (mmol/L)						
HDL-C	0.87 ± 0.11	0.66 ± 0.06	0.130	3.93 ± 0.27	4.73 ± 0.16	0.023
LDL-C	0.37 ± 0.01	0.31 ± 0.04	0.170	1.63 ± 0.25	3.31 ± 0.23	<0.001
Total CHO	1.03 ± 0.09	1.06 ± 0.14	0.829	4.20 ± 0.3	7.51 ± 0.44	<0.001
Triglyceride	0.23 ± 0.03	0.226 ± 0.03	0.975	0.87 ± 0.05	0.95 ± 0.07	0.349
In the liver (mmol/g protein)						
HDL-C	0.31 ± 0.03	0.36 ± 0.105	0.633	1.49 ± 0.22	2.85 ± 0.51	0.049
LDL-C	6.03 ± 0.58	6.01 ± 0.43	0.980	13.38 ± 0.49	17.01 ± 1.17	0.007
Total CHO	1.33 ± 0.28	1.22 ± 0.23	0.840	3.58 ± 0.32	5.51 ± 0.51	0.006
Triglyceride	8.93 ± 1.04	8.26 ± 1.09	0.661	4.77 ± 0.40	7.85 ± 1.31	0.067

Note: Dietary treatment: Control=basal diet, COS=basal diet + COS (30 mg COS/kg basal diet); Data are expressed as mean SEM, n = 8

levels of circadian clock genes, and hepatic cholesterol concentration in the newborn piglets on day 14 (Xie et al. 2016c). Collectively, these results provided new knowledge in regulating hepatic lipid metabolism in the offspring through utero and postnatal manipulation of the maternal diet. In addition, more studies need to carry out on the molecular mechanisms which COS affected hepatic cholesterol accumulation regulated by circadian clock in the experimental piglets.

From the above discussion, the survival rate of lactating piglets directly affects the production level and benefit of the whole stage of pig farm. There are many factors causing the death of lactating piglets, such as external factors, nutritional factors, self-factors, and disease factors. Generally, the mortality rate of lactating piglets is 10–20%. COS can improve the growth rate and survival rate of suckling piglets (Fig. 11.7). Supplementing COS in sow diet improved villus length and ratio of villus length to crypt depth in the jejunum and ileum to promote the development of the small intestine in suckling piglets. Moreover, COS increased SOD and GPx1 in the colon and duodenum, plasma glutathione peroxidase activity to improve the intestinal antioxidant capacity. In addition, dietary COS supplement during late gestation and lactation reduced piglet hypoglycemia by stimulating hepatic gluconeogenesis and improved the growth rate of suckling piglets, because of the G6Pase, PEPCK-C and PEPCK-M in the liver of piglets were increased through COS intervention.

11.2.3 The Application of COS on Weaned Piglets

Weaning stress has been proved to disturb intestinal health balance, which is related to physiological and morphological changes that include villous atrophy and crypt hyperplasia, intestinal inflammation, and reduced epithelial brush border activity (Zhu et al. 2012). In addition to this, a recent study demonstrated that weaning could

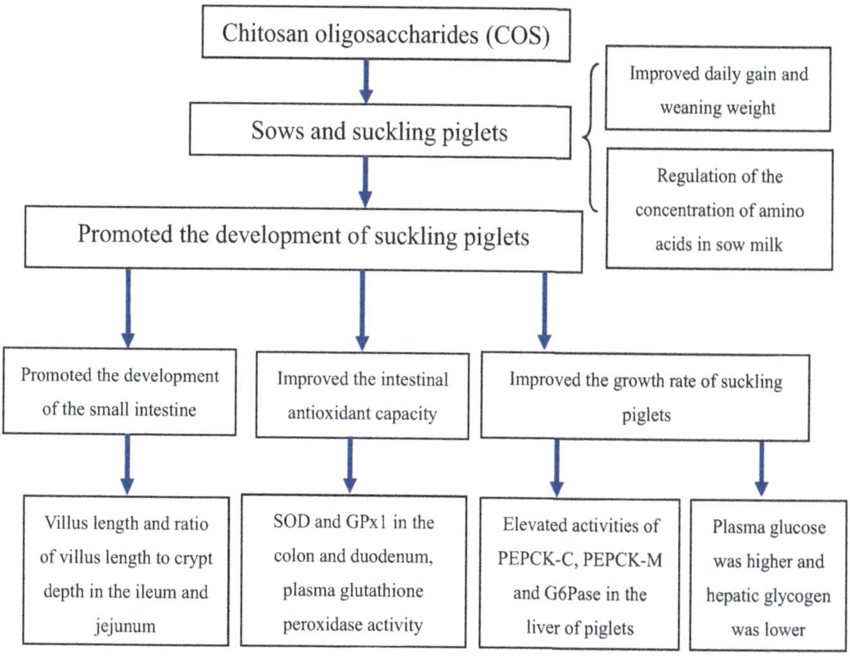

Fig. 11.7 Possible mechanism of chitosan oligosaccharides regulating growth ability of suckling piglets

also induce oxidative stress and then resulted in oxidative damage in piglets (Xu et al. 2018b). Moreover, these factors favored diarrhoea, translocation of bacteria and reduced digestive and absorptive capacity of small-intestinal enterocytes, which resulted in a decline in daily gain and then cause serious economic losses in the swine industry. Over the past decades, the post weaning disorders were kept under control using antibiotic growth promoters in weaned piglet diets. At present, specific dietary interventions could offer a practical and viable approach to alleviate physiological disorder after weaning. Some nutrients such as COS can play a positive effect on improving the intestinal development as well as in down-regulated oxidative stress.

11.2.3.1 Effect of COS on the Growth Performance of Weaned Piglets

To investigate the effects of COS on the growth performance and serum immune parameters, Wan et al. demonstrated that upon COS supplementation, during the overall trial period, piglets had greater ($P < 0.05$) ADG compared with those in the control group. However, no differences ($P > 0.05$) were found in the ADFI and F/G throughout the whole experimental period between the two groups (Table 11.8) (Wan et al. 2017a). The similar results were demonstrated by other earlier research, Liu et al. (2008) founded that dietary supplementation with 0.01% or 0.02% of COS positively improved BWG, FI and FCR, promoted the digestibility of nutrients and

Table 11.8 Effects of COS on the growth performance of weaned pigs (Wan et al. 2017a, b)

Items	Treatments[a]		*P*-value
	CON	COS	
Days 1–21			
ADG[b] (g)	476.05 ± 15.67	516.83 ± 10.37*	0.038
ADFI[c] (g)	704.99 ± 20.00	748.81 ± 13.40	0.079
F/G[d]	1.49 ± 0.03	1.45 ± 0.01	0.221

*$P < 0.05$ versus the CON group. **$P < 0.01$ versus the CON group
[a]CON, a corn-soybean basal diet; COS, chitosan oligosaccharides (the basal diet supplemented with 100 mg kg^{-1} COS)
[b]*ADG* average daily body weight gain
[c]*ADFI* average daily feed intake
[d]*F/G* the ratio of feed to gain

increased villus height and the villus/crypt ratio in the jejuna and ilea of weaned piglets (Liu et al. 2008). The goal of a research by Yang et al. (2012) was to investigate the effect of dietary COS (0.02%, 0.04%, or 0.06%) on the growth performance and intestinal morphology of weaned piglets. Pigs supplementation with 0.04% or 0.06% COS had increased FCR and BWG from day 1 to 14 post-weaning, but dietary COS did not alter villous height or crypt depth of the duodenum, ileum or jejunum (Yang et al. 2012).

11.2.3.2 Effect of COS on the Amino Acid Profiles, Fatty Acids and Serum Biochemical Indicators in Weaned Piglets

Yang et al. (2016) investigated the effect of COS on serum biochemical variables, and intestinal mucosal amino acid profiles and alkaline phosphatase (ALP) activities in weaned piglets (Yang et al. 2016). A total of 24 piglets (BW = 7.82 ± 0.21 kg) were weaned at 25 day of age and for a 14 days period, and the dosage of COS supplemented was 30 mg/kg in COS group. The results demonstrated that fed diet containing COS elevated urea nitrogen contents and serum IgG, and tended to elevate serum calcium (Yang et al. 2016). The ileal mucosal ALP activity in piglet fed with COS diet was greater than that in control group. Dietary supplementation with COS elevated the contents of Cys and Asn, and tended to elevate the contents of Orn and Asp in the small intestinal mucosa of weaned pigs (Yang et al. 2016). Furthermore, the contents of short chain fatty acid (SCFA) in colonic and caecal digesta of weaned pigs were regulated by dietary COS supplementation (Table 11.9). There were interactions between dietary COS and intestinal section for digesta SCFA contents and mucosa amino acid contents in weaned pigs. Collectively, the results of the present investigation suggested that fed diet containing COS affected immune and intestine functions of weaned pigs.

11.2.3.3 Effect of COS on the Growth Factors Expression in Weaned Piglets

Insulin-like growth factor (IGF) is a group of growth-promoting polypeptide substances, which can promote animal growth. The level of IGF-I in vivo is regulated

Table 11.9 Effects of dietary supplementation with COS on colonic and caecal short chain fatty acid (mmol/kg wet digesta) contents of weaned piglets (Yang et al. 2016)

	CON		COS		*P*-value		
Item	Caecum	Colon	Caecum	Colon	T	S	T×S
Acetate	50.4 ± 7.1	63.8 ± 4.7	54.8 ± 11.4	53.6 ± 1.3	0.353	0.055	0.026
Propionate	17.8 ± 1.3	24.6 ± 1.9	21.6 ± 3.7	22.0 ± 2.5	0.353	0.055	0.026
Isobutyrate	1.5 ± 0.2	1.9 ± 0.3	1.8 ± 0.5	2.2 ± 0.6	0.214	0.079	0.995
Butyrate	7.2 ± 1.9	11.1 ± 2.9	10.2 ± 1.7	14.2 ± 1.7	0.005	0.001	0.990
Isovalerate	2.5 ± 0.4	3.2 ± 0.4	3.1 ± 0.3	3.9 ± 0.6	0.005	0.001	0.846
Valerate	1.9 ± 0.3	2.7 ± 0.8	2.5 ± 0.5	3.0 ± 0.3	0.059	0.012	0.445
Total SCFA[a]	81.2 ± 7.8	107.4 ± 3.9	94.1 ± 10.1	9.0 ± 9.9	0.539	0.001	0.009

[a]*SCFA* short chain fatty acid

T effects of dietary treatment, *S* effects of intestinal section, *T×S* effects of interaction between dietary treatment and intestinal section

by growth hormone, and the secretion of growth hormone is also negatively regulated by IGF-I. Tang et al. (2005) studied the effects of dietary supplementation of COS on serum IGF-I and growth hormone levels, and hepatic and muscle IGF-I mRNA expression in early-weaned piglets (Tang et al. 2005). The results also showed that blood urea nitrogen (BUN) level was decreased whereas serum total protein concentration was elevated in COS supplementation (Tang et al. 2005). Dietary supplementation of COS also elevated the serum IGF-I and growth hormone levels along with increasing muscle and hepatic IGF-I mRNA abundance. From the above research results, it can be concluded that dietary supplementation of COS may increase growth and feed conversion efficiency by elevating plasma IGF-I and growth hormone levels in early-weaned pigs.

11.2.3.4 Effect of COS Supplementation on Immune Response, Antioxidant Capacity, and Barrier Function in Weaned Piglets

Xiong et al. (2015) studied the effect of dietary supplementation of a low dosage of COS on immune response, antioxidant capacity, intestinal morphology and barrier function in weaned piglets (Xiong et al. 2015). A total of 120 weaned piglets (21 days of age, 7.86 ± 0.22 Kg average BW) were fed with dietary treatments. COS group supplemented 30 mg/kg COS. Piglets fed diet containing COS had greater stomach pH than those fed the control diet on day 14 post weaning. The concentrations of IL-10 (jejunum, duodenum, and ileum) and secretory immunoglobulin (SIgA; ileum and duodenum) were higher in pigs fed diet containing COS compared with control pigs (Xiong et al. 2015). These results further indicated that supplementation COS at this level could induce an oxidative stress and immune response in small intestine and have compromised the intestine barrier integrity in weaned pigs (Xiong et al. 2015). These researches will provide some guidance on the low dosage of COS application on weaned piglets.

11.2.3.5 Effect of COS on Small Intestinal Morphology of Weaned Piglets

Suthongsa et al. (2017) examined the effect of COS supplementation on growth performance, small intestinal functions and nutrient digestibility in weaned piglets as an effective alternative to antibiotic addition in post-weaning diets. In the diets, the COS dosage were 75, 150, 225 mg/kg, and antibiotics dosage was 110 mg/kg lincomycin (Suthongsa et al. 2017). These results demonstrated that pigs in 150 mg/kg COS group or lincomycin group showed more active cell division (as indicated by Ki-67 marker of jejunal and duodenal crypt cells) on day 56 of the experiment (Fig. 11.8). These results suggested that 150 mg/kg COS could be helpful dietary supplement to improve digestibility efficiency and nutrient absorption.

11.2.3.6 Effect of Dietary COS on Diarrhea in Weaned Piglets

Early weaned piglets often possess an underdeveloped immune system, digestive disorders and post-weaning diarrhea. Enterotoxigenic *E. coli* can colonize the small intestine, and release enterotoxins to regulate epithelial cells to secrete fluid into the lumen of the gut to induce diarrhea. Xiao et al. performed experiments to verify the effect of dietary COS on diarrhea in weaned piglets. The diarrhea gradually declined as the experiment progressed, and four pigs had diarrhea at the end of the feeding trial in the control group. By contrast, in the chlortetracycline group, five pigs had diarrhea on the 5th day, and two pigs had diarrhea from day 17 to the end. However, in the COS group, from day 14 to the end, none of the pigs had diarrhea. These results demonstrated that 300 mg/kg COS had a similar role on decreasing diarrhea as 50 mg/kg chlortetracycline (Xiao et al. 2014). Sun et al. also investigated the effect of dietary COS on diarrhea in weaned piglets challenged with pathogenic *E.coli*. The results showed that dietary COS reduced diarrhea days and diarrhea rate, which may reflect that COS improve disease resistance (Sun et al. 2009).

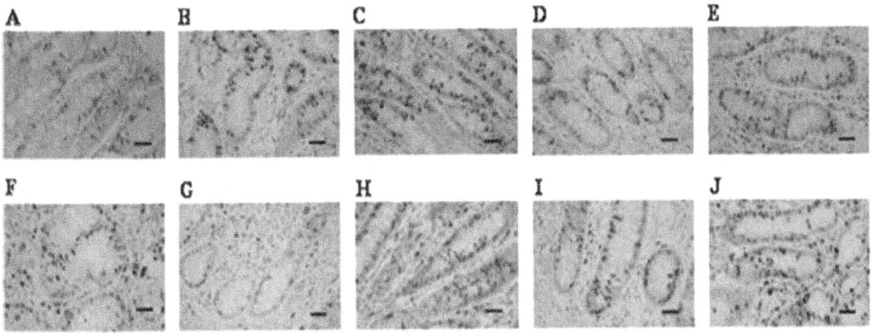

Fig. 11.8 Immunohistochemical staining of Ki-67 depicted from jejunal mucosa on day 28 (**a–e**) and day 56 (**f–j**) of the experiment. (**a**, **f**) Basal diet; (**b**, **g**) Basal diet plus 75 mg/kg of COS; (**c**, **h**) Basal diet plus 150 mg/kg of COS; (**d**, **i**) Basal diet plus 225 mg/kg of COS; (**e**, **j**) Basal diet plus 110 mg/kg of lincomycin. The positive crypt cell nuclei were indicated by reddish-brown nuclear stain (scale bar 10 μm). (Suthongsa et al. 2017)

11.2.3.7 Effect of COS on Intestinal Inflammation in LPS Challenged-Piglets

The intestinal epithelium is continuously exposed to potentially harmful pathogens, antigens, toxins and air pollutants, which will induce intestinal inflammation. Huang et al. (2016) investigated the effects of dietary COS on the intestinal inflammatory response using a lipopolysaccharide (LPS)-challenged pig model. The main factors were dietary treatment (basal or 300 μg/kg COS) and inflammatory challenge (saline or LPS). The results indicated that pigs challenged with LPS had an obvious decline in average daily gain and histopathological injury in the ileum and jejunum, but dietary supplementation with COS dramatically prevented intestinal injury induced by LPS. Pigs supplementation with COS had lower serum concentrations of interleukin (IL) 6, IL-8, and tumor necrosis factor alpha (TNF-α) compared with pigs fed the basal diet among LPS-challenged pigs ($P < 0.05$). Dietary COS promoted intestinal calcium-sensing receptor (CaSR) and PLCβ2 protein expressions in both LPS and saline treated pigs, but inhibited IKKα/β, p-NF-κB p65, and IκB protein expressions in LPS-challenged pigs ($P < 0.05$). The results demonstrated that COS had the potential to alleviate the intestinal inflammatory response, which was concomitant with the stimulation of CaSR and the prevention of NF-κB signaling pathways under an inflammatory response (Huang et al. 2016).

11.2.3.8 Effect of COS on Intestinal Flora and Antibiotics Substitution in Weaned Piglets

Antibiotics can improve disease resistance, increase feed utilization efficiency and promote growth performance. However, the antibiotics lead to the decline in product quality and drug resistance problem. Based on the reported antibacterial and anti-inflammatory effects, COS has a promising future as an alternative to antibiotics thereby improving the productivity of swine. Yang et al. (2012) investigated the effect of COS on growth performance, intestinal barrier function, intestinal morphology, and cecal microflora in 180 weanling pigs (Yang et al. 2012). The dosage of COS supplement were 200, 400, 600 mg/kg in the diet, and the antibiotics group contained colistin sulfate (CSE) 20 mg/kg supplementation. From day 1 to 7 post weaning, pig fed CSE or COS had greater ADFI and ADG compared with the control groups. From day 1 to 14, diet with either 600 or 400 mg/kg COS, or 20 mg/kg CSE promoted ADG and G:F compared with the control diet. These results indicated that dietary supplementation of COS at 600 or 400 mg/kg promoted growth performance and improved gut barrier function, increased the population of *Lactobacilli* and *Bifidobacteria*, and decreased *S. aureus* in the cecum of weaned piglets (Table 11.10). Therefore, COS can be an alternative to antibiotics in feed to improve growth performance and health of pigs.

What are the factors affecting the performance of weaned piglets? These factors include the following: age, body weight, feed nutrition, disease status, quality of managers, environmental conditions, etc. These factors together affect the performance and survival rate of piglets. The mechanism of COS promoting growth of weaned piglets can be explained from the following aspects (Fig. 11.9): (1) Dietary COS supplementation promoted the growth performance, nutrition digestibility and

Table 11.10 Effect of dietary treatments on cecal microflora in weaned pigs[1,2] (Yang et al. 2012)

Item	CON[3]	Dietary COS, mg/kg			CSE	SEM	*P*-value
		200	400	600			
Bifidobacteria							
Day 7	7.57[bc]	8.22[ab]	8.45[a]	8.31[ab]	7.38[c]	0.374	0.028
Day 14	7.94[bc]	8.62[ab]	8.71[ab]	8.90[a]	7.80[c]	0.392	0.032
Lactobacilli							
Day 7	7.94[b]	8.48[ab]	8.80[a]	9.11[a]	7.79[b]	0.388	0.042
Day 14	8.07	8.28	8.41	8.24	8.12	0.397	0.856
E. coli							
Day 7	7.07	6.70	6.46	6.59	6.26	0.381	0.445
Day 14	6.97	6.79	6.57	6.36	6.62	0.371	0.402
S.aureus							
Day 7	6.64[a]	5.85[b]	5.70[b]	5.44[b]	5.36[b]	0.407	0.044
Day 14	7.42	6.95	6.49	6.34	6.30	0.442	0.094

[a–c]Means in the same row with different superscript letters differ ($P < 0.05$)
[1]Bacterial numbers are expressed as log10 colony forming units per gram. [2]n = 6. [3]*CON* control diet, *COS* chito-oligosaccharides, *CSE* colistin sulfate

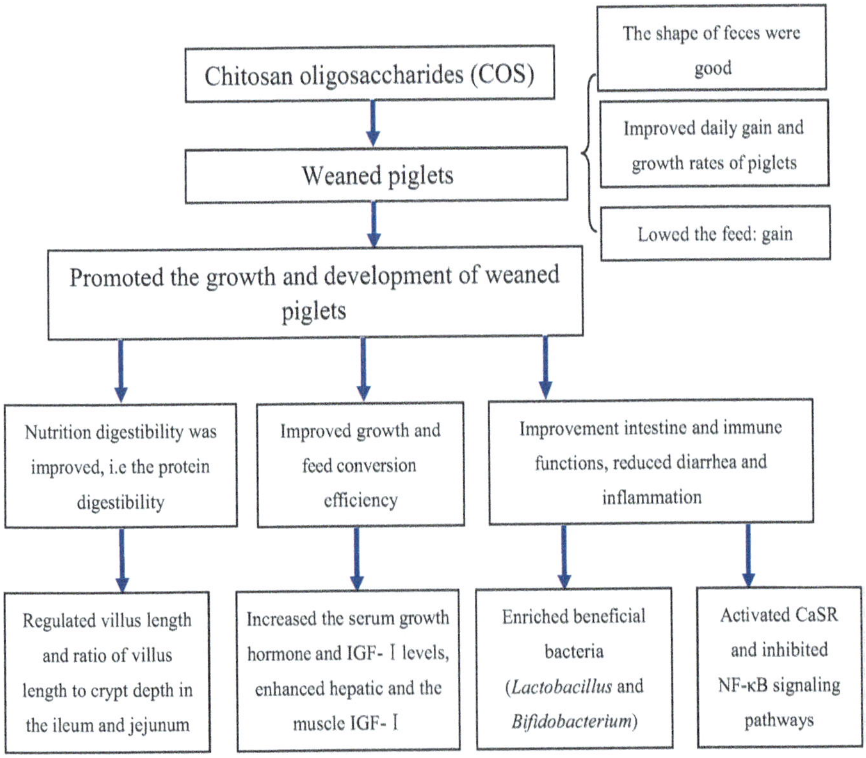

Fig. 11.9 Possible mechanism of COS regulating growth ability of weaned piglets

health status of weaned piglets. (2) Dietary COS supplementation improved intestinal flora by enriched beneficial bacteria (*Lactobacillus* and *Bifidobacterium*) and suppressed the harmful bacteria (*Peptostreptococcus* and *S. aureus*). (3) Dietary supplementation of COS might promote growth and feed conversion efficiency via increasing plasma IGF-I and growth hormone levels. (4) Dietary supplementation with COS affected immune and intestine functions of weaned pigs. (5) Dietary COS reduced diarrhea days and diarrhea rate, which may reflect that COS improve disease resistance. (6) Dietary COS had the potential to decrease the intestinal inflammatory response.

11.3 The Application of Oligosaccharides in Poultry Industry

In modern poultry industry, intensive genetic selection for faster growth rates and greater feed conversion of broiler chickens are usually accompanied by more mortality rates, mainly because of metabolic disorders, and finally results in the susceptibility to some infectious diseases. Some oligosaccharides have been selected to be added to poultry farming to alleviate these problems, and COS is one of the important oligosaccharides additives. Compared with the research and application of COS in livestock, the studies on the efficacy of dietary COS in the nutrition of poultry were much lower, and were mainly focused on broilers and hens.

11.3.1 The Application of COS in Broiler Chickens

A lot of different feed additives are routinely investigated to maintain metabolic status and good health and improved the performance indices of intensively produced broiler industry. Some studies have indicated that appropriate supplementation of COS may improve nutrient digestibility and daily weight gain in broiler chickens, enhance immunity, regulate lipid metabolism, and promote growth and development (Swiatkiewicz et al. 2014).

11.3.1.1 Effect of COS on Nutrient Digestibility and Daily Weight Gain in Broiler Chickens

As early as 2005, Huang et al. compared the effect of diet supplementation with increasing levels of COS (0, 50, 100, 150 mg/kg) and with an antibiotic (flavomycin) on digestibility of nutrients and performance indices in broiler chickens. These results indicated that dietary supplementation with COS was useful in increasing the ileal digestibility of nutrients and feeding efficiency in broilers (Huang et al. 2005). A later study by Li et al. detailed investigated the effects of COS on nutrient digestibility and daily weight gain in broiler chickens. Diet supplementation with COS significantly improved the growth performance in broilers (Table 11.11). The results indicated that the effect was related to increasing average daily feed intake (FI) and nutrient (dry matter, calcium, phosphorus, and energy) digestibility (Li et al. 2007). The recent study by Swiatkiewicz et al. also demonstrated that dietary

Table 11.11 Effects of dietary COS supplementation on the performance of broiler chickens (Li et al. 2007)

Diet						
Item	Control	Chlortetracycline	COS, 50 mg/kg	COS, 100 mg/kg	SEM	*P*-value
Starter period (1–21 day)						
Average daily gain	34.9^{c}	36.3^{b}	37.6^{a}	37.9^{a}	0.30	<0.01
Average daily feed intake	50.4^{c}	50.9^{bc}	52.3^{a}	52.0^{ab}	0.25	<0.01
Feed conversion	1.44^{a}	1.40^{ab}	1.39^{b}	1.38^{b}	0.01	0.04
Grower period (22–42 day)						
Average daily gain	59.7^{b}	62.7^{a}	63.4^{a}	61.9^{ab}	0.52	<0.01
Average daily feed intake	144.2	148.5	149.8	149.4	0.97	0.73
Feed conversion	2.42	2.37	2.37	2.42	0.00	0.84
Overall experiment (1–42 day)						
Average daily gain	47.3^{c}	49.5^{b}	50.8^{a}	50.1^{ab}	0.32	<0.01
Average daily feed intake	97.8^{b}	99.7^{ab}	100.6^{a}	100.2^{a}	0.37	0.04
Feed conversion	2.07^{a}	2.02^{ab}	1.98^{b}	2.00^{b}	0.00	0.02

$^{a–c}$Means within a row with the same or no letter do not differ ($P > 0.05$)

COS (0.01%) increased growth performance during either the first or second phases or the whole feeding period, along with the digestibility of nutrients (Swiatkiewicz et al. 2014).

11.3.1.2 Effect of COS on Immunity of Broiler Chickens

Epidemics or plagues in the immunosuppressive state can induce severe morbidity and mortality, which will increase economic loss in broiler production. The application of nutritive ingredients such as COS has become promising one preventive strategy. Deng et al. (2008) evaluated the different mechanisms of the immune system of broiler chickens fed diets containing COS (100 mg/kg) or feed antibiotic chlortetracycline (80 mg/kg). The results indicated that dietary COS had a more significantly effect on immune function than chlortetracycline and promoted the immunity of broilers by increasing the weight of the main immune organs (such as thymus, spleen and bursa of fabricius), stimulating IgM secretion, optimizing macrophage function by increasing the production of cytokines (IL-1β, IL-6, IFN-γ and TNF-α) and regulating inducible nitric oxide synthase to release NO. As a result, the authors suggested that COS was a hopeful alternative to the use of antibiotics in poultry production (Deng et al. 2008). A recent study by Chi et al. also demonstrated that the relative weights of spleen, bursa of fabricius and thymus in the COS group were elevated when compared with those in the control group. The results demonstrated that when COS (350 mg/kg) were supplemented to the broiler, some cells in the immune organs accelerated entering into the division stage, which could lead to the increase of relative weights of the three immune organs, and could eventually contribute to the improvement of immune function (Chi et al. 2017).

11.3.1.3 Effect of COS on Hematological and Biochemical Blood Characteristics in Broiler Chickens

The effects of supplementation with COS on blood characteristics in broilers were investigated by Zhou et al. COS supplementation had no obvious effect on the total protein, white blood cells, lymphocyte, albumin, and total cholesterol. However, the red blood cell counts were 18.2% greater ($P < 0.05$) in broilers in the COS group than that in the control group (Zhou et al. 2009). Nuengjamnong and Angkanaporn also comprehensively evaluated the function of COS on hematological parameters in broilers, including haemoglobin, haematocrit, white blood cell, heterophil, basophil, eosinophil, lymphocyte, monocyte, heterophil/lymphocyte (H/L) ratio. Most of hematological parameters in all treatments were not different except the white blood cell (WBC) count. Chickens had higher WBC count in COS group than those in the control group and amoxicillin group ($P < 0.05$). There was no significant difference in H/L ratio of chickens among different groups (Nuengjamnong and Angkanaporn 2017). In addition, COS supplementation did not obviously affect the serum IGF-I level, but the serum growth hormone level of broilers supplemented 50 mg/kg of COS was higher ($P < 0.05$) than the other treatments (Li et al. 2007).

11.3.1.4 Effect of COS on Lipid Metabolism of Broiler Chickens

Excessive body fatness in broiler chickens is widely deemed to a problem in the poultry industry. In chickens, the lipogenic activities in liver are much greater than that in adipose tissue, and most of the fats were accumulated in the adipose tissues. Li et al. (2007) reported that COS had a positive effect on serum lipid. The broilers fed COS (100 mg/kg) had declined ($P < 0.05$) serum triglyceride, increased ($P < 0.05$) serum high-density lipoprotein cholesterol, and were higher serum total protein content than broilers in control treatments (Table 11.12). It was probably through the influence of lipid absorption in the intestine by binding bile acids, which would elevate cholesterol elimination (Li et al. 2007). Correspondingly, Li et al. recent report investigated the effects of COS on serum levels of lipid metabolism-related biomarkers in broiler chickens. Low-dose COS supplementation (20 mg/kg) for 21 days provoked a significant reduction of triglycerides (TG) (−19.05%; $P < 0.05$), an effect also mediated by supplementation of 40 mg/kg COS since TG decreased by 17.46% ($P < 0.05$) in the respective groups (Li et al. 2017b).

11.3.1.5 Effect of COS on Intestinal Function of Broiler Chickens

Normal morphology of the intestinal crypt and villus, with adequate improvement of intestinal nutrient transporters, is important to a healthy gut. Ibitoye et al. demonstrated that dietary cricket COS and shrimp COS obviously ($P < 0.05$) increased jejunal villus height and decreased crypt depth (Ibitoye et al. 2018). Furthermore, Li et al. showed that the activities of superoxide dismutase and glutathione peroxidase, and the ability of total antioxidant capacity and eliminate hydroxy radical as well as the contents of glutathione in the COS (350 and 500 mg/kg) groups were obviously elevated, but the levels of malonedialdehyde were obviously reduced. The percentages of S and gap 2/mitosis (G2M) phases and proliferating index of ileum mucosal lymphocytes in the COS (350 and 500 mg/kg) groups were also elevated compared with the control group. The authors suggested dietary COS supplements might increase the antioxidant function and promote lymphocytes proliferation in the ileum mucosa of chicken (Li et al. 2017a). In addition, Li et al. found that broilers fed diet containing COS (100 mg/kg) obviously elevated concentrations of *Lactobacillus* in the cecum compared with those in the other groups. Correspondingly, the concentrations of *Escherichia coli* in the cecum were obviously declined by dietary supplementation of COS (100 mg/kg) in comparison with the control group (Li et al. 2007).

What are the factors affecting the performance of broiler chickens? These factors include the following: strain, feed nutrition, disease status, quality of managers, environmental conditions, etc. These factors together affect the performance and survival rate of chickens. The mechanism of COS promoting growth of chickens can be summarized as the following aspects (Fig. 11.10): (1) Dietary COS supplementation promoted the growth performance, nutrition digestibility and health status of chickens. (2) Dietary COS supplementation enhanced the immunity of broilers, such as elevating the weight of the some immune organs, improving of humoral and cellular immunity, and activating the immune effector molecule. (3) Dietary supplementation of COS influenced the hematological and biochemical

Table 11.12 Effect of dietary COS supplementation on total protein, triglyceride, total cholesterol, high-density lipoprotein (HDL) cholesterol, and low-density lipoprotein (LDL) cholesterol concentrations in the serum of broiler chickens (Li et al. 2007)

Diet						
Item	Control	Chlortetracycline	COS, 50 mg/kg	COS, 100 mg/kg	SEM	P-value
Day 21						
Total protein, g/dL	2.8	3.0	2.9	3.0	0.04	0.69
Triglyceride, mg/dL	88.2^{ab}	89.8^{a}	76.6^{bc}	74.0^{c}	2.53	0.04
Total cholesterol, mg/dL	111.0^{a}	111.2^{a}	110.6^{a}	96.3^{b}	1.78	<0.01
HDL cholesterol, mg/dL	68.1	74.3	65.8	65.5	1.54	0.13
LDL cholesterol, mg/dL	16.9^{b}	24.6^{a}	17.7^{b}	21.5^{ab}	1.01	0.02
Day 42						
Total protein, g/dL	3.3^{c}	3.5b^{c}	3.5^{b}	3.7^{a}	0.05	<0.01
Triglyceride, mg/dL	86.7^{a}	76.0^{a}	78.1^{a}	63.4^{b}	2.53	<0.01
Total cholesterol, mg/dL	110.5^{b}	116.6^{b}	110.4^{b}	131.2^{a}	2.85	0.02
HDL cholesterol, mg/dL	72.2^{b}	69.0^{b}	73.3^{b}	81.1^{a}	1.14	<0.01
LDL cholesterol, mg/dL	26.2^{b}	34.9^{ab}	38.1^{a}	41.8^{a}	1.98	0.02

$^{a-c}$Means within a row with the same or no letter do not differ ($P > 0.05$)

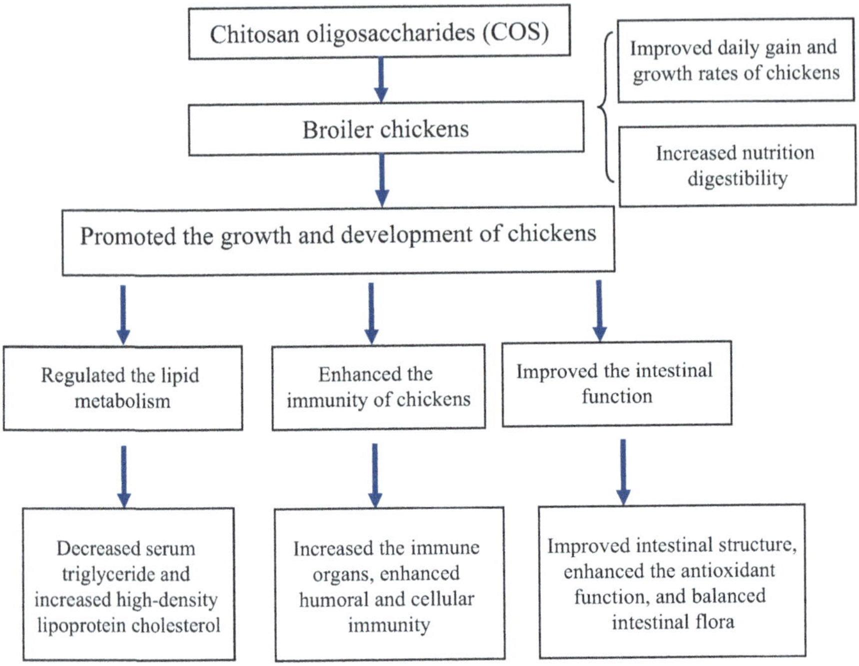

Fig. 11.10 Possible mechanism of COS regulating growth ability of broiler chickens

blood characteristics in broiler chickens. (4) Dietary supplementation of COS regulated the lipid metabolism of broiler chickens by decreasing serum triglyceride and increasing high-density lipoprotein cholesterol. (5) Dietary supplementation of COS improved the intestinal function of broiler chickens, such as improving jejunal villus height and reducing crypt depth, enhancing the antioxidant function of ileum, and balancing intestinal flora.

11.3.2 The Application of COS in Laying Hens

The number of reports on the effect of dietary COS in the nutrition of laying hens is much less than that of broiler reports. Besides the effects and mechanisms of COS found in broilers, more attention has been paid to the effects of COS on egg production and quality. An application effect of dietary COS on the dietetic value of eggs was investigated by Nogueira et al. (2003), who demonstrated that supplementation with 30 g/kg of COS decreased egg yolk contents of cholesterol, stearic and palmitic acids, and elevated the content of oleic acid (Nogueira et al. 2003). The goal of an experiment by Yan et al. (2010) was to evaluate dietary supplementation of COS on the production performance, hematological characteristics and egg quality in laying hens. The results indicated that dietary supplementation of COS (100 mg/kg, 200 mg/kg) linearly increased yolk color and egg weight, and haugh units (Yan

Table 11.13 Effects of COS for 45 days on performance and lipid profile of egg yolk of laying hens (Ghada et al. 2016)

Item	Control group	COS group	*P* value
Daily feed intake (g)	107.38 ± 4.99	113.67 ± 7.19	0.668
Daily egg production (%)	76.67 ± 2.50[a]	90.35 ± 2.67[b]	0.008*
Daily egg weight in g	45.00 ± 1.54[a]	52.78 ± 1.71[b]	0.007*
Daily egg mass production	34.68 ± 2.17[a]	47.90 ± 2.83[b]	0.005*
Feed conversion ratio	4.45 ± 0.14[a]	3.38 ± 0.17[b]	0.001*
Total lipid %	13.06 ± 0.51[a]	10.5 ± 0.50[b]	0.004*
Total cholesterol (mg/g)	73.2 ± 2.92[a]	64.2 ± 1.62[b]	0.034*
LDL (mg/g)	39.8 ± 0.66[a]	35.6 ± 1.08[b]	0.013*
Triglyceride (mg/g)	122 ± 2.59[a]	109.8 ± 1.28[b]	0.002*
HDL (mg/g)	28 ± 1.73[a]	35.8 ± 0.80[b]	0.009*

Data were represented as means ± SE.[a,b] Mean in the same row are significantly different
*Significantly difference using ANOVA test at $P < 0.05$

et al. 2010). Similarly, Meng et al. (2010) demonstrated that dietary inclusion of COS appeared to elevate egg quality and production by increasing nutrient digestibility of hens (Meng et al. 2010).

In a recent report on hens fed a diet containing a high level of distillers dried grains with solubles (20%), the inclusion of COS elevated the number of eggs produced and daily egg mass, and decreased concentration of cholesterol in egg yolks of hens (Swiatkiewicz et al. 2013). Ghada et al. investigated the effect of COS on egg production and performance of laying hens. The results showed that addition of COS to drinking water improved feed intake, egg production, average egg weight, egg mass production and feed conversion ratio, and a significant decrease in total lipid, total cholesterol, low density lipoprotein (LDL), triglyceride, but an obvious increase in high density lipoprotein (HDL) in egg yolk. These results clarified that the use of COS improved the performance and lipid profile in egg yolk of laying hens (Table 11.13) (Ghada et al. 2016).

11.4 The Application of Oligosaccharides in Ruminant Farming

Compared with the research of COS in livestock and poultry, the studies of COS in ruminants are very limited, and only few reports have been found in cattle.

11.4.1 Effect of COS on Nutrient Digestibility and Blood Metabolites in Dairy Cows

Feed additives have been investigated to meet high productive dairy cow energy requirements. Mingoti et al. investigated the effects of COS on dry matter intake (DMI) and nutrient digestibility in lactating dairy cows. COS supplementation did not alter DMI, but elevated the digestibility of organic matter (OM), dry matter (DM), crude protein (CP) and neutral detergent fiber (NDF) ($P < 0.05$). Blood urea nitrogen concentration was elevated ($P = 0.01$) with COS supplementation. COS reduced ($P = 0.02$) nitrogen (N) fecal excretion without influence nitrogen balance (Mingoti et al. 2016). Valle et al. also evaluated dietary COS effects on mid-lactation dairy cow intake, digestibility, metabolism and productive performance. COS increased dry matter and crude protein digestibility ($P < 0.05$), promoted feed efficiency, and increased milk unsaturated fatty acids concentration (Valle et al. 2017).

11.4.2 Effect of COS on Antioxidative and Immune Function in Beef Cattle

Li et al. conducted experiments to investigate the effects of COS on antioxidative and immune function in beef cattle. A total of 24 fattening Simmental cattle with similar age and body weight were divided randomly into different trial groups, and the experimental diets contained 0, 500, and 1000 mg/kg COS, respectively. The whole feeding trial lasted for 84 days. It was concluded that: (1) the inclusion of COS in diets increased ($P < 0.05$) the levels of interleukin-1 and IgA in serum at middle stage. The levels of IgA and IgM tended to be elevated ($P < 0.1$) by dietary COS at later stage of the trial; (2) the inclusion of 500 mg/kg COS in diets elevated ($P < 0.1$) total superoxide dismutase activity and reduced ($P < 0.05$) malondialdehyde content in serum at early and later stages, respectively. Collectively, these findings suggested that inclusion of 500 mg/kg COS improved cellular and humoral immune responses, and promoted the antioxidative function of beef cattle (Li et al. 2015).

11.4.3 Effect of COS on Diarrhea in Hanwoo Calves

The neonatal calves' diarrhoea is a major reason of breeding loss to the cattle industry, which can lead the death in dairy heifer and beef calves. In hence, it is important to explore alternative therapeutic agents for effective treatment of the disease. Alam et al. (2012) investigated the therapeutic effects of COS on diarrhea in Korean native (Hanwoo) calves (Alam et al. 2012). The dosage of COS in the treated group was 50 ml/day. Among the COS experimental group 41 (out of 46) calves recovered completely from diarrhoea and in the control group 5 calves (out of 40) spontaneously recovered after 5 days (Table 11.14). Because COS was found to effectively inhibit diarrhoea bacterial (*Escherichia coli* and *Salmonella typhimurium*) growth

Table 11.14 Therapeutic effects of COS on diarrhoea in Hanwoo calves (Alam et al. 2012)

	Routine diet group (n = 40)		Treated group (n = 46)	
Parameters	Control	5 days after normal diet	Control	5 days after COS treatment
Faeces				
Normal	–	5	–	41
Loose	31	5	36	2
Very loose	9	32	10	2
Runny	–	3	–	1
Dehydration (skin tenting test)				
Normal (<2 s)	28	5	31	40
Mild (2–3 s)	12	11	15	1
Moderate (3–6 s)	–	20	–	3
Severe (>6 s)	–	4	–	2

and pathogenicity, the authors suggested that COS could be promising applied for the clinical treatment of diarrhoea in Hanwoo calves (Alam et al. 2012).

11.4.4 The Antibacterial Effect of COS Against Infection by *Staphylococcus aureus* Isolated from Bovine Mastitis

Mastitis is induced primarily by microbial intramammary infection (IMI) and is a costly disease for dairy cattle breeding. Bovine mastitis induced by *S. aureus* infection constitutes a serious challenge to dairy producers due to the frequent inability of both the immune response and antibiotics to destroy the pathogen in the intramammary environment. Moon et al. (2007) studied the antibacterial effect of COS. The growth of *Staphylococcus aureus* isolated from bovine mastitis was declined within 10 min of incubation with COS in concentrations ranging from 0.0001% to 0.5%. In addition, electron microscopic results demonstrated that the surface of the COS-treated bacteria was distorted, expanded and lysed compared to that of the control bacteria. This indicated that COS was a hopeful agent for the treatment and improvement of bovine mastitis because of its strong antibacterial activity against *S. aureus* (Moon et al., 2007).

Though few studies have been reported on the effects of COS on ruminants, these results indicated that COS could be used as promising feed additive in bovine. COS increased dry matter and crude protein digestibility, and improved feed efficiency. In addition, COS had immunostimulative effect and antibacterial activity against infection by *Staphylococcus aureus* isolated from bovine mastitis. COS also could be potentially applied for the clinical treatment of diarrhoea in Hanwoo calves. Therefore, COS can improve the health level of bovine and prevent the incidence of diseases (Fig. 11.11).

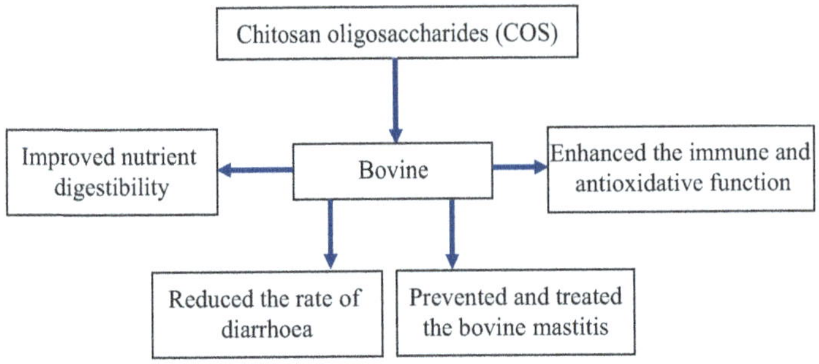

Fig. 11.11 Possible mechanism of COS regulating performance of bovine

11.5 Conclusion and Perspective

11.5.1 The Conclusion of COS Feed Additives

From the above reports, we found that dietary inclusion COS improved the growth performance and reproductive performance of sows, piglets, chicken, hens and bovine. COS regulated animal physiological function as follows (Fig. 11.12): (1) Dietary supplementation COS increased the reproductive performance of sows (including total number of piglets born, total number of piglets born alive, the average weights for piglets born alive and the litter weighs for piglets born alive. (2) Dietary COS promoted the growth performance of suckling piglets. COS could improve the growth rate and the survival rate of suckling piglets. Supplementation of the maternal diet with COS elevated weaning weight and daily gain and the concentration of amino acids in sow milk. Supplementing COS in the sow diet improved villus length and ratio of villus length to crypt depth in the jejunum and ileum to promote the development of the small intestine in suckling pigs. (3) Dietary COS promoted the growth performance of weaned piglets. Dietary supplementation COS improved the growth performance, nutrition digestibility and health status of weaned piglets. Dietary COS increased growth rates of piglets and the feed conversion rate by increasing plasma growth hormone and IGF-I levels. COS supplementation improved intestinal flora by enriched beneficial bacteria (*Lactobacillus* and *Bifidobacterium*) and suppressed the harmful bacteria (*Peptostreptococcus* and *Escherichia coli*). (4) Dietary COS supplementation promoted the growth performance, nutrition digestibility and health status of chickens. Dietary COS supplementation enhanced the immunity, regulated the lipid metabolism and improved the intestinal function of broiler chickens. In addition, Dietary COS supplementation improved the production and quality of eggs. (5) COS increased dry matter and crude protein digestibility, and improved the feed efficiency in bovine. As well as COS can improve the health level of bovine and prevent the incidence of diseases.

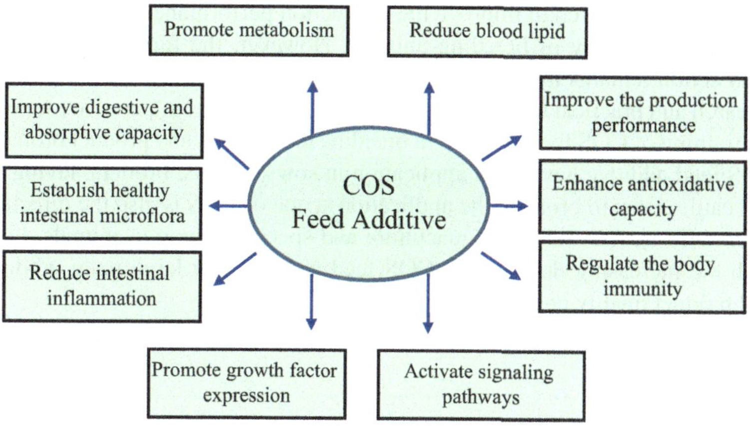

Fig. 11.12 Mechanism of COS as feed additives

11.5.2 The Perspective of COS Feed Additive

As a new kind of functional oligosaccharide feed additive, COS has good feed effects and application prospects. However, there is still a lot of work to further carry out and vigorously promote in some aspects.

1. Through the continuous efforts of research scholars, some progresses have been made in animal epidemiological research, but the breeding epidemics have always been harmful to the breeding industry, such as African swine fever and avian influenza. The situation of animal epidemic prevention and control is still grim. It not only affects consumption, but also breeding enterprises confidence. As an effective immunopotentiator, COS can effectively improve the immunity of breeding animals and enhance their resistance to external environment and disease infection. Recently, COS have been shown to be the potential as adjuvants to some vaccines in livestock and poultry farming. Collectively, COS is expected to play a greater role in the field of breeding epidemiology prevention and control.
2. It is well known that antibiotics supplementation in the diet usually improve the growth rate and feed efficiency of breeding animals. However, antibiotics have serious safety problems for human beings due to drug residues and drug resistance. It is necessary to identify alternatives to antibiotics to maintain growth performance benefits. Many researchers, based on their experimental data, indicated that COS can improve growth performance, immune function and feed utilization efficiency. Compared with the effects of some antibiotics, the effects of COS are basically the same. At least within a certain range, COS has good prospects for the application of alternative antibiotics. However, there are still many problems to be solved, such as the choice of substitution time and dose.

3. COS has been proved to improve the production performance, immune function and product quality of breeding animals. However, the further mechanism of COS action remains to be explored. In addition, how to combine the theoretical research and practical industry of COS feed additive? It is imperative to speed up the training of COS feed additive in breeding technology and product promotion. COS feed additive have been application in sows, piglets, broilers, laying hens, and cattle. How to broaden the application scope of COS is also the direction of future efforts, for example, aquaculture and special economic animals. Finally, with the increasing demand for COS feed additive, the large-scale production and product quality control are particularly important.

References

Alam MR, Kim WI, Kim JW, Na CS, Kim NS (2012) Effects of Chitosan-oligosaccharide on diarrhoea in Hanwoo calves. Vet Med 57(8):385–393

Cheng L, Wang L, Xu Q, Huang L, Zhou D, Li Z, Li S, Du Y, Yin H (2015) Chitooligosaccharide supplementation improves the reproductive performance and milk composition of sows. Livest Sci 174:74–81

Chi X, Ding X, Peng X, Li X, Fang J (2017) Effects of chitosan oligosaccharides supplementation on the cell cycle of immune organs in broilers. Kafkas Un Vet Fak Der 23(6):1003–1006

Deng XZ, Li XJ, Liu P, Yuan SL, Zang JJ, Li SY, Piao XS (2008) Effect of chito-oligosaccharide supplementation on immunity in broiler chickens. Asian Australas J Anim Sci 21(11):1651–1658

Ghada I, Amal F, Nabila I, Aml SH (2016) Effect of chito-oligosaccharide as feed additives on egg production and performance of laying hens. Egypt J Chem Environ Health 2(2):183–194

Huang RL, Yin YL, Wu GY, Zhang YG, Li TJ, Li LL, Li MX, Tang ZR, Zhang J, Wang B, He JH, Nie XZ (2005) Effect of dietary oligochitosan supplementation on heal digestibility of nutrients and performance in broilers. Poult Sci 84(9):1383–1388

Huang B, Xiao D, Tan B, Xiao H, Wang J, Yin J, Duan J, Huang R, Yang C, Yin Y (2016) Chitosan oligosaccharide reduces intestinal inflammation that involves calcium-sensing receptor (CaSR) activation in lipopolysaccharide (LPS)-challenged piglets. J Agric Food Chem 64(1):245–252

Ibitoye EB, Lokman IH, Hezmee MNM, Goh YM, Zuki ABZ, Jimoh AA, Danmaigoro A, Nicholas NP (2018) Gut health and serum growth hormone levels of broiler chickens fed dietary chitin and chitosan from cricket and shrimp. Poult Sci:pey419-pey419

Li XJ, Piao XS, Kim SW, Liu P, Wang L, Shen YB, Jung SC, Lee HS (2007) Effects of chito-oligosaccharide supplementation on performance, nutrient digestibility, and serum composition in broiler chickens. Poult Sci 86(6):1107–1114

Li T, Na R, Yu P, Shi B, Yan S, Zhao Y, Xu Y (2015) Effects of dietary supplementation of chitosan on immune and antioxidative function in beef cattle. Czeh J Anim Sci 60(1):38–44

Li X, Ding X, Peng X, Chi X, Cui H, Zuo Z, Fang J (2017a) Effect of chitosan oligosaccharides on antioxidant function, lymphocyte cycle and apoptosis in ileum mucosa of broiler. Kafkas Univ Vet Fak Derg 23(4):571–577

Li S, Jin E, Qiao E, Wu G, Li K (2017b) Chitooligosaccharide promotes immune organ development in broiler chickens and reduces serum lipid levels. Histol Histopathol 32(9):951–961

Liaqat F, Eltem R (2018) Chitooligosaccharides and their biological activities: a comprehensive review. Carbohydr Polym 184:243–259

Liu P, Piao XS, Kim SW, Wang L, Shen YB, Lee HS, Li SY (2008) Effects of chito-oligosaccharide supplementation on the growth performance, nutrient digestibility, intestinal morphology, and fecal shedding of Escherichia coli and Lactobacillus in weaning pigs. J Anim Sci 86(10):2609–2618

Meng QW, Yan L, Ao X, Jang HD, Cho JH, Kim IH (2010) Effects of chito-oligosaccharide supplementation on egg production, nutrient digestibility, egg quality and blood profiles in laying hens. Asian Australas J Anim Sci 23(11):1476–1481

Mingoti RD, Freitas JE, Gandra JR, Gardinal R, Calomeni GD, Barletta RV, Vendramini THA, Paiva PG, Rennó FP (2016) Dose response of chitosan on nutrient digestibility, blood metabolites and lactation performance in holstein dairy cows. Livest Sci 187:35–39

Moon JS, Kim HK, Koo HC, Joo YS, Nam HM, Park YH, Kang MI (2007) The antibacterial and immunostimulative effect of chitosan-oligosaccharides against infection by *Staphylococcus aureus* isolated from bovine mastitis. Appl Microbiol Biotechnol 75(5):989–998

Nogueira CM, Zapata JFF, Fuentes MFF, Freitas ER, Craveiro AA, Aguiar CM (2003) The effect of supplementing layer diets with shark cartilage or chitosan on egg components and yolk lipids. Br Poult Sci 44(2):218–223

Nuengjamnong C, Angkanaporn K (2017) Efficacy of dietary chitosan on growth performance, haematological parameters and gut function in broilers 17(2):1–8

Sun ZH, Tang ZR, Yin YL, Huang RL, Li TJ, Tang SX, Tan ZL (2009) Effect of dietary supplementation of galacto-mannan-oligosaccharides and chitosan on performance and serum immune parameters of 28-day weaned piglets challenged with pathogenic E.coli. J Appl Anim Res 36(2):207–211

Suthongsa S, Pichyangkura R, Kalandakanond-Thongsong S, Thongsong B (2017) Effects of dietary levels of chito-oligosaccharide on ileal digestibility of nutrients, small intestinal morphology and crypt cell proliferation in weaned pigs. Livest Sci 198:37–44

Swiatkiewicz S, Arczewska-Wlosek A, Krawczyk J, Puchala M, Jozefiak D (2013) Effects of selected feed additives on the performance of laying hens given a diet rich in maize dried distiller's grains with solubles (DDGS). Br Poult Sci 54(4):478–485

Swiatkiewicz S, Arczewska-Wlosek A, Jozefiak D (2014) Feed enzymes, probiotic, or chitosan can improve the nutritional efficacy of broiler chicken diets containing a high level of distillers dried grains with solubles. Livest Sci 163:110–119

Swiatkiewicz S, Swiatkiewicz M, Arczewska-Wlosek A, Jozefiak D (2015) Chitosan and its oligosaccharide derivatives (chito-oligosaccharides) as feed supplements in poultry and swine nutrition. J Anim Physiol Anim Nutr 99(1):1–12

Tang ZR, Yin YL, Nyachoti CM, Huang RL, Li TJ, Yang CB, Yang XJ, Gong J, Peng J, Qi DS, Xing JJ, Sun ZH, Fan MZ (2005) Effect of dietary supplementation of chitosan and galacto-mannan-oligosaccharide on serum parameters and the insulin-like growth factor-I mRNA expression in early-weaned piglets. Domest Anim Endocrinol 28(4):430–441

Valle TAD, Paiva PGD, Jesus EFD, Almeida GFD, Zanferari F, Costa AGBVB, Bueno ICS, Renn, Oacute FP (2017) Dietary chitosan improves nitrogen use and feed conversion in diets for mid-lactation dairy cows. Livest Sci 201:22–29

Wan J, Yang K, Xu Q, Chen D, Yu B, Luo Y, He J (2016) Dietary chitosan oligosaccharide supplementation improves foetal survival and reproductive performance in multiparous sows. RSC Adv 6(74):70715–70722

Wan J, Jiang F, Xu Q, Chen D, Yu B, Huang Z, Mao X, Yu J, He J (2017a) New insights into the role of chitosan oligosaccharide in enhancing growth performance, antioxidant capacity, immunity and intestinal development of weaned pigs. RSC Adv 7(16):9669–9679

Wan J, Jiang F, Zhang J, Xu Q, Chen D, Yu B, Mao X, Yu J, Luo Y, He J (2017b) Amniotic fluid metabolomics and biochemistry analysis provides novel insights into the diet-regulated foetal growth in a pig model. Sci Rep 7:44782

Xiao DF, Wang YF, Liu G, He JH, Qiu W, Hu XG, Feng ZM, Ran ML, Nyachoti CM, Kim SW, Tang ZR, Yin YL (2014) Effects of chitosan on intestinal inflammation in weaned pigs challenged byenterotoxigenic *escherichia coli*. Plos One 9(8):e104192

Xie C, Guo X, Long C, Fan Z, Xiao D, Ruan Z, Deng Z-y, Wu X, Yin Y (2015) Supplementation of the sow diet with chitosan oligosaccharide during late gestation and lactation affects hepatic gluconeogenesis of suckling piglets. Anim Reprod Sci 159:109–117

Xie C, Long C, Wu X, Yang H, Fan Z, Xiao D, Wang Y, Yin Y (2016a) Effect of maternal supplementation with chitosan oligosaccharide on the antioxidant capacity of suckling piglets. J Anim Sci 94:453–456

Xie C, Wu X, Guo X, Long C, Li S, Hu C-AA, Yin Y (2016b) Maternal chitosan oligosaccharide supplementation affecting expression of circadian clock genes, and possible association with hepatic cholesterol accumulation in suckling piglets. Biol Rhythm Res 47(2):253–265

Xie C, Xin W, Long C, Wang Q, Fan Z, Li S, Yin Y (2016c) Chitosan oligosaccharide affects antioxidant defense capacity and placental amino acids transport of sows. BMC Vet Res 12(1):243

Xiong X, Yang HS, Wang XC, Hu Q, Liu CX, Wu X, Deng D, Hou YQ, Nyachoti CM, Xiao DF, Yin YL (2015) Effect of low dosage of chito-oligosaccharide supplementation on intestinal morphology, immune response, antioxidant capacity, and barrier function in weaned piglets. J Anim Sci 93(3):1089–1097

Xu Q, Qu C, Wan J, Cheng G, Yang W, Gong C, He J, Du Y (2018a) Effect of dietary chitosan oligosaccharide supplementation on the pig ovary transcriptome. RSC Adv 8(24):13266–13273

Xu Y, Wang Z, Wang Y, Yan S, Shi B (2018b) Effects of chitosan as growth promoter on diarrhea, nutrient apparent digestibility, fecal microbiota and immune response in weaned piglets. J Appl Anim Res 46(1):1437–1442

Yan L, Lee JH, Meng QW, Ao X, Kim IH (2010) Evaluation of dietary supplementation of delta-aminolevulinic acid and chito-oligosaccharide on production performance, egg quality and hematological characteristics in laying hens. Asian Australas J Anim Sci 23(8):1028–1033

Yang C, Ferket P, Hong Q, Zhou J, Cao G, Zhou L, Chen A (2012) Effect of chito-oligosaccharide on growth performance, intestinal barrier function, intestinal morphology and cecal microflora in weaned pigs. J Anim Sci 90(8):2671–2676

Yang H, Xiong X, Li J, Yin Y (2016) Effects of chito-oligosaccharide on intestinal mucosal amino acid profiles and alkaline phosphatase activities, and serum biochemical variables in weaned piglets. Livest Sci 190:141–146

Zhou TX, Chen YJ, Yoo JS, Huang Y, Lee JH, Jang HD, Shin SO, Kim HJ, Cho JH, Kim IH (2009) Effects of chitooligosaccharide supplementation on performance, blood characteristics, relative organ weight, and meat quality in broiler chickens. Poult Sci 88(3):593–600

Zhu LH, Zhao KL, Chen XL, Xu JX (2012) Impact of weaning and an antioxidant blend on intestinal barrier function and antioxidant status in pigs. J Anim Sci 90(8):2581

The manufacturer's authorised representative in the EU is Springer Nature Customer Service Centre GmbH, Europaplatz 3, 69115 Heidelberg, Germany. If you have any concerns regarding our products, please contact ProductSafety@springernature.com

Printed and bound by CPI Group (UK) Ltd, Croydon, CR0 4YY

23/07/2026

02175005-0009